T0296503

It is usually the case that scientists examine either ecological systems or social systems, yet the need for an interdisciplinary approach to the problems of environmental management and sustainable development is becoming increasingly obvious. Developed under the auspices of the Beijer Institute in Stockholm, this new book analyses social and ecological linkages in selected ecosystems using an international and interdisciplinary case-study approach. The chapters provide detailed information on a variety of management practices for dealing with environmental change. Taken as a whole, the book will contribute to the greater understanding of essential social responses to changes in ecosystems, including the generation, accumulation and transmission of ecological knowledge, structure and dynamics of institutions, and the cultural values underlying these responses. A set of new (or rediscovered) principles for sustainable ecosystem management is also presented.

Linking Social and Ecological Systems will be of value to natural and social scientists interested in sustainability.

LINKING SOCIAL AND ECOLOGICAL SYSTEMS: MANAGEMENT PRACTICES AND SOCIAL MECHANISMS FOR BUILDING RESILIENCE

LINKING SOCIAL AND ECOLOGICAL SYSTEMS

Management practices and social mechanisms
for building resilience

Edited by

FIKRET BERKES

Natural Resources Institute, The University of Manitoba,
Winnipeg, Canada

and

CARL FOLKE

Department of Systems Ecology and Centre for Research on Natural Resources
and the Environment, Stockholm University Sweden,
and The Beijer International Institute of Ecological Economics,
Royal Swedish Academy of Sciences, Stockholm, Sweden

and with the editorial assistance of

JOHAN COLDING

The Beijer International Institute of Ecological Economics,
Royal Swedish Academy of Sciences, Stockholm, and
Department of Systems Ecology, Stockholm University, Sweden

CAMBRIDGE
UNIVERSITY PRESS

PUBLISHED BY THE PRESS SYNDICATE OF THE UNIVERSITY OF CAMBRIDGE
The Pitt Building, Trumpington Street, Cambridge, United Kingdom

CAMBRIDGE UNIVERSITY PRESS
The Edinburgh Building, Cambridge CB2 2RU, UK   http://www.cup.cam.ac.uk
40 West 20th Street, New York, NY 10011–4211, USA   http://www.cup.org
10 Stamford Road, Oakleigh, Melbourne 3166, Australia
Ruiz de Alarcón 13, 28014 Madrid, Spain

First published 1998
Reprinted 2000
First paperback edition 2000

A catalogue record for this book is available from the British Library

Library of Congress Cataloguing in Publication data
Linking social and ecological systems : management and practices and social
mechanisms / edited by Fikret Berkes and Carl Folke; with the
assistance of Johan Colding.
p.   cm.
Includes index.
ISBN 0 521 59140 6 (hardbound)
1. Human ecology. 2. Social ecology. 3. Social systems.
4. Biotic communities–Management. 5. Sustainable development.
I. Berkes, Fikret. II. Folke, Carl. III. Colding, Johan.
GF21.L55 1997
304.2–dc21   97-6082 CIP

ISBN 0 521 59140 6 hardback
ISBN 0 521 785626 paperback

Transferred to digital printing 2002

Contents

Contributors

James M. Acheson
Department of Anthropology, University of Maine, Orono, Maine 04469, USA

Janis B. Alcorn
Biodiversity Support Program, World Wildlife Fund, 1250 24th Street NW, Washington DC 200 37, USA

Alpina Begossi
Núcico de Estudos e Pesquisas Ambientais, Center of Environmental Studies & Research, Universidade Estadudual de CP M6166 Campinas, 13081-970, SP Brazil

Fikret Berkes
Natural Resources Institute, The University of Manitoba, Winnipeg, Manitoba, Canada R3T 2N2

Johan Colding
The Beijer International Institute of Ecological Economics, The Royal Swedish Academy of Sciences, Box 50005, S-104 05 Stockholm, Sweden, and Department of Systems Ecology, Stockholm University, S-106 91 Stockholm, Sweden

A. Christopher Finlayson
Department of Human Ecology, Cook College, Rutgers University, PO Box 231, New Brunswick NJ 08903, USA

Carl Folke
The Beijer International Institute of Ecological Economics, The Royal Swedish Academy of Sciences, Box 50005, S-104 05 Stockholm, Sweden and The Department of Systems Ecology, Stockholm University, S-106 91 Stockholm, Sweden

Madhav Gadgil
Centre for Ecological Sciences, Indian Institute of Science, Bangalore 560 012, India

Susan S. Hanna
Department of Agricultural & Resource Economics, Oregon State University, Ballard Extension Hall 213, Corvallis, Oregon 97331-3601, USA

Natabar Shyam Hemam
Anthropometry and Human Genetics Unit, Indian Statistical Institute, Calcutta 700035, India

C. S. Holling
Department of Zoology, University of Florida, 111 Bartram Hall, P.O. Box 118525, Gainesville, Florida 32611-8525, USA

Narpat S. Jodha
Mountain Enterprises and Infrastructure Division, International Centre for Integrated Mountain Development, P.O. Box 3226, Kathmandu, Nepal

Bonnie J. McCay
Department of Human Ecology, Cook College, Rutgers University, P.O. Box 231, New Brunswick, New Jersey 08903, USA

Maryam Niamir-Fuller
SHIS QI 25, Conjunto 1, Casa 21, CEP 70660-300, Brasilia, DF Brazil

Gísli Pálsson
Faculty of Social Science, University of Iceland, 101 Reykjavik, Iceland

Evelyn Pinkerton
School of Resource and Environmental Management, Simon Fraser University, Burnaby, British Columbia V5A 1S6, Canada

Jennifer Pinkston
CIKARD, Iowa State University, 324 Curtiss Hall, Ames, Iowa 50011-1050 USA

B. Mohan Reddy
Anthropometry and Human Genetics Unit, Indian Statistical Institute, Calcutta 700035, India

Ulf Sporrong
Department of Human Geography, Stockholm University, 106 91 Stockholm, Sweden

Robert S. Steneck
Department of Oceanography, University of Maine, Orono, Maine 04469, USA

Victor M. Toledo
Instituto de Ecologia, Universitad Autonoma de Mexico, Apdo 41-4, Sta. Maria Guido, Morelia, Michoacan 58090, Mexico

D. Michael Warren
CIKARD, Iowa State University, 324 Curtiss Hall, Ames, Iowa 50011-1050, USA

James A. Wilson
Agricultural and Resource Economics, University of Maine, 5782 Winslow Hall, Orono, Maine 04469-5782, USA

Preface

The Beijer Institute of the Royal Swedish Academy of Sciences is an international institute of ecological economics. Its major objectives are to carry out research and to stimulate collaboration and training between scientists, university departments, and institutes working at the interface of natural and human systems.

In 1993 the Beijer Institute initiated a major research programme on *Property Rights and the Performance of Natural Resource Systems*, in recognition of the crucial importance of property rights issues in resource management. Within this programme, we co-ordinated a subproject on *Linking Social and Ecological Systems for Resilience and Sustainability*. This book is the main product of that subproject. It is an edited volume, but it is different from most edited volumes. We have used a common framework; each chapter, written by scholars from a diversity of disciplines, has been developed on the basis of that framework. The authors were also invited to participate in a series of three workshops of the Beijer Institute at which the framework and progress of cases were discussed. The introductory chapter describes the framework, and the objectives and content of the book.

It is more the rule than the exception that scientists examine either ecological systems or social systems. In this volume, social and ecological systems are analysed as one system. The volume is deliberately interdisciplinary because the problems addressed do not fit into any one discipline such as ecology, anthropology, economics or political science.

The focus of the book is on social and ecological linkages in selected ecosystems, and there is a systematic treatment of the mechanisms behind these linkages. The overall purpose has been to learn from cases which show successful adaptations or learning, and to try to unravel management practices and social mechanisms that can cope with resource and ecosystem change. Such mechanisms are scarce in modern society, and hence we can hope to learn from some traditional systems, and also some recently evolved common-property systems in contemporary society. Many of these social–ecological linkages represent a range of historical human experience with environmental management, and provide a reservoir of active adaptations which may be of universal importance in designing for sustainability. This is not to romanticize about traditional societies and the commons being the best solution; that is not at all the purpose of the book. Many of the chapters certainly show both successes and failures, and they focus on the dynamics of change and the evolution of adaptations through learning.

The research questions posed in the book explicitly link ecology, economics and social science. They are analysed through an interdisciplinary, international, case-study approach. The case studies provide lessons to be learned from resource management in environments such as northern coastal ecosystems, arid/semi-arid land and temperate land ecosystems, mountain ecosystems, temperate and tropical forest ecosystems, and subarctic ecosystems.

The framework of the book provides a checklist of a number of factors which we hypothesize as being important, and individual chapters do not necessarily address every issue raised in it. Each chapter analyses key aspects of the framework and provides important information on essential social responses to changes in ecological systems, contributing to an understanding of social mechanisms for dealing with resource and environmental fluctuations. For example, several chapters indicate the importance of mechanisms which contribute to enhancing the buffering capacity, the resilience of social and ecological systems to stress. All chapters deal with both social and ecological resilience, but vary in their degree of emphasis on each.

In the last chapter we summarize lessons that can be learned to assist in the designing of more sustainable resource management systems, and we discuss how adaptiveness and flexibility can be built into institutions so that they are capable of responding to processes that contribute to the resilience of ecosystems. Drawing on the insights from the case studies, a list of working hypotheses or 'rediscovered' principles for resource and environmental management is also presented.

It is our hope that the book *Linking Social and Ecological Systems* will stimulate collaborative research and policy on how to improve management and build resilience in interdependent social–ecological systems, so urgently needed in modern society

Stockholm, October 1996   *Fikret Berkes*
                                                              Carl Folke

Acknowledgements

We are grateful to many people who have supported us during the work with this book. First we would like to acknowledge the referees. All chapters have been reviewed by three referees. In addition to the help of those who are also chapter authors, we are grateful for the constructive comments by Jean Ensminger, Milton Freeman, David Feeny, Arturo Gómez-Pompa, Monica Hammer, Torsten Hägerstrand, Edvard Hviding, Svein Jentoft, Robin Mearns, Russ McGoodwin, Margaret McKean, Jeffrey McNeely, Barbara Neis, Peter Parks, and Nancy Peluso. We are also grateful to Manoshi Mitra, Elinor Ostrom and Kenneth Ruddle for providing input to the framework. Susan Hanna has been a key player in her capacity as the director of the Beijer Institute's research programme on *Property Rights and the Performance of Natural Resource Systems*, as well as being a chapter author. We are grateful to Icongraph Johan Hultenheim for preparing the illustrations, to Christina Leijonhufvud and Astrid Auraldsson for excellent organization and co-ordination of the Beijer workshops, and to the chapter authors for their patience with two (and in some cases three) rounds of chapter drafts and with the editors' inquiries and review comments on the chapters.

This book is part of the output of the property rights programme[1] of the Beijer International Institute of Ecological Economics, The Royal Swedish Academy of Sciences, with support from the World Environment and Resources Program of the John D. and Catherine T. MacArthur Foundation. Additional financial support has been provided by the

---

[1] Publication list of other output from the programme can be attained from The Beijer Institute, The International Institute of Ecological Economics, The Royal Swedish Academy of Sciences, Box 50005, S-104 05 Stockhom, Sweden. Fax: 46 8 15 24 64. E-mail: beijer @ beijer.kva.se

Swedish Council for Planning and Coordination of Research (FRN), the Swedish Council for Forestry and Agricultural Research (SJFR), and the Pew Scholars Program in Conservation and the Environment of the Pew Charitable Trusts.

1

Linking social and ecological systems for resilience and sustainability

FIKRET BERKES & CARL FOLKE

Introduction

There is much evidence of poor management of ecosystems; the conventional prescriptions of resource management are in many cases not resulting in sustainability. In fact, some of the resource crashes of recent years are of greater magnitude than those observed historically. Some authors attribute this fact to human 'shortsightedness and greed', and question whether resources could ever be managed sustainably (Ludwig, Hilborn and Walters 1993). Others argue that resource management science may be fundamentally flawed as a system of thought and practice in that its premises are based on the *laissez-faire* ideology which still persists in neo-classical economics (Daly and Cobb, 1989). Gadgil and Berkes (1991) and McNeely (1991), among others, have pointed out that scientific resource management has its roots in the utilitarian and exploitative worldview which assumes that humans have dominion over nature. In the historical process of converting the world's life-support systems into mere commodities, resource management science was geared for the efficient utilization of resources as if they were limitless. Methods of resource development and management, in both biological and economic areas, have treated the environment as discrete boxes of 'resources', the yields from which could be individually maximized. The field has relied on the use of fixed rules for achieving constant yields, as in fixed carrying capacity of animals and fixed maximum sustainable yields (MSY) of fish and forest products.

Much of the development in resource management science since around the 1970s has sought to deal with the environmental and social problems created by resource mismanagement and depletion. Many of the new approaches have been reformist in nature, seeking to alleviate the excesses of classical resource management. An example would be the creation of multi-species models in fisheries, as opposed to the modelling of single species populations and calculating fixed MSYs as if these species were not affected by changes in the populations of their competitors, predators, prey, and fluctuations of their physical environment (Larkin, 1977). Other approaches have been more radical, rejecting resource management altogether as a valid objective. One example would be deep ecology, as the preservation of ecosystems as a public good independently of their utility as resources; an extreme example would be the animal rights movement.

The basic assumption behind the work that led to the present volume is that resource management is necessary but that it requires fundamentally different approaches, not mere tinkering with current models and practices. The volume seeks to integrate two streams of resource management thought that fundamentally differ from the classic utilitarian approach. The first is the use of systems approach and adaptive management, with their emphasis on linkages and feedback controls (Holling, 1978; Walters, 1986). The systems approach is replacing the view that resources can be treated as discrete entities in isolation from the rest of the ecosystem and the social system. For example, the volume *Investing in Natural Capital* (Jansson *et al.*, 1994) explored in some detail the necessity of a systems-oriented, wide-scope ecological economics approach to sustainability.

The second stream of thought is that improving the performance of natural resource systems requires an emphasis on institutions and property rights. A people-oriented approach which focuses on the resource user rather than on the resource itself is not a new idea; many have pointed out that 'resource management is people management'. However, tools and approaches for such people management are poorly developed, and the importance of a *social science* of resource management has not generally been recognized. The present volume follows from and extends the findings of a number of books that have tried to fill this gap, including Clark and Munn (1986) on the various dimensions of sustainability; Ostrom (1990) on institutions and collective action; Bromley (1991), Hanna and Munasinghe (1995) and Hanna, Folke and Maler (1996) on property rights; McCay and Acheson (1987), Berkes (1989) and Baland and Platteau (1996) on community-based resource management; and Lee (1993) and Gunderson, Holling and Light (1995) on institutional learning and resource management.

Following the statement of objectives, this introductory chapter will review some definitions that will be widely used in the book, and will then cover five concepts or themes on which our arguments hinge: property rights, the systems approach, adaptive management, ecological resilience, and traditional resource management systems. These sections explain the basis for the analytical framework used in the book, followed by a description of the framework itself.

Objectives

The general objective of the book is to investigate how the management of selected ecosystems can be improved by learning from a variety of management systems and their dynamics. The essential feature of this inquiry is to mobilize a *wider range* of considerations and sources of information than those used in conventional resource management. Many of the cases in the book investigate a mix of systems and their change over time, and some of the cases focus on traditional and newly emergent local systems. All chapters address the question of sustainability, and seek principles that may assist in successful resource management or help restore degraded ecosystems to generate a sustainable flow of services. To accomplish this task, social and ecological linkages in selected ecosystem types are investigated systematically, using a common analytical framework. Specifically, in each of the case studies, the authors address two objectives:

- how the local social system has developed management practices based on ecological knowledge for dealing with the dynamics of the ecosystem(s) in which it is located; and
- social mechanisms behind these management practices.

The volume is organized into four parts. Part I includes three chapters that take very different approaches to deal with the question of learning from locally devised systems. One of these is a 'traditional' system from India; the other two are from Europe, one historical and one contemporary. Part II includes three chapters from the Americas and one from Africa dealing with the emergence of new adaptive systems. These case studies show how institutions can adapt to local ecosystem characteristics and provide an understanding of some of the dynamic processes in social – ecological systems undergoing change. Part III, also with four chapters, is concerned with regional experiences as well as local experiences, and with generalizations that emerge from the accumulated body of literature from four different regions of the world. These four chapters help emphasize the point that local social systems are not isolated but are subject to national and regional influences. The four chapters in Part IV address the question of designing new approaches to management. Three of them explore the ways to combine local and scientific knowledge, or to combine traditional and conventional resource management systems.

Many previous studies have analysed the impact of human activities on the ecosystem, but few have studied the interdependence of social systems and ecological systems. Depending on the discipline base of the author(s), either the social system or the ecological system tends to be taken as a 'given'. In many volumes on resource management and environmental studies, the human system has been treated as external to the ecosystem. By contrast, studies of institutions have mainly investigated processes *within* the social system, treating the ecosystem largely as a 'black box'. Only a few studies (including some of those cited in the previous section) have explicitly analysed linkages between social systems and ecological systems. The present volume addresses this issue of linkage through its objective to *relate* management practices based on ecological understanding, *to* the social mechanisms behind these practices, in a variety of geographical settings, cultures, and ecosystems.

Definitions

Some definitions are needed to establish a common vocabulary. Our objective is *sustainability,* defined by WCED (1987) as 'development that meets the needs of the present without compromising the ability of future generations to meet their own needs'. Sustainability, as used here, is a process and includes ecological, social and economic dimensions. We recognize that the question of 'what is to be sustained?' has to be addressed on a case by case basis (Costanza and Patten, 1995). For our general purposes, sustainability implies not challenging ecological thresholds on temporal and spatial scales that will negatively affect ecological systems and social systems. *Social systems* that are of primary concern for this volume deal with property rights, land and resource tenure systems, systems of knowledge pertinent to environment and resources, and world views and ethics concerning environment and resources. The term *ecological system* (ecosystem) is used in the conventional ecological sense to refer to the natural environment. We hold the view that social and ecological systems are in fact linked, and that the delineation between social and natural systems is artificial and arbitrary. Such views, however, are not yet accepted in conventional ecology and social science. When we wish to emphasize the integrated concept of humans-in-nature, we use the terms *social – ecological system* and *social – ecological linkages.*

The term *indigenous knowledge* (IK) is used to mean local knowledge held by indigenous peoples, or local knowledge unique to a given culture or society, consistent with Warren, Slikkerveer and Brokensha, (1995). The term can be used interchangeably with traditional knowledge, but we prefer to use *traditional ecological knowledge* (TEK) more specifically to refer to a cumulative body of knowledge and beliefs, handed down through generations by cultural transmission, about the relationship of living beings (including humans) with one another and with their environment (Berkes, Folke and Gadgil, 1995). The word *traditional* is used to refer to historical and cultural continuity, recognizing that societies are constantly redefining what is considered 'traditional'. Some chapters refer to *neo-traditional resource management systems*, defined here as local resource management which does not have historical continuity but which is based on observations, experience and local knowledge of resource users themselves (as opposed to government scientists and managers). It is used here interchangeably with *newly emergent resource management systems*.

Traditional and local management is contrasted with *Western resource management science*, defined as resource management based on Newtonian science and on the expertise of government resource managers. We use the term interchangeably with *scientific resource management* and *conventional resource management*. We recognize that all societies have their own science, but identify Western science and scientific method to represent a particular brand of science which is used as the basis of resource management by centralized bureaucracies in all parts of the world.

Institutions are defined as 'humanly devised constraints that structure human interaction. They are made up of formal constraints (rules, laws, constitutions), informal constraints (norms of behavior, conventions and self-imposed codes of conduct), and their enforcement characteristics' (North, 1993). Institutions are 'the set of rules actually used (the working rules or rules-in-use) by a set of individuals to organize repetitive activities that produce outcomes affecting those individuals and potentially affecting others' (Ostrom, 1992). The emphasis in the book is on institutions that deal with property rights and common property resources. Here we define *property* as the rights and obligations of individuals or groups to use the resource base; a bundle of entitlements defining owner's rights, duties, and responsibilities for the use of the resource, or 'a claim to a benefit (or income) stream, and a property right is a claim to a benefit stream that some higher body – usually the state – will agree to protect through the assignment of duty to others who may covet, or somehow interfere with, the benefit stream' (Bromley, 1992). *Common-property (common-pool) resources* are defined as a class of resources for which exclusion is difficult and joint use involves subtractability (Berkes, 1989; Feeny *et al.*, 1990). Institutions have to deal with the two fundamental management problems that arise from the two basic characteristics of all such resources: how to control access to the resource (the exclusion problem), and how to institute rules among users to solve the potential divergence between individual and collective rationality (the subtractability problem).

The term *feedback* is used in the conventional systems sense to refer to the result of any behaviour which may reinforce (positive feedback) or modify (negative feedback) subsequent behaviour. More specifically, the book is concerned with the recognition of environmental feedbacks (e.g. depletion of particular resources, decline of catch per unit of effort) that signal for changes in management responses, and the ability of resource management institutions to receive and to respond to these signals. *Resilience* is the buffer capacity or the ability of a system to absorb per-turbations; the magnitude of disturbance that can be absorbed before a system changes its structure by changing the variables and processes that control behavior (Holling *et al.*, 1995). *Threshold* is the point where a system flips from one equilibrium state to another. *Surprise* denotes the condition when perceived reality departs *qualitatively* from expectation. Surprises occur when causes turn out to be sharply different than was con-ceived, when behaviours are profoundly unexpected, and when action pro-duces a result opposite to that intended (Holling, 1986).

Capital is a stock resource with value embedded in its ability to produce a flow of benefits. We make a distinction among three kinds of capital: (a) *human-made capital,* which is generated through economic activity through human ingenuity and technological change, the produced means of produc-tion; (b) *natural capital,* which consists of non-renewable resources extracted from ecosystems, renewable resources produced by the processes and functions of ecosystems and environmental services sustained by the workings of ecosystems; and (c) *cultural capital,* which refers to the factors that provide human societies with the means and adaptations to deal with the natural environment and actively to modify it (Berkes and Folke, 1994). Coleman (1990: 300–21) used *social capital* to refer to features of social organization such as trust, norms and networks. Ostrom (1990: 190, 211) used social capital to refer to the richness of social organization, and *institu-tional capital* to refer to the supply of organizational ability and social struc-tures, literally the 'capital' of institutions that a society has at its disposal.

Property rights institutions

Recent advances in common-property theory have shown why institutions and property rights are important considerations for resource management

(McCay, 1995). It used to be popularly believed that users of common-property resources were always trapped in an inexorable 'tragedy of the commons' (Hardin, 1968). However, many studies, especially since the mid-1980s, have shown Hardin's generalization does not hold. *If* the resource is freely open to access by any user, a tragedy of the commons does eventually follow. However, many resources used by rural communities are not open-access but are used under communal property rights arrangements. That is, more often than not, rules exist regarding access and joint use, as shown by many of the case studies in this book, as well as in the volumes by McCay and Acheson (1987), Berkes (1989) and Bromley (1992).

Property rights arrangements in a given area may be complex because resource tenure often involves 'bundles of rights', including use rights, rights to exclude others, rights to manage, and the right to sell (Schlager and Ostrom, 1992). Determining the actual rights is often a challenge, as in many marine resources (Palsson, 1991). Even within an administrative area with common legal and fiscal interventions, the actual status of local property rights to resources may vary from village to village (Jodha, 1986). Also, different resources within a given area may be held under different property rights regimes. For example, in the case of forest resource management in mountainous areas in Asia, patches of privately owned cropland may alternate with state-controlled and managed forest land, common grazing land, and common grass and bush land from which users may be obtaining a diversity of products (Messerschmidt, 1993).

Generally speaking, local social systems of rights and responsibilities develop for any resource deemed important for a community. Even under rapidly changing conditions, there are usually incipient property rights; rules arise and evolve according to local needs (Berkes, 1986; 1989). Ostrom (1990) has reviewed six commons cases in depth, and a number of others in less detail, to formulate eight design principles for successful common-property regimes. Most of these design principles fall into two clusters: those dealing with access, group boundary and resource boundary issues; and those dealing with decision-making for joint use, including issues of representation, monitoring, sanctions, conflict resolution and legal recognition, consistent with the definition of common property resources.

Hardin's tragedy often results, not from any inherent failure of common property, but from institutional failure to control access to the resource, and to make and enforce internal decisions for collective use. Institutional failure could be due to internal reasons, as in the inability of the users to manage themselves, or it could be due to external reasons, as in the incursion of outsiders (Dove, 1993). Failure could also occur as a result of such factors as population growth and technology change, especially if the changes occur too rapidly for the ability of the local social system to absorb them (Berkes, 1989; Ostrom, 1990).

The likelihood of users designing successful common property institutions will be improved if the group is relatively small and stable; if it is relatively homogenous, with the members using similar technologies and having similar values and discount rates; if there is reciprocity and trust; and if the transaction cost for making and enforcing rules is low. Not all of these conditions are necessary for success. For example, much larger numbers of users can be accommodated if users are organized in nested enterprises or multiple layers, as in the *huerta* irrigation systems in Spain (Ostrom, 1990).

The analysis of institutions also needs to include questions of jurisdiction and the respective roles of local groups and government agencies. Often the user community is dependent on the enforcement and protection of local rights by higher levels of government. Even those indigenous groups with well-functioning local management systems are dependent on the central government for the legal recognition of their rights and their protection against outsiders. Many systems of property rights show a mix of local jurisdiction and government jurisdiction. The sharing of resource management responsibility and authority between users and government agencies (co-management or collaborative management) has been receiving increased attention (Pinkerton, 1989; Jentoft and McCay, 1995; Chapter 8).

Perhaps the most striking finding of the common property literature is the rich diversity of common-property institutions and property rights arrangements, especially in the older, historically rooted resource management systems (Feeny *et al.,* 1990). Examples include Swiss alpine meadows (Netting, 1981), and the reef and lagoon tenure systems of Oceania with their diverse array of rules from island group to island group (Ruddle and Johannes, 1990; Freeman, Matsuda and Ruddle, 1991). As compared to the rather narrow set of management prescriptions based on scientific resource management, some of which may inadvertently act to reduce ecosystem resilience (Holling *et al.,* 1995), traditional common-property systems tend to be locally diverse and operate under systems of knowledge which may differ substantially from Western knowledge systems (Banuri and Apffel Marglin, 1993).

Systems approach and social–ecological linkages

The systems approach broadly refers to a holistic view of the components and the interrelationships among the components of a system. The systems view most relevant to our discussion is the ecosystem view or the ecosystem perspective (e.g. Odum, 1989). But unlike biological ecology, which tends to view humans as external to ecosystems (e.g. Pomeroy and Alberts, 1988; Likens, 1992), we use an ecosystem perspective that does explicitly include humans or, more specifically, the *social system*. The analysis is consistent with the classical human ecology literature from Park (1936) onwards which emphasized the interactions of population, technology, organization and culture. Also, the analysis is consistent with the way many traditional societies see their relationships with the environment. With a few exceptions, including the Western industrial societies of the last 400 years or so, human societies have generally regarded themselves as *part of nature* and not separate from it. Of particular interest are cases of traditional integrated human–nature concepts of the environment, such as the *vanua* concept in Fiji which regards the land, water and human environment as a unit, one and indivisible (Ruddle, Hviding and Johannes, 1992). Several such pre-scientific ecosystem concepts are known from Europe, North America and Asia as well as throughout Oceania where they have been well documented (Costa-Pierce, 1987; Gadgil and Berkes, 1991).

It is perhaps significant that scientific concepts of ecosystem are deficient in the description and analysis of such human-in-nature systems. There is no single, universally accepted way of formulating the linkage between social systems and natural systems. Findings of the common-property literature in recent years stress the importance of social, political and economic organization, with institutions as the mediating factor that governs the relationship between a social group and the life-support ecosystems on which it depends. In the ecological economics literature, the emphasis is on the sustainable use of *natural capital* (natural resources and ecological services generated and sustained by ecosystems and their biodiversity) by the use of economic incentives and other tools, and by the use of appropriate economic institutions. To make the analysis more complete, Berkes and Folke (1994) suggested that a third kind of capital needs to be considered, *cultural capital*, by which societies convert *natural capital* into *human-made capital* or the produced means of production. In this volume, institutions are considered to be a part of this cultural capital.

It is probably true that the unity of humans and nature is an easier concept to accept in many non-Western societies than in Western ones, although a shift in worldview is well underway. However, it is also true that environmental degradation and resource depletion are an even larger problem in many non-Western societies than in Western ones, for various reasons. In this volume, we approach this dilemma by glorifying neither

Western-style nor 'traditional' non-Western style resource management. Rather, we focus on the *ability of the management system to respond to feedbacks from the environment*. One scientific management approach that explicitly emphasizes feedback learning is adaptive management.

Adaptive management

Resource management, as a branch of applied ecology, is a difficult field in which to carry out scientific research. As Hilborn and Ludwig (1993) put it, the difficulty is easy enough to explain: 'experiments take longer, replication, control, and randomness are harder to achieve, and ecological systems have the nasty habit of changing over time.' The authors do not think that the problem is the inherent complexity of the system under study. Single cells are very complex systems too, and yet research progress in molecular biology has been spectacular in providing applications based on predictive models. By comparison, predictive models in ecology are hard to come by, and this is certainly true for the various areas of resource management.

Of the various areas of difficulty mentioned by Hilborn and Ludwig (1993), recent conceptual work has focused on the propensity of ecosystems to change over time in an unpredictable manner. Further, stressed ecosystems, as in resource overexploitation, tend to change not gradually but in lurches, through threshold effects and in surprises, whereby outcomes differ from predictive models not only quantitatively but *qualitatively* (Holling, 1986; Gunderson *et al.*, 1995; Holling *et al.*, 1995).

Adaptive management deals with the unpredictable interactions between people and ecosystems as they evolve together. It takes the view that resource management polices can be treated as 'experiments' from which managers can learn (Holling, 1978; Walters, 1986). Organizations and institutions can 'learn' as individuals do, and hence adaptive management is based on social and institutional learning (Lee, 1993). Adaptive management differs from the conventional practice of resource management by emphasizing the importance of *feedbacks* from the environment in shaping policy, followed by further systematic (i.e. non-random) experimentation to shape subsequent policy, and so on. The process is iterative; it is feedback and learning-based. It is co-evolutionary (Norgaard, 1994) in the sense that it involves two-way feedback between management policy and the state of the resource. Hence, adaptive management is an inductive approach, relying on comparative studies that combine ecological theories with observation, and with active human interventions in nature, based on an understanding of human response processes (Gunderson *et al.,* 1995). More recent work in adaptive management has focused on the importance of scale in time and space (see Chapter 13).

Adaptive management is a relatively new approach in resource management science, but its common-sense logic that emphasizes learning by doing and its elimination of the barrier between research and management resemble traditional resource management systems. Both rely on feedback and learning, and on the progressive accumulation of knowledge, often over many generations in the case of traditional systems. Adaptive management has the advantage of systematic experimentation and the incorporation of scientific research into the overall management scheme.

Ecological resilience

Environmental surprises, such as those that arise from the construction of large dams, have been discussed for a long time but often anecdotally (e.g. Farvar and Milton, 1972). The beginnings of a new 'science of surprise' can be traced to Holling (1986), who pointed out that there was a general pattern to unexpected changes and resource crises. Typically, there is a sequence of events which starts with efficient exploitation that eventually leads to inadvertent loss of ecosystem resilience. To supply markets, resource management tries to control a target resource (e.g. supply of fish or timber) by reducing the *variability* of the target resource. This helps meet production targets and economic objectives (e.g. revenue and employment). The management policy is successful in the short term, but its very success causes inadvertent changes in the functioning and resilience of the ecosystem. Over a period of time, management emphasis shifts to improving the efficiency of the methods of resource utilization, and the need for other ecosystem support and services (e.g. water regulation capacity of a forest) and the loss of resilience are not perceived. The ecosystem and the target resource have become more vulnerable to surprise, while the resource management institution, devoted to production efficiency, has become more rigid and less responsive to environmental feedbacks, thus setting the stage for a resource management crisis (Holling, 1986).

This general pattern of unforeseen effects and nasty surprises is thought to occur through a mechanism involving the loss of ecological resilience. The very success of management, effective in the short term, 'freezes' the ecosystem at a certain stage of natural change by actively blocking out environmental variability and feedbacks that govern change. Instead of allowing smaller perturbations to act on the system, management causes the accumulation of perturbations, inviting larger and less predictable feedbacks at a level and scale that threaten the functional performance of the whole ecosystem, and thereby also the flow of resources and services that it generates. Holling (1986) used the examples of budworm control in Canadian forests (more and more control seems to lead to larger and larger infestations when they do finally occur) and forest fire suppression (following a century of fire suppression, nearly half of Yellowstone National Park in the USA burned down in one major fire).

Holling's concept of a four-stage ecosystem renewal cycle, consisting of exploitation–conservation–release–reorganization stages, is key to this argument (Holling, 1986; Holling *et al.*, 1995). So is the resilience concept. Resilience has been defined in two very different ways in the ecological literature (Holling *et al.*, 1995). The first definition concentrates on stability at a presumed steady-state, and stresses resistance to a disturbance and the speed of return to the equilibrium point. This is the conventional, equilibrium-centred, linear, cause-and-effect view of a predictive science as used by many in ecology, economics and other disciplines. In resource management science, this view leads to the assumption that resources are manageable and yields predictable. Discrete yield levels, such as maximum sustained yields of fish or timber, can be calculated, and perturbations (such as fire and pest outbreaks) can be controlled and excluded from the system.

By contrast, the second definition of resilience, and the one used in the present volume, emphasizes conditions in which disturbances (or perturbations) can flip a system from one equilibrium state to another. In this case, the important measure of resilience is the magnitude or scale of *disturbance that can be absorbed* before the system changes in structure by the change of variables and processes that control system behaviour. This is a fundamentally different view of science, in which determining causal effects and making predictions are not simple matters at all. Rather, systems are seen to be complex, non-linear, multi-equilibrium and self-organizing; they are permeated by uncertainty and discontinuities. Resilience in this context is a measure of robustness and buffering capacity of the system to changing conditions (more detail in Chapter 13).

The kind of science implied by the second definition of resilience represents a move away from the positivist emphasis on objectivity and towards a recognition that fundamental uncertainty is large, yields are unpredictable, certain processes are irreversible, and qualitative judgments do matter. This kind of science is in many ways symphatetic to 'savage thought' of Levi-Strauss (1962) and indigenous systems of environmental knowledge. The parallels between traditional ecological wisdom and this multi-equilibrium paradigm of ecological systems show promise for further inquiry.

Local and traditional resource management systems

The similarities between the ecosystem perspective and traditional ecosystem-like concepts, and the similarities between adaptive management and traditional management compelled us to ask the question: can resource management be improved by supplementing scientific data with local and traditional knowledge? Can information from resource users themselves broaden the base of knowledge necessary for decision making for sustainable resource use?

Traditional resource management systems or other local-level systems, which are based on the knowledge and experience of the resource users themselves, may have potential that has hardly been tapped. We assume that every society has its own means and adaptations to deal with its natural environment, its own *cultural capital* (Berkes and Folke 1994). In some cases, the capital of local knowledge may be used and organized in such a way that it, in effect, amounts to a management system. Such is the case with some shifting cultivating systems in the tropical forest (Alcorn, 1984; Ramakrishnan, 1992), island ecosystems (Costa-Pierce, 1987), inshore fisheries of Pacific islands (Johannes, 1978; Ruddle and Johannes, 1990), and grazing systems in semi-arid lands (Niamir, 1990). Given that Western resource management has not been all that successful in managing many of these environments sustainably, perhaps there are lessons to be learned from the cultural capital of the people who are the local experts, a view which gained legitimacy after it was expressed in *Our Common Future* (WCED, 1987: 12).

The issue does not only concern *traditional* peoples; more generally, the issue is the significance and legitimacy of local expertise. Ancient cultures and indigenous peoples often have a longer-term relationship with their environment than do others, but they certainly do not have monopoly over local ecological wisdom. There are cases of local, newly emergent or 'neo-traditional' resource management systems which cannot claim historical continuity over thousands of years, but which are nevertheless based on local knowledge and practice appropriately adapted to the ecological systems in which they occur (e.g. Smith and Berkes, 1993).

These considerations suggest that resource managers need to be cautious about assuming that 'our' Western system of acquiring scientific knowledge is a universal epistemology (Funtowicz and Ravetz, 1990). Some field ecologists have known for a long time that the 'non-scientific' knowledge of local experts can be both substantial and essential for management (e.g. Johannes, 1981). Nevertheless, non-Western knowledge systems have received relatively little attention from a resource management point of view. Feyerabend (1987) distinguished two different traditions of thought. The first is the *abstract tradition* which corresponds to Western scientific epistemology. It allows the scientist to formulate scientific statements, in accordance with the rules and conventions of science, without necessarily having met a single one of the objects described.

The kind of knowledge possessed by traditional or small-scale societies is of a different kind, labelled by Feyerabend as the *historical tradition*. Here, the knowledge held by the observer is based on his/her personal experience with the object; it is concrete rather than abstract (Levi-Strauss, 1962). The knowledge is often encoded in the cultural practices of everyday life. Culture and folk-science are not distinct, and religion often serves to code local ecological knowledge (Rappaport, 1971).

For an observer schooled in the abstract tradition, local knowledge is easy to ignore, and local management systems may at first make little sense because '. . . folk beliefs are a melange of truth and inaccuracy. Much of the world's ancient belief systems seems like preposterous nonsense to the modern scientist' (Anderson, 1996: 101). It is a major task, therefore, to seek out ecologically sensible practices and knowledge from the mixture of superstition, beliefs and folk-science. Local resource management strategies pose problems because they incorporate 'multiple epistemologies, possessed by different groups of people', as Redclift (1994) points out. It is not possible, therefore, to make sense of local management systems without understanding the context of local knowledge and the mixture of epistemologies.

There is more than one possible way to organize environmental knowledge, and the diversity of systems of knowledge and environmental world views deserves a re-examination. Cultural diversity may be related to biodiversity (Gadgil and Berkes, 1991), and both may be important for improving the sustainability of the world's ecological systems, as well as for their own sake. There is potential to involve local knowledge for the improvement of resource management in environments such as northern coastal ecosystems, arid lands, mountain ecosystems, tropical forests, and subarctic ecosystems. Traditional systems in these areas represent many millennia of human experience with environmental management, and

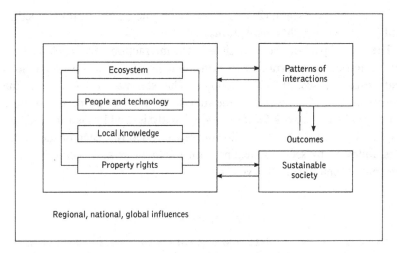

Figure 1.1.   A framework for analysing the link between social and ecological
systems for resilience and sustainability.

provide a reservoir of active adaptations which may be of universal importance in designing for sustainability (Gadgil, Berkes and Folke, 1993;
Berkes *et al.*, 1995).

The analytical framework

'Conceptual frameworks are neither models nor theories', Rapoport (1985:
256) wrote. 'Models describe how things work, whereas theories explain
phenomena. Conceptual frameworks do neither; rather they help to think
about phenomena, to order material, revealing patterns – and pattern
recognition typically leads to models and theories' (Rapoport, 1985).

The research questions posed by the present study explicitly link
ecology, economics and social science. They require an interdisciplinary,
international, case-study approach. To help keep focus, provide direction, and assist in the synthesis, a common framework is needed for the
case studies. The framework is designed to help identify the relevant
characteristics of the ecosystem, people and technology, local knowledge, and property rights institutions that characterize the case study.
The framework in Figure 1.1 is only one way to represent embedded relationships diagrammatically. It is a heuristic device and not meant to
imply that the various elements in the overall system are discrete boxes.
The framework in Figure 1.1 distinguishes four sets of elements which
can be used to describe social and ecological system characteristics and

linkages: (1) ecosystem, (2) people and technology, (3) local knowledge, and (4) property rights institutions.

The crucial part of the analysis is in the interactions (feedbacks) of the four components. The framework is meant to help focus on key interactions that result in sustainable outcomes. The schema borrows from the Oakerson (1992) framework for the analysis of common property resources and from the framework for institutional analysis used by Ostrom and colleagues (Ostrom, 1990). The following sections describe each of the attributes, followed by sections dealing with interactions and outcomes, with reference to chapters in this volume.

The ecosystem

Ecosystems may be characterized in a variety of ways, focusing on structure or function or both. In terms of physical attributes, for example, subarctic ecosystems can be characterized by highly variable temperatures and short growing seasons. In terms of biological characteristics, biodiversity and rates of nutrient cycling are low; stored energy (and biomass) in forests and lakes is high relative to rates of biological productivity (see Chapter 5). Not all characteristics of ecosystems are equally significant. Recent research has made it possible to suggest that the diversity and complexity of ecosystems can be traced to a relatively small number of biotic and abiotic variables and physical processes (Holling, 1992; Holling *et al.,* 1995). It seems that a relatively few species, or groups of species, run these processes, thereby contributing to the functional performance of the ecosystem. Remaining organisms occupy niches in the system shaped by these processes. These organisms may seem to be redundant in the short term, but they are crucial in maintaining system resilience, and serve as a system insurance for unpredictable events (Walker, 1992; Schulze and Mooney, 1993; Solbrig, 1993; Barbier, Burgess and Folke, 1994; Folke, Holling and Perrings, 1996) Hence, biodiversity is important, and biodiversity-conserving local management practices receive particular attention in this volume (e.g. Chapters 2 and 9). Also singled out for special attention is the inherent unpredictability and non-linear nature of ecosystems (Chapters 13 and 15).

People and technology

The level of analysis is not the individual or the household but the social group, which could be a small community (e.g. Chapter 5), a district

(Chapter 4), a tribal group (Chapter 7), or a regional population (Chapters 10 and 11). The description of the social system starts with the people organized as user communities, and the technology employed by them. Even within the smaller, geographically bounded case study areas, there will be considerable complexity in the user communities, the resources they pursue and the technology they use, for example, the smaller-scale inshore and the larger-scale offshore fisheries of Iceland (Chapter 3). The type of technology available to potential users for exploiting a resource will be important; for example, the stationary gear and small boats used by the small-scale fishers of Newfoundland limit their areas of use, whereas the trawlers of the offshore fleet are by necessity more mobile and exploit a larger area (Chapter 12), thereby impacting on resources and ecosystems in different ways.

The use or choice of technology may also provide clues to distinguish user communities and perhaps also the sustainability of their practices. Examples include the limitation of the clam harvest by the nature of the technology used (Chapter 8), the different users of the Brazilian Amazon Forest (Chapter 6), Himalayan villagers at different levels of market integration (Chapter 11), and the various groups of herders of the African Sahel (Chapter 10). Chapter 15 builds on an argument for the adaptive match of technology, social organization, and resource management practice (through qualitative, behavioural controls) in traditional fishing societies.

Local knowledge

Any resource user will have a certain amount of local environmental knowledge that will allow him/her to carry out a particular activity. This local knowledge may be very substantial, especially if it includes culturally transmitted knowledge accumulated over generations. Many indigenous groups as well as other historically continuous communities, such as some North Atlantic fishers, will possess traditional knowledge (Chapter 3). In some cases, local knowledge may be organized and used in a way which, in effect, amounts to a traditional management system. Such is the case with certain shifting cultivators (Chapter 9), Amerindian hunter–trappers (Chapter 5), and tribal groups in Asia which maintain sacred groves (Chapter 2). Especially important for the present volume is knowledge related to the maintenance of ecosystem resilience, as in traditional agricultural and aquacultural systems that use a multiplicity of crop varieties and species (Chapters 9 and 11), as opposed to monocultures that tend to predominate industrial agricultural systems.

It is important to note that this volume is not dealing with isolated groups and anthropological curiosities. Many of the chapters are about groups that operate in the context of conventional resource management science, and in integrated economic systems. This is the case with the Maine clammers in Chapter 8, forest users in Chapter 14, and North Atlantic fishers in Chapters 3 and 12. Other chapters deal with non-Western societies, but each of these is also embedded in a market economy. The sacred forests in Chapter 2 became protected areas in the modern conservation idiom. The knowledge systems of the Ara (Chapter 7) and the James Bay Cree (Chapter 5), for example, bear the marks of mixing with Western knowledge systems prior to the twentieth century.

Property rights institutions

Useful management lessons come from societies that have survived resource scarcities. These are societies that adapt to changes and learn to interpret signals from the resource stock through a dynamic social–ecological process, thus developing flexible institutions to deal with resource management crises. Chapters in this volume show that these institutions involve a range of property rights regimes. But Western resource management science often assumes a very limited set of property rights: state property (regime based on government regulation), private property (market-based regime), or else a 'tragedy of the commons'. However, no fewer than nine of the chapters in this volume (Chapters 2, 4–7, 9, 10, 14, 15) show that the real world also contains many working examples of common-property (or communal property) systems in which an identifiable group of users holds the rights and responsibilities for the use of a resource. This is not to say that pure common-property systems are the norm or that they ought to replace state-property and private-property regimes. The nine chapters mentioned above in fact encompass a great diversity of institutional arrangements, many in combination with state-property and private-property regimes, and include cases in which government intervention has caused the disruption of local institutions (Chapter 4).

Of the remaining chapters, several discuss explicit combinations of local property rights and government management regimes. These include the Iceland case (Chapter 3) with the 'trawling rally', a weak co-management system involving minimal local participation in management, and the Maine case (Chapter 8), a strong co-management system with a hierarchy of institutions involving local and state management. Chapter 6 explores why some groups are more successful than others in making themselves heard by policy makers.

Several chapters deal with the failure of management and the redesign of institutions. These include Chapter 5, which analyses episodes of resource depletion in subarctic Canada, Chapter 14 which deals with re-emergent neo-traditional institutions, and Chapter 12 which describes the recent spectacular crash of a major resource (the Newfoundland cod) which can potentially trigger institutional redesign. Three of the chapters (Chapters 9, 10 and 11) deal with meso-scale or regional disruptions that create complex problems in resource management.

Patterns of interaction

Patterns of interaction address the question of dynamic change and are the key to the analysis of case studies. The essential questions to be addressed are those that arise from the objectives of the volume: how the local social system has developed management practices based on ecological knowledge for dealing with the dynamics of the ecosystem(s) in which it is located; and the social mechanisms behind these management practices.

The analysis of patterns of interactions requires an evolutionary focus, as both social systems and natural systems have an evolutionary character (Chapter 13) There are, however, no ready guidelines for the study of interactions, and each case is different. Key factors driving patterns of interaction may be externally imposed perturbations in one case, population growth in another, and market factors in yet another. We distinguish between local-scale disturbances, meso-scale (or regional and national-scale) disturbances, and global-scale disturbances. The box enclosing Figure 1.1 is meant to depict the embeddedness of local systems in meso-scale and global-scale influences.

The analysis of interactions requires a focus on feedback mechanisms. Some societies adapt to changing conditions better than others, and there is very little agreement on what accounts for such variability (Kuran, 1988). The key factor in successful adaptation may be the presence of appropriate feedback mechanisms which enable consequences of earlier decisions to influence the next set of decisions which make adaptation possible. Conversely, factors that obliterate feedbacks may result in loss of cultural adaptations. For example, the development of market economy in a previously isolated area may 'free' local people from traditional ecological constraints, triggering a change in agricultural practices in favour of cash crops (Chapter 11) and monocultures (Chapter 9), or a change in the kinds of resources used in a tropical rain forest (Chapter 6). These changes may be fine if such transitions improve the well-being of the combined social, economic, and ecological system. However, the result is often the opposite because of loss of cultural adaptations.

The analysis also requires a focus on social–ecological linkages. For example, in the Delacarlia region of Sweden, the land use pattern was conditioned by the rules of partial inheritance (Chapter 4), just as the shifting cultivation system in tropical Mexico was conditioned by *milpa* as a social system (Chapter 9). The agroforestry system in Ara, Nigeria, shapes the land tenure system of the Yoruba, and in turn is shaped by it (Chapter 7). The fishers of Newfoundland cod were responding to market opportunities and government incentives; the overfishing that resulted affected the viability of fishing companies and communities (Chapter 12).

Outcomes

Patterns of interaction produce certain outcomes. The biophysical environment may or may not be used sustainably; the functional performance of the ecosystem may or may not be damaged; and benefits may or may not be shared equitably or fairly. The question of performance of natural resource systems begs the question of evaluative criteria. Oakerson (1992) suggested two criteria, efficiency (defined as Pareto Optimality) and equity. Other criteria include empowerment and livelihood security, as suggested for example by some development professionals (Pomeroy, 1994).

In seeking a criterion which is both human-centric and resource-centric, and not exclusively one or the other, Feeny *et al.* (1990) suggested sustainability (*sensu* WCED, 1987), and this is the criterion used in this volume. However, there are operational problems with this concept (Costanza and Patten, 1995). Whereas the criteria for ecological sustainability are relatively well known, there are no agreed-upon criteria for economic and social/cultural sustainability. In this study, our working assumption will be that social–ecological systems which have survived over extended periods of time are sustainable. This assumption is consistent with Ostrom (1990), and will facilitate the search for *mechanisms* for the resilience of the integrated social–ecological system.

The framework can be used as a guide for identifying social practices based on local ecological knowledge, and the social mechanisms behind these practices. It can assist in identifying similarities, general patterns and principles that can be drawn from the case studies, and the lessons that can be learned to assist in the designing of more sustainable resource management systems. There are additional possibilities raised by the analysis in this volume towards a long-term research programme.

The framework can be used as a policy guide for designing more sustainable systems. For example, depending on the outcome, the interaction of ecological and social systems may be modified. One mechanism by which such a modification may come about is co-evolution (Norgaard, 1994). There is some evidence in Chapters 5, 7 and 14 that society and the natural environment mutually modified one another over a period of time, and that local knowledge systems and property rights institutions became attuned to the resources used. How can adaptiveness and resilience be built into institutions so that they are capable of responding to the processes that contribute to the resilience of ecosystems? *We hypothesize that maintaining resilience may be important for both resources and social institutions – that the well-being of social and ecological systems is thus closely linked.*

The interaction of social and ecological systems may be modified through adaptations for maintaining resilience. Chapters 2 and 15, among others, make this argument. The resource management systems of special interest are those that allow less intensive use and greater biological diversity, and thus help maintain resilience. These are systems in which ecosystem processes (as well as populations of target species) may be maintained and environmental feedbacks managed for sustainability. It is possible that traditional and neo-traditional knowledge and resource management systems may escape some of the limitations of conventional scientific resource management with its assumptions of controllable nature, predictable yields, and exclusion of environmental perturbations. *We hypothesize that successful knowledge and resource management systems will allow disturbances to enter on a scale which does not disrupt the structure and functional performance of the ecosystem and the services it provides.*

Such resource management systems have to be able to recognize the feedbacks that signal these disturbances. Thus, it would seem that they require *mechanisms* by which information from the environment may be received, processed and interpreted. If the resource management institution is a government agency, the mechanism in question will involve scientific/managerial bodies. However, if the resource management institution is a local or traditional body, then we expect to find tangible evidence of *social* mechanisms behind social–ecological practices that deal with disturbances and maintain system resilience. Thus, *we hypothesize that there will be social mechanisms behind management practices based on local ecological knowledge, as evidence of a co-evolutionary relationship between local institutions and the ecosystem in which they are located.*

References

Alcorn, J.B. 1984. Development policy, forest and peasant farms: reflections on Huastec-managed forests' contributions to commercial production and resource conservation, *Economic Botany 38*: 389–406.

Anderson, E.N. 1996. *Ecologies of the Heart.* New York and Oxford: Oxford University Press.

Baland, I.-M. and Platteau, J.-P. 1996. *Halting Degradation of Natural Resources: Is There a Role for Rural Communities?* Oxford: FAO/Clarendon Press, Oxford.

Banuri, T. and Apffel Marglin, F., eds. 1993. *Who Will Save the Forests?* London: The United Nations University/Zed Books.

Barbier, E.B., Burgess, J. and Folke, C. 1994. *Paradise Lost? The Ecological Economics of Biodiversity.* London: Earthscan.

Berkes, F. 1986. Local-level management and the commons problem: A comparative study of Turkish coastal fisheries, *Marine Policy 10*: 215–29.

Berkes, F., ed. 1989. *Common Property Resources. Ecology and Community-Based Sustainable Development.* London: Belhaven.

Berkes, F. and Folke, C. 1994. Investing in cultural capital for the sustainable use of natural capital. In *Investing in Natural Capital: The Ecological Economics Approach to Sustainability*, pp. 128–49, ed. A.M. Jansson, M. Hammer, C. Folke, and R. Costanza. Washington DC: Island Press.

Berkes, F., Folke, C. and Gadgil, M. 1995. Traditional ecological knowledge, biodiversity, resilience and sustainability. In *Biodiversity Conservation: Policy Issues and Options*, pp. 281–99, ed. C. Perrings, K.-G. Mäler, C. Folke, C.S. Holling, and B.-O. Jansson, Dordrecht: Kluwer Academic Publishers.

Bromley, D.W. 1991. *Environment and Economy: Property Rights and Public Policy.* Oxford: Basil Blackwell.

Bromley, D.W. 1992. The commons, property, and common-property regimes. In: *Making the Commons Work: Theory, Practice, and Policy*, pp. 3–15, ed. D.W. Bromley. San Francisco: Institute for Contemporary Studies.

Clark, W.C. and Munn, R.E., eds. 1986. *Sustainable Development of the Biosphere.* Cambridge: Cambridge University Press.

Coleman, J.S. 1990. *Foundations of Social Theory.* Cambridge, Mass.: Harvard University Press.

Costanza, R. and Patten, B.C. 1995. Defining and predicting sustainability. *Ecological Economics 15*: 193–6.

Costa-Pierce, B.A. 1987. Aquaculture in ancient Hawaii. *BioScience 37*: 320–31.

Daly, H.E. and Cobb, J.B. 1989. *For the Common Good: Redirecting the Economy toward Community, the Environment, and a Sustainable Future*, Boston: Beacon Press.

Dove, M.R. 1993. A revisionist view of tropical deforestation and development, *Environmental Conservation 20*: 17–24.

Farvar, M.T. and Milton, J.P., eds. 1972. *Careless Technology: Ecology and International Development.* Garden City, New York: Natural History Press.

Feeny, D., Berkes, F., McCay, B.J. and Acheson, J.M. 1990. The tragedy of the commons: Twenty-two years later. *Human Ecology 18*: 1–19.

Feyerabend, P. 1987. *Farewell to Reason*. London: Verso.

Folke, C., Holling, C.S. and Perrings, C. 1996. Ecosystems, biological diversity and the human scale, *Ecological Applications* 6: 1018–24.

Freeman, M.M.R., Matsuda,Y. and Ruddle, K., eds. 1991. *Adaptive Marine Resource Management Systems in the Pacific,* Special Issue of *Resource Management and Optimization, 8,* No. 3/4.

Funtowicz, S.O. and Ravetz, J.R. 1990. *Uncertainty and Quality in Science for Policy*. Dordrecht: Kluwer Academic Publishers.

Gadgil, M. and Berkes, F. 1991. Traditional resource management systems. *Resource Management and Optimization* 8: 127–41.

Gadgil, M., Berkes, F. and Folke, C. 1993. Indigenous knowledge for biodiversity conservation. *Ambio 22*: 151–6.

Gunderson, L., Holling, C.S. and Light, S., eds. 1995. *Barriers and Bridges to the Renewal of Ecosystems and Institutions*. New York: Columbia University Press.

Hanna, S., Folke, C. and Mäler, K.-G., eds. 1996. *Rights to Nature*. Washington, DC: Island Press.

Hanna, S. and Munasinghe, M., eds. 1995. *Property Rights and the Environment. Social and Ecological Issues.* Washington DC: Beijer International Institute of Ecological Economics and the World Bank.

Hardin, G. 1968. The tragedy of the commons. *Science 162*: 1243–8.

Hilborn, R. and Ludwig, D. 1993. The limits of applied ecological research. *Ecological Applications 3: 550–2.*

Holling, C.S. 1978. *Adaptive Environmental Assessment and Management.* London: Wiley.

Holling, C.S. 1986. The resilience of terrestrial ecosystems: local surprise and global change. In *Sustainable Development of the Biosphere,* pp.292–317, ed. W.C. Clark and R.E. Munn. Cambridge: Cambridge University Press.

Holling, C.S. 1992. Cross-scale morphology, geometry and dynamics of ecosystems, *Ecological Monographs* 62: 447–502.

Holling, C.S., Schindler, D.W., Walker, B.W. and Roughgarden, J. 1995. Biodiversity in the functioning of ecosystems: An ecological synthesis. In *Biodiversity Loss: Economic and Ecological Issues,* pp.44–83, ed. C. Perrings, K.-G. Mäler, C. Folke, C.S. Holling, and B.-O. Jansson, Cambridge: Cambridge University Press.

Jansson, A.M., Hammer, M., Folke, C. and Costanza, R., eds. 1994. *Investing in Natural Capital: The Ecological Economics Approach to Sustainability.* Washington DC: ISEE/Island Press.

Jentoft, S. and McCay, B.J. 1995. User participation in fisheries management. Lessons drawn from international experiences, *Marine Policy 19*: 227–46.

Jodha, N.S. 1986. Common property resources and rural poor in dry regions of India, *Economic and Political Weekly 21*: 1169–81.

Johannes, R.E. 1978. Traditional marine conservation methods in Oceania and their demise, *Annual Review of Ecology and Systematics 9*: 349–64.

Johannes, R.E. 1981. *Words of the Lagoon. Fishing and Marine Lore in the Paulau District of Micronesia.* Berkeley: University of California Press.

Kuran, T. 1988. The tenacious past: theories of personal and collective conservation. *Journal of Economic Behaviour and Organization 10*: 143–71.

Larkin, P.A. 1977. An epitaph for the concept of maximum sustained yield. *Transactions of the American Fisheries Society 106*: 1–11.

Lee, K.N. 1993. *Compass and Gyroscope: Integrating Science and Politics for the Environment.* Washington DC: Island Press.

Levi-Strauss, C. 1962. *La Pensee Sauvage.* Paris: Librarie Plon.

Likens, G.E. 1992. *The Ecosystem Approach: Its Use and Abuse.* Oldendorf/Luhe, Germany: Excellence in Ecology, Ecology Institute.

Ludwig, D., Hilborn, R. and Walters, C. 1993. Uncertainty, resource exploitation, and conservation: lessons from history. *Science 260*: 17, 36.

McCay, B.J. 1995. Common and private concerns. *Advances in Human Ecology 4*: 89–116.

McCay, B.J. and Acheson, J.M. eds. 1987. *The Question of the Commons. The Culture and Ecology of Communal Resources.* Tucson: University of Arizona Press.

McNeely, J.A. 1991. Common property resource management or government ownership: Improving the conservation of biological resources, *International Relations 1991*: 211–25.

Messerschmidt, D.A., ed. 1993. *Common Resource Management. Annotated Bibliography of Asia, Africa and Latin America.* Community Forestry Note 11. Rome: FAO.

Netting, R. 1981. *Balancing on an Alp.* Cambridge: Cambridge University Press.

Niamir, M. 1990. *Herder's Decision-making in Natural Resources Management in Arid and Semi-arid Africa.* Community Forestry Note No. 4. Rome: FAO.

Norgaard, R.B. 1994. *Development Betrayed: The End of Progress and a Coevolutionary Revisioning of the Future.* Routledge, New York.

North, D.C., 1993. *Economic Performance through Time.* Les Prix Nobel. The Nobel Prizes 1993. Stockholm: The Nobel Foundation and the Royal Swedish Academy of Sciences.

Oakerson, R.J. 1992. Analyzing the commons: a framework. In *Making the Commons Work: Theory, Practice, and Policy*, pp. 41–59, ed. D.W. Bromley. San Francisco: Institute for Contemporary Studies.

Odum, E.P. 1989. *Ecology and Our Endangered Life-Support Systems.* Sunderland, Mass.: Sinuaer.

Ostrom, E. 1990. *Governing the Commons: The Evolution of Institutions for Collective Action.* Cambridge: Cambridge University Press.

Ostrom, E. 1992. *Crafting Institutions for Self-Governing Irrigation Systems.* San Francisco: Institute for Contemporary Studies Press.

Pálsson, G. 1991. *Coastal Economies, Cultural Accounts. Human Ecology and Icelandic Discourse.* Manchester and New York: Manchester University Press.

Park, R.E. 1936. Human ecology. *American Journal of Sociology 42*: 1–15.

Pinkerton, E., ed. 1989. *Co-operative Management of Local Fisheries: New Directions for improved Management and Community Development.* Vancouver: University of British Columbia Press.

Pomeroy, L.R. and Alberts, J.J. 1988. *Concepts of Ecosystem Ecology.* New York: Springer-Verlag.

Pomeroy, R.S., ed. 1994. *Community Management and Common Property of Coastal Fisheries in Asia and the Pacific.* Manila: ICLARM.

Ramakrishnan, P.S. 1992. *Shifting Agriculture and Sustainable Development. An Interdisciplinary Study from North-Eastern India.* Paris: Unesco/Parthenon.

Rappaport, R.A. 1971. The sacred in human evolution, *Annual Review of Ecology and Systematics 2*: 23–44.

Rapoport, A. 1985. Thinking about home environments: A conceptual framework. In *Home Environments,* pp.255–86, ed. I. Altman and C.M.Werner. New York: Plenum Press.

Redclift. M. 1994. Reflections on the 'sustainable development' debate. *International Journal of Sustainable Development and World Ecology 1*: 3–21.

Ruddle, K., Hviding, E. and Johannes, R.E. 1992. Marine resources management in the context of customary marine tenure, *Marine Resource Economics 7*: 249–73.

Ruddle, K. and Johannes, R.E., eds. 1990. *Traditional Marine Resource Management in the Pacific Basin: An Anthology.* Jakarta: Unesco/ROSTSEA.

Schlager, E. and Ostrom, E. 1992. Property-rights regimes and natural resources: a conceptual analysis. *Land Economics 68*: 249–62.

Schulze, E.-D. and Mooney, H.A. eds. 1993. *Biodiversity and Ecosystem Function.* Heidelberg: Springer-Verlag.

Smith, A.H. and Berkes, F. 1993. Community-based use of mangrove resources in St. Lucia, *International Journal of Environmental Studies 43*: 123–32.

Solbrig, O.T. 1993. Plant traits and adaptive strategies: Their role in ecosystem function. In *Biodiversity and Eco-system Function,* pp.97–116, eds. E.-D. Schulze and H.A. Mooney. Heidelberg: Springer-Verlag.

Walker, B.H. 1992. Biodiversity and ecological redundancy. *Conservation Biology 6*: 18–23.

Walters, C.J. 1986. *Adaptive Management of Renwable Resources.* New York: McGraw-Hill.

Warren, D.M., Slikkerveer, L.J. and Brokensha, D., eds. 1995. *The Cultural Dimension of Development. Indigenous Knowledge Systems.* London: Intermediate Technology Publications.

WCED 1987. *Our Common Future. The Report of the World Commission on Environment and Development.* Oxford: Oxford University Press.

Part I

Learning from locally devised systems

Introduction

Until recently, few people thought that resource management based on local practices had any lasting contributions to make. But this view has been changing, and the ecological sense of traditional practices has now become a topic fit for scholars to investigate. The three chapters in this section deal with three very different areas, resource types and social systems. How are these systems linked with ecological systems? Chapter 2, by Gadgil and colleagues, offers a model of ecological prudence that can develop by trial and error in a traditional society, and a discussion with reference to locally developed ecological refugia. Chapter 3, by Pálsson, concentrates on the importance of the practical knowledge of Icelandic fishermen, and compares the experts' discourse of fishing to that of the fishers. Chapter 4, by Sporrong, provides a historical geographic assessment of a land-use system that survived in central Sweden for several centuries in an apparently sustainable manner, both socially and ecologically.

Chapter 2 starts with a discussion of traditional sacred groves in northeastern India. Social–institutional change is explored through a model of trial-and-error innovation followed by diffusion through imitation. The starting point of the authors' argument is that certain small-scale societies are intimately dependent on, and in control of, biological resources of their immediate environments. Such societies often exhibit many practices of restrained resource use. These practices, often rooted in religious beliefs and social conventions, are enforced through systems of community-based social sanctions. Many such practices promote sustainable use of biological resources, although community members rarely attribute such a function to them. On the basis of the model, the authors suggest that it is nevertheless likely that practices such as the total protection of the resource population in parts of its range set aside as refugia may have arisen through a trial-and-error process based on simple 'rules of thumb'. Practices that turned out to be advantageous may then have spread through imitation by other communities. Pre-British India harboured an extensive network of such refugia in the form of sacred groves and ponds. Such practices, albeit greatly reduced, still persist in many areas. The authors present a case study of revival of protection to forest refugia on the basis of explicit recognition of their value in providing a variety of ecosystem services by communities of shifting cultivators in northeastern India. The case study of sacred groves turned into safety forest illustrates the idea that some traditional resource use practices may have been designed to fulfill conservation functions. The continuation of these ecologically desirable practices would have to be grounded in a system of state-sponsored positive incentives for local communities.

Chapter 3 is an analysis of local and practical knowledge held by Icelandic fishers, as opposed to what Pálsson considers 'modernist' scientific fishery management. The author argues that there is increasing evidence and a growing body of theoretical scholarship to suggest grounds for questioning some of the assumptions of 'modernist' resource management, including its restrictive emphasis on disembedded knowledge, linearity and control. The chapter discusses practical knowledge in Icelandic fishing and its relevance for fisheries management, emphasizing the theory of practice and the notions of situated action and mutual engagement. Pálsson argues that while current attempts to collect and store practical knowledge are seriously misguided, it is essential, given the uncertainty of many marine ecosystems, to pay attention to how practical knowledge is acquired and what that process entails. Also, it is important to explore the ways in which it could be brought more systematically into the process of resource management. As an illustration of the kind of joint management that may be possible, the chapter examines one collaborative attempt involving both fishers and professional marine biologists, the so-called 'trawling rally'. A kind of 'index fishing' for monitoring of the population, it involves repeated trawling under controlled conditions, along predetermined paths. The trawling rally can also be considered an institution for social learning that collectively engages fishers and managers towards an objective of mutual interest. The chapter concludes that in trying to link social and ecological systems, it is important to look for institutional frameworks which allow for both democratic participation and the development of mutual trust among users and scientists.

Chapter 4 is about the peculiar fragmented land use system in the Dalecarlia district in central Sweden, and the struggle between the local people who wanted to maintain it and the central government which wanted to institute land reform. The land-use system allowed for the diversified use of resources in this forest-and-mountain district. Individual families owned parcels of land spread over large areas, and every family used a variety of ecological zones in the landscape. Was the system sustainable? The author admits that there are few reliable historical ecological data from the district. But on the basis of civil servants' reports of high crop yields and cattle productivity in the eighteenth century, he concludes that land use must have been ecologically sustainable. Socially, the most striking feature of the district was the lack of an upper and a lower class, unlike the rest of Sweden at the time. There were no landless people. Land management involved a mix of individual rights and communal control. Village elders exercised leadership, but everyone took part in all socio-economic decisions. The author identifies 'partial inheritance' and a high degree of flexibility as the mechanisms that drove the system, and argues that the system must have conferred a degree of social and ecological resilience, pointing out that the mean size of farms was constant during a time of great expansion of population and arable land, 1734–1819. Enclosure was introduced fairly late in the district. Few asked for land reform, and for the social – ecological system of Dalecarlia, enclosure was a catastrophe. Starting in the early nineteenth century, ever decreasing farm sizes led to poverty and emigration from the area. In recent years, the author points out, partial inheritance and the holding of scattered parcels of land are enjoying a revival.

2

People, refugia and resilience

MADHAV GADGIL, NATABAR SHYAM HEMAM & B. MOHAN REDDY

Introduction

Thirty years ago, Slobodkin (1968) asked if there were any examples of prudent predators in the world of animals. By prudent he meant predators which exhibit restraints on harvests such that long-term yields from the prey populations are enhanced at the cost of some immediate harvests. His answer was in the negative. Animals always seem to behave as optimal foragers, concentrating at any time on prey that maximizes the energy or nutrient returns per unit time, or minimizes the risks incurred. If animals leave some species, or age class, or patches of prey alone, it is only because they have better options. Humans, too, behave as optimal foragers much of the time (Borgerhoff-Mulder, 1988). In the Torres Strait, fishing may be stopped in localities where fish yields are observed to have declined; in parts of New Guinea, the hunting of birds of paradise may be temporarily abandoned if their population declines (Eaton, 1985; Nietschmann, 1985). These responses may merely indicate that the returns from that prey species or those localities are lower than the returns possible from alternative species or localities. But in other cases humans seem to refrain from harvesting resources which might provide higher returns than the alternatives exploited. Thus residents of the village Kokre Bellur near Bangalore in the south of India strictly protect painted storks and grey pelicans breeding in the midst of their village, although the same birds may be hunted outside the breeding season. Obviously the nesting birds are far easier prey, but they are nevertheless left alone, probably enhancing the long term availability of the prey population. Modern resource management practices, too, include examples of deliberate restraints on resource harvests, whether these be mesh-size regulations, closed seasons or protection to endangered species (Gadgil and Berkes, 1991).

Modern resource management practices are based on explicitly stated rationale; they are implemented on the basis of written prescriptions, rules and regulations. It is, of course, possible that the motivations underlying particular practices may be different from the stated ones, and that the prescriptions and rules may not be adhered to in practice. But such practices are accompanied by explanations cast in an idiom acceptable to modern science-based societies. Traditional resource management practices, on the other hand, are not supported by such explanations. The concerned community members may accept certain restraints on the use of biological resources so as not to offend deities, or because their violation would attract social sanctions. It is unlikely that they would state that a given practice is agreed to by all members of the community because it fulfills some secular purpose, such as the provision of an ecosystem service. It has therefore often been argued that the conservation consequences of traditional resource use practices are totally unintended, merely incidental consequences (Diamond, 1993). It is, of course, possible that this is so in certain cases. But it is also plausible that on occasions, traditional societies might have arrived at practices that promoted sustainable resource use through a trial-and-error process based on some simple rules of thumb. If sustainability of resource use conferred an advantage on the concerned community, other communities might also acquire the same practices through imitation. Such practices of ecological prudence may then spread as a part of a system of religious beliefs or social conventions without their secular function being explicitly recognized.

We consider such a possibility in two different contexts. Firstly, with the help of a simple model, we demonstrate that practices leading to sustainable use may be arrived at through a trial-and-error process based on very simple rules of thumb. Secondly, we present a case study of a resource conservation practice (sacred groves) grounded in religious beliefs, that was abandoned and then revived essentially in the original form when the community realized its value in the provision of ecosystem services. This natural experiment suggests that pre-scientific societies can and do adopt conservation practices on the basis of their experience; practices that in the past were implemented through the medium of religious beliefs.

Negative feedbacks

Our model is based on the postulate that many kin based, small scale societies, intimately dependent on the natural resources of their own restricted resource catchments, would be sensitive to signs of depletion of their resource base. As a group, they would be aware of levels of harvesting pressure exerted by group members, as well as levels of resource harvests realized by them. The simplest, most general rule of thumb that such societies may employ to adjust harvesting effort would be: reduce the pressure of harvest on a resource if there are signs of difficulties, permit the harvesting pressures to increase if there are no signs of problems. There are many ways in which the harvesting pressure may be regulated. It may involve a uniform lowering of harvesting pressure at all times and over the entire range. It may take the form of a closed season during which all harvests are suspended, or immunity from harvests to certain life-history stages such as pregnant females. We focus here on one particular form of restraint, namely complete or near-complete immunity from harvests of all species in specific localities. Such localities then serve as refugia for all biological populations. Sacred groves, sacred ponds, sacred lagoons on coral islands are examples of such refugia established by societies outside the ambit of modern Western resource management systems; core areas of national parks are examples of refugia set up under the latter regime.

Harvesting entire resource populations

We summarize here the results of the simplest of hierarchy of models explored by Joshi and Gadgil (1991); the original paper discusses others which lead to similar conclusions. Consider a homogeneous social group in full control of a resource population, attempting a trial-and error-process to arrive at some desirable level of harvesting effort. This desirable level is presumed to be equivalent to the maximum sustainable yield (MSY) as defined in the modern fishery management theory (Clark, 1990). The dynamics of the resource population are specified in terms of a discrete-time deterministic model without any age or sex structure. Each time step consists of a growth phase followed by a harvesting phase. We assume logistic growth, i.e.

$$B_t' = rB_t \cdot (1 - B_t)$$

where B is the resource biomass, B' the biomass resulting from logistic growth, r the intrinsic rate of growth, and the subscript t denotes the time step. This growth is assumed to be followed by a harvesting phase, where harvesting effort E_t results in the harvest H_t. The effort – harvest relationship is assumed to fetch diminishing returns, such that:

$$H_t = B_t'[1 - \exp(-K \cdot E_t)]$$

In such a case, under constant E the resource population would reach an equilibrium value

$$1-\frac{1}{r\cdot\exp(-E)}$$

so long as $E<lnr$. When $E>lnr$ the resource population would be eliminated. When $\hat{E}=ln[(1+r)/2]$, the harvest attains a maximum value of:

$$\hat{H}=\frac{(r-1)^2}{4r}$$

Our interest then is in exploring the nature of the harvesting strategy that would allow a community to zero in on $\hat{E}$, given that it possesses information on harvesting effort put in and harvests realized at all times. The basic decision to be made at every point in time is whether to decrease, maintain at the same level, or enhance the harvesting effort from one time interval to the next. The simplest procedure involving a negative feedback would be to operate on the following set of rules:

1. If an increase in effort has led to an increase in harvest, then increase the effort further.
2. If an increase in effort has led to a decrease in harvest, then decrease the effort.
3. If a decrease in effort has led to an increase in harvest, then decrease the effort further.
4. If a decrease in effort has led to a decrease in harvest, then increase the effort.

In a discrete time model, if $D(t)$ is the change in effort so that $E(t+1)=E(t)+D(t)$, then the prescription would be:

$$D(t)=X\cdot D(t-1) \text{ if } H(t)>H(t-1)$$

and

$$D(t)=-X\cdot D(t-1) \text{ if } H(t)<H(t-1) \tag{1}$$

where X is some positive scale factor. Here we confine to the simple case of X being a constant.

It turns out that such a decision rule always leads to an unchecked increase in harvesting effort till the resource is wiped out. This is because when harvesting effort can change at each step, the harvest obtained by enhancing the effort is always higher, even when the harvesting effort exceeds the optimal level, $\hat{E}$ (Figure 2.1). This results from the fact that a decrease in harvesting effort exceeding $\hat{E}$ never leads to an immediate

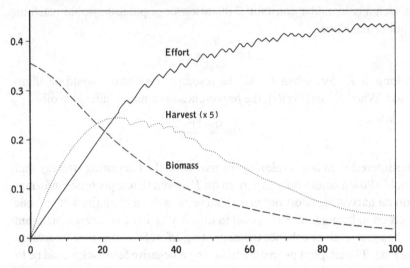

Figure 2.1. Variation of harvesting effort, biomass level, and the harvest obtained as a function of time when the harvesting effort is changed at every time step of a fixed increment or decrement of 0.01. (r=1.5.)

enhancement of the harvest, but might do so only if the harvesting effort is held steady for long periods of time. We do not expect societies to organize a complicated trial-and-error procedure involving very long waits holding the effort constant, followed by comparisons of the harvests and efforts at the end of such a period with those at the end of one or more previous periods of steady effort. It is then to be expected that societies following the simple procedure sketched above would inevitably overharvest and often exterminate the resource, unless higher levels of harvests were not desired for some other reason.

Harvesting with refugium

This model may be extended to include a refugium. We now assume that regardless of the total biomass, a fraction Z is always immune from harvesting. Only the remaining fraction $(1-Z)$ lying outside the refugium patches is exploited. Then the harvest H_t is given by:

$$H_t = (1-Z)B_t'[1-\exp(-E_t)]$$

and the biomass remaining after exploitation equals:

$$(1-Z)B_t'\exp(-E_t)+ZB_t'$$

We assume the refugia to be numerous, small and well dispersed so that there is complete mixing between the non-harvested population in the refugia and the harvested population outside, during every phase of biomass growth. It can be shown that when the refugium is sufficiently large, i.e. $Z>1/r$, the resource population will persist however high the harvesting effort. Furthermore, the maximum sustained harvest in the presence of a refugium still equals the value of:

$$\hat{H}=\frac{(r-1)^2}{4r}$$

in its absence. This value may be realized for a whole series of combinations of Z and E which satisfy the relation:

$$Z=\frac{\dfrac{2}{r-1}-\exp(-E)}{1-\exp(-E)}$$

We may then enquire whether there are reasonable harvesting strategies that would enable a community to zero in on such a combination of Z and E that would result in maximal sustainable yields. Since the feedback strategy should help reduce the risk of resource extermination, the procedure should involve an increase in the refugium size whenever the harvesting effort is stepped up. The decision rule (1) above may then be amplified by adding to a change in harvesting effort, a change in refugium size as:

$$Z=p \cdot D \tag{2}$$

where p is a positive constant. Then if E_0 and Z_0 are the initial harvesting effort and refugium size, they would at any time be related by:

$$Z_t=Z_0+p(E_t-E_0)$$

given that $0 \leq Z \leq 1$

The question then is: does the amplified decision rule (2) reduce the risk of resource decimation, and does it render it more likely that the harvest stabilizes close to the maximal sustainable yield? It turns out that this procedure does, indeed, reduce the risk of harvesting effort drifting to higher and higher values. This is because higher values of harvesting effort are no longer necessarily rewarded by non-sustainable higher harvests, thanks to the protection to the resource capital afforded by the accompanying increases in refugium size. In consequence, E and Z do tend to stabilize at

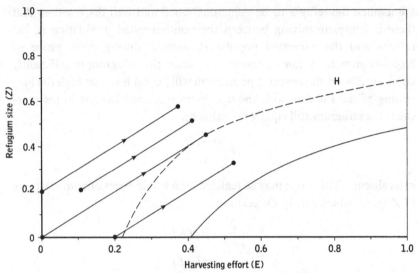

Figure 2.2.   Trajectories corresponding to simultaneous changes in the harvesting effort (*E*) and the refugium size (*Z*). The initial and final equilibrium point of each trajectory is designated by filled circles. The two middle trajectories end close to the combination of *Z* and *E*, corresponding to maximum sustainable harvest ($\hat{H}$), depicted by a dashed line. ($r=1.5$, $p=1$.)

moderate levels. Furthermore, as Figure 2.2 shows these do stabilize at a harvest level not far from the optimal under a wide range of parameters. This is, however, not guaranteed. Refugia such as sacred groves may thus have evolved as components of resource use systems combining conservation and sustainable harvests and operated on the basis of simple decision rules involving negative feedbacks. Such practices, originating in a few communities, may have conferred on them an advantage in avoiding resource depletion in their limited resource catchments. Other societies may then have adopted such practices through imitation of practices of communities which appeared to be successful.

Social and ecological systems
Societies practising prudence

Refugia in the form of sacred groves persist to this day in many parts of Asia, Africa and Mexico. They also covered much of the Middle East and Europe before the spread of Christianity and Islam. The famous 'Epic of Gilgamesh' describes the destruction of a sacred cedar grove protected by a wild giant to build the king's palace in the city of Uruk in Mesopotamia.

The Greek and Roman landscape was dotted like a leopard skin with hundreds of sacred spaces, which usually contained groves of trees and springs of water. The groves varied in size from small plots with a temple and a few trees to those covering several square kilometres. The third sacred war in Greece (355–347 BC) was fought over the issue of illegal cultivation of Apollo's sacred ground at Crisa (Hughes, 1994). As Christianity spread, these groves became objects of religious zeal. The emperor Theodosius II (fifth century AD) issued an edict directing that the pagan groves be cut down unless they had already been appropriated for some purpose compatible with Christianity. A few of them became monastery gardens and churchyards. The others were razed to the ground (Hughes, 1984). But even today, of the thousands of sacred sites to which Christian pilgrims are attracted in Western Europe, nearly a sixth are groves and rock formations, a third are springs and the rest are high places in the hills (Nolan 1981).

Sacred groves continue to this day in Mexico (Gomez-Pompa, Flores and Fernandez 1990), Zimbabwe and Ghana in Africa (Isaac, 1961–62; Dorm-Adzobu and Veit, 1991), in China, Thailand, Nepal and many parts of India (Risley, 1908; Gadgil and Vartak, 1976; Seeland 1986; Brockleman, 1987; Buri, 1987; Watchel, 1993). Dietrich Brandis (1897), the first Inspector General of Forests under the British regime in India, had this to say of the sacred groves:

Very little has been published regarding sacred groves in India, but they are, or rather were very numerous. I have found them in nearly all provinces. As instances I may mention the Garo and Khasia hills . . . the Devara kadus or sacred groves of Coorg Presidency . . . Well known are the Swami shola on the Yelagiris, the sacred forests on the Shevaroys. These are situated in the moister parts of the country. In the dry region sacred groves are particularly numerous in Rajaputana. . . . In the southernmost states of Rajaputana, in Partabgarh and Banswara, in a somewhat moister climate, the sacred groves consist of a variety of trees, teak among the number. These sacred forests, as a rule, are never touched by the axe except when wood is wanted for the repair of religious buildings, or in special cases for other purposes.

In spite of the radical social, economic, religious and ideological changes that have taken place since the beginning of British rule and following independence, sacred groves continue to be protected in India, especially in the more remote regions populated by relatively autonomous, small scale tribal and peasant societies. Figure 2.3 attempts to depict the nature of the interacting social and ecological systems that may have employed sacred groves and other refugia as elements of their resource management systems.

We believe such societies to have been motivated to use their biological resource base in a prudent fashion because restrained resource use may

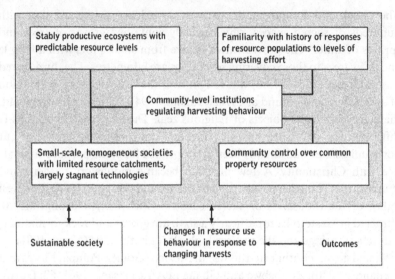

Figure 2.3. Operation of a system of sustainable use of biological resources coupled to conservation by a small-scale homogeneous society in control of its resource base.

have benefited the often tightly knit kin-based group. For this to operate, the ecosystems must have been stably productive, and not excessively influenced by migrant animal populations such as the great herds of herbivores on the East African savannas. Tropical forest ecosystems which once covered most of India would be such relatively stable ecosystems. For kin or group interests to play a significant role in moulding behaviour, the concerned societies must have been relatively small and homogeneous, as tribal and peasant societies dependent on subsistence agriculture coupled to hunting – gathering have been in many parts of India. For long term benefits of prudence to flow to the communities, they must have an effective control over their local resource base, as tribal and rural societies seem largely to have enjoyed in pre-British India. To evolve effective systems of resource management based on negative feedbacks, the societies must have been rooted to particular localities, as was the case with many tribal and peasant societies of pre-British India. Such societies were organized as largely autonomous, self-regulating, tribal-caste-village communities, and were therefore equipped with community-level institutions that could deal with resource management, along with many other functions (Gadgil and Guha, 1992). The case study to be detailed below suggests that such societies may respond to changes in resource level by appropriately adjusting their patterns of harvest. We believe that, given this context, these societies had evolved a variety of cultural traditions such as refugia that promoted sustainable use of their natural resource base.

Societies in flux

Beginning with the British rule in the late eighteenth to early nineteenth century, the small-scale societies with autonomous control of the local resource base have been ever-more effectively absorbed in the large-scale, stratified society of the newly constituted Indian state. First the colonial, and then the independent, Indian state has tended either to privatize or, more often, bring under state control the common property resources earlier under de facto community control. The forest and other resources of their privatized or state-controlled lands have then been mobilized to support the larger economy. The local communities, no longer in control over their resource base, and having to deal with rapidly changing levels of resource stocks influenced much more by outside demands, have tended to lose their motivation for sustainable use, and along with that many systems of conservation of biodiversity, including protection to refugia such as sacred groves. The larger commercial interests working in league with the state apparatus have no stake in the sustainable use of resource stocks from any particular locality and have therefore promoted a process of sequential exploitation. During such a process, resource harvests at any one time are focused on the most easily accessible, profitably exploitable elements. When these are exhausted, the focus shifts to the next most-readily accessible, most profitably exploitable element. When the resulting resource exhaustion leads to serious resource shortages, technological innovations permit tapping of new kinds of resources, or opening up of entirely new localities for resource exploitation (Gadgil, 1991). In such a system the resources and the society are in permanent flux (Figure 2.4).

Sacred groves and safety forests

Churachandpur district in the state of Manipur in northeastern India provides an interesting setting in which to investigate the fate of refugia in the context of a rapidly transforming social and ecological system. This case study suggests that traditional conservation practices might have originated to serve secular functions, although they were implemented through systems of social sanctions based on religious beliefs and social conventions. A breakdown of these beliefs, coupled to major changes in the socio-economic setting of the community, has led to widespread abandonment of

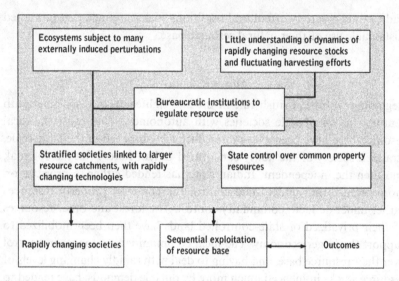

Figure 2.4.   Operation of the system of rapidly changing patterns of resource use accomplished by sequential exploitation by a large-scale, stratified society with a very diffuse control of the resource base.

many traditional practices, including protection to refugia. In several cases the conservation practices have subsequently been revived by the communities, with an explicit understanding that they serve certain secular functions. Notably, the practices are implemented through the same system of social sanctions as before.

Churachandapur is a hilly, remote region not far from the borders of India, Myanmar and China (Figure 2.5). Its population primarily consists of shifting cultivators belonging to the Kuki group of tribal people. Our studies involving field work extending over a total period of eight months in 1992–94 have focused on one particular group of Kukis known as Gangtes, today numbering about 10000 people distributed over a tract of 400–500 km^2 in the northern part of the district. The Gangte population has probably increased by a factor of four or so over the last four decades; it might earlier have been relatively stable at a size of around 2500 individuals. Traditionally, the tribe comprised some five bands, members of each band living together in a hamlet and carrying out shifting cultivation over an area of 50–100 km^2. Each band was an egalitarian group led by a chief who controlled all the material wealth of the members of the band. The chief was also gifted 16 kg of grain produced through shifting cultivation by each household; this was used by the chief to entertain visitors. The Gangte territory was completely outside the fold of any larger state or

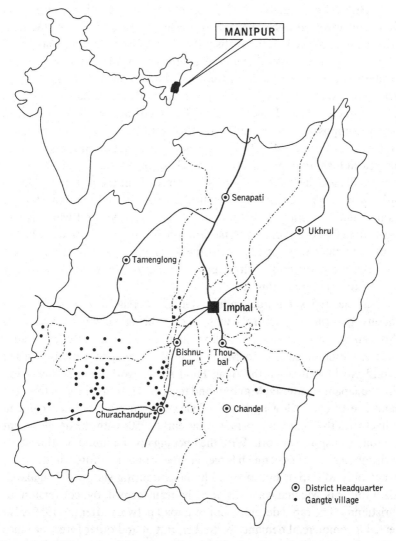

Figure 2.5. The state of Manipur, showing the location of district headquarters and Gangte villages.

economy till the early years of the twentieth century. At this time their only contact with larger markets was for the acquisition of a few highly valued objects such as pieces of iron. But journeys to the markets were hazardous since they entailed passing clandestinely through territories of alien tribal groups who had a tradition of head hunting, possibly till the middle of nineteenth century.

The British first established a measure of control over this region around 1908. This involved assignment of ownership over land. Under the land settlement introduced by the British, the chief of each band was accorded the status of owner of all the land controlled by the band, with other members converted to the legal status of sharecroppers. The British had earlier created a similar system of a few landlords controlling huge tracts of lands in the eastern Indian states of Bihar and Bengal. But that system was superimposed on an already stratified caste society in which the landlords began to extract 40–50% of the share of agricultural produce from the peasant sharecroppers. Amongst Gangtes, especially in the more remote villages, the chiefs remain only nominally in the position of landlords, with the sharecroppers giving them the traditional share of 16 kg of grain, which is around 2% of the agricultural produce and which is still largely used for entertaining guests. However, in villages more closely linked to markets, the chiefs tend to assert their ownership over land and forest resources more forcefully, although the society does retain a basically egalitarian structure to this date.

Until the 1950s Churachandpur remained relatively isolated, with tenuous government control, with people being slowly converted to Christianity yet continuing to follow their traditional patterns of subsistence. Communications to Manipur really developed only during the Second World War when the Azad Hind Army, working in co-operation with the Japanese, entered Manipur and reached Moirang, 110 km from the Gangte territory. Following independence, Churachandpur itself came in contact with the mainland society in the early 1950s with the development of a road transport network. With the opening up of communication came Christian missionaries in much larger numbers, and the entire tribal society of not only Manipur, but also of the neighbouring states of Nagaland, Mizoram and Meghalaya, was rapidly transformed by conversion to Christianity. The rapid development of a road network after the 1950s also created a commercial demand for timber, rattan and other forest produce for the first time. Indeed, by the 1970s, the forest resources of mainland India were largely depleted, so that Manipur and other northeastern states have come to play an increasingly significant role in the supply of timber and plywood to the rest of the country over the last 25 years. At the same time the near-complete cessation of the intertribal warfare, and control of diseases such as malaria, have led to a rapid increase in the population of Gangtes: their numbers increased from 4856 in 1961 to 7891 in 1981.

The traditional land-use system of Gangtes involved leaving aside extensive tracts of land not subject to shifting cultivation as sacred groves, believed to be abodes of nature spirits who would be offended if the groves were interfered with. Our extensive field interviews suggest that anywhere between 10% and 30% of the land was thus maintained as sacred groves. Furthermore, all of the trees felled during the shifting cultivation cycle were burnt on the spot, leading to substantial return of nutrients to the soil. Today much of the valuable wood is transported to the market, leading to lower levels of nutrient return to the soil. How much of the wood is so marketed depends on the proximity of the shifting cultivation fields to the roads and main markets. Thus all the wood felled in the villages of Khoushabung and Bunglon near the district market town of Churachandpur is marketed as fuelwood, the more valuable timber having been exhausted by the early 1970s. On the other hand, only the most valuable timber trees are transported from Santing, 92 km from Churachandpur. This diversion of nutrients from the fields forces the cultivators to abandon them sooner, thus shortening the fallow cycle. The fallow cycle is further shortened because of increasing demand for land closer to the more permanent settlement near the roadside, introduction of longer duration commercial crops such as bananas and population growth. While shifting cultivation cycles are thus being shortened, primeval sacred groves that used to protect the watersheds, provide shade and serve as fire breaks have been encroached for cultivation or felled for the timber. Effects of this elimination of refugia in terms of impairment of their ecosystem services, in particular as firebreaks, have been very noticeable. In the absence of these moist forests, fires started during the slash-and-burn operations can more readily spread to the hamlets and burn them down. As a result, a few years after the complete elimination of all the sacred groves, protection to forest patches has been reinstituted in several Gangte villages. A sacred groves is termed 'gamkhap' in Gangte language; the protected forest patches continue to be called gamkhaps. When speaking to outsiders, however, Gangtes term them as 'forest reserves', while in the neighbouring state of Mizoram, such newly protected forest patches are termed 'safety forests' (Malhotra, 1990). While these refugia are no longer considered to be inviolable as abodes of spiritual beings, the system of community-based vigilance and protection is identical to that prevailing with the sacred groves. Notably, taboos against extraction of any plant material from these refugia are often total and may extend even to the extraction of rattan which is in much commercial demand.

All the larger refugia in present-day Gangte villages surround the habitation and serve as effective firebreaks. Thus the village of Santing has a 20-year-old refugium of over 100 hectares in extent, which is today a closed canopy evergreen forest patch. In addition, Saichang – a neighbouring village – has a smaller refugium away from habitation. This was apparently established when the wife of a farmer died during the course of a slash-and-burn cycle in that spot. Thus, Gangtes seem to react to perceived misfortunes by setting up refugia. Saichang village also has a large patch rich in bamboo, called Mauhak or bamboo reserve. Extraction of bamboo shoots (a much relished food item) from this patch is totally prohibited, and bamboos are permitted to be extracted only for domestic use.

Today, Gangtes are fully convinced that the refugia did perform useful services and would like to continue protecting them in spite of commercial pressures. We have had detailed discussions with them on this score; they would like to guarantee the revival of some areas as refugia and continued protection of other areas that still retain natural vegetation in return for help in developing commercial horticultural crops and marketing facilities for the produce of such crops.

Prospects

Remnants of the system of refugia, sacred groves and sacred ponds continue to survive in many parts of India. They have been treated as a hindrance to development by resource managers. Thus, in the early 1970s Karnataka Forest Department took up felling of the excellent system of sacred groves of Coorg on the exhaustion of plywood resources of the state controlled forests. Indeed, at this time, a senior forest official asked one of us why we were attempting to persuade them to protect these 'stands of overmature timber'. Fishery managers have been equally active in poisoning sacred ponds, eliminating the indigenous fish fauna, and converting them into carp aquaculture sites. Hindu priests have also helped in the elimination of sacred groves by persuading the local villagers that they would perform religious rites propitiating the deities of sacred groves, so that timber contractors could be permitted to fell them (Gadgil and Chandran, 1992; Chandran and Gadgil, 1993; Roy Burman, 1994).

Yet in many places, largely in the more remote localities away from the pull of market forces, people continue to recognize the manifold ecosystem services provided by the sacred groves. Sacred groves also continue to be newly established in many localities, some as large as one of 25 hectares near Udupi in Dakshina Kannada district of Karnataka. Nevertheless, the current system of protection of refugia on the basis of religious beliefs is not likely to remain operational for any length of time. It must be replaced by a newer system of incentives. These could most appropriately be developmental or technical assistance or straightforward financial rewards.

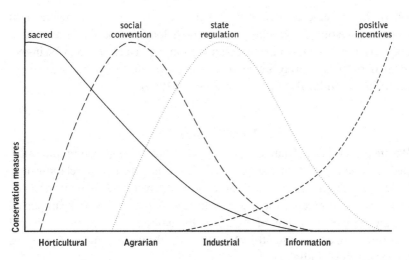

Figure 2.6. Relative significance of different social mechanisms for promoting sustainable use of biological resources at different stages of social development.

Thus Gangtes are willing to guarantee maintenance of sacred groves in return for assistance in developing commercial horticulture. Others, such as farmers on the thickly settled coast of Kerala who are by now integrated into the larger market yet continue to protect a number of sacred groves dedicated to serpent deities, may be more inclined to accept monetary incentives. Their village councils could then be given some special annual grants to be used at their discretion for community work provided they continue to protect the sacred groves (Gadgil and Rao, 1994; 1995).

Human societies at different stages of development thus employ different mechanisms to promote sustainable use of biological resources and ecological resilience (Fig. 2.6; Holling, 1986). Small-scale horticultural societies grounded in hunting – gathering – shifting cultivation – subsistence agriculture do so on the basis of a feeling of respect for the sacred or for social conventions. As these give way to agrarian societies, state sponsored regulation assumes greater significance, especially at the cost of respect for the sacred. Indeed, the latter was almost totally discarded in the societies that came under the sway of Christianity and Islam (White, 1967). The hunting preserves of the nobility have constituted a significant component of the state sponsored regulation of agrarian societies; this has given way to national parks and other protected areas in the industrial societies. Industrial societies have also instituted a series of supposedly science-based regimes of regulation of harvests of biological resources; these have, however, often failed to serve their purpose (Ludwig, Hilborn and Walters,

1993). In the emerging information age, decentralized management of ecological resources, grounded in detailed locality specific-information, may turn out to be the most desirable option. Such a regime may be implemented most effectively by providing appropriate positive incentives to individuals and local communities (Gadgil, 1996).

Acknowledgements

We are grateful to our numerous friends amongst Gangte communities for their hospitality and for the many insights in people–nature relationships. M.D. Subash Chandran, two anonymous referees and the editors made many useful comments and suggestions. This research has been supported by the Indian Institute of Science, Bangalore, the Indian Statistical Institute, Calcutta and the Ministry of Environment and Forests, Government of India.

References

Borgerhoff-Mulder, M. 1988. Behavioural ecology in traditional societies. *Trends in Ecology and Evolution* 3(10): 260–4.

Brandis, D. 1897. *Indian Forestry*. Oriental Institute, Woking.

Brockleman, W. 1987. Nature conservation. In *Thailand Natural Resources Profile*, pp. 00–00, ed. Arbhabhirama *et al.* Bangkok: Thailand Development Research.

Buri, R. 1987. Wildlife in Thai culture. In *Culture and Environment in Thailand*. Symposium Proceedings of the Siam Society.

Chandran, M.D.S. and Gadgil, M. 1993. *Kans-safety Forests of Uttara Kannada*. Proceedings of IUFRO Forest History Group Meeting on Peasant Forestry, Freiburg, Germany, No. 40, pp 49–57.

Clark, C.W. 1990. *Mathematical Bioeconomics. The Optimal Management of Renewable Resources*, 2nd ed. New York: John Wiley.

Diamond, J. 1993. New Guineans and their natural world. In *The Biophilia Hypothesis*, pp. 251–71, S. Kellert and E.O. Wilson, eds. Washington DC: Island Press.

Dorm-Adzobu, C. and Veit, P.G. 1991. Religious beliefs and environmental protection: the Malshegu sacred grove in Northern Ghana. In *From the Ground Up*, Case Study No. 4, pp. 1–34. Center for International Development and Environment. World Resources Institute, USA.

Eaton, P. 1985. Customary land tenure and conservation in Papua New Guinea. In *Culture and Conservation: The Human Dimension in Environmental Planning*, pp. 181–91, ed. J.A. McNeely and D. Pitt. Dublin: Croom Helm.

Gadgil, M. 1991. Restoring India's forest cover: the human context. *Nature and Resources* 27(2): 12–20.

Gadgil, M. 1996. Managing biodiversity. In *Biodiversity: A Biology of Numbers and Differences*, pp. 345–66, ed. K.J. Gaston, Oxford: Blackwell.

Gadgil, M. and Berkes, F. 1991. Traditional resource management systems. *Resource Management and Optimization* 18(3–4): 127–141.

Gadgil, M. and Chandran, M.D.S. 1992. Sacred groves. In *Indigenous Vision. People of India. Attitudes to the Environment*, pp. 183–7, ed. G. Sen. New Delhi: Sage Publications and India International Centre.

Gadgil, M. and Guha, R. 1992. *This Fissured Land: An Ecological History of India*. Delhi : Oxford University Press; Berkeley: University of California Press.

Gadgil, M. and Rao, P.R.S. 1994. A system of positive incentives to conserve biodiversity. *Economic and Political Weekly*, Aug. 6: 2103–7.

Gadgil, M. and Rao, P.R.S. 1995. Designing incentives to conserve India's biodiversity. In *Property Rights in a Social and Ecological Context*, pp. 53–62, ed. S. Hanna and M. Munasinghe. Washington DC: The Beijer International Institute of Ecological Economics and The World Bank.

Gadgil, M. and Vartak, V.D. 1976. The sacred groves of Western Ghats in India. *Economic Botany* 30: 152–60.

Gomez-Pompa, A., Flores, J.S. and Fernandez, M.A. 1990. The sacred cacao groves of the Maya. *Latin American Antiquity* 1: 247–57.

Holling, C.S. 1986. Resilience of ecosystem: local surprise and global change. In *Sustainable Development of the Biosphere*, pp. 292–317, ed. E.C. Clark and R.E. Munn. Cambridge: Cambridge University Press.

Hughes, J.D. 1984. Sacred groves: the gods, forest protection and sustained yield in the ancient world. In *History of Sustained Yield Forestry*, ed. H.K. Steen. Durham: Forest History Society.

Hughes, J.D. 1994. *Pan's Travail: Environmental Problems of the Ancient Greeks and Romans*. Baltimore: Johns Hopkins University Press.

Isaac, E. 1961–62. The act and the covenant. *Landscape* 11(2): 12–17.

Joshi, N.V. and Gadgil, M. 1991. On the role of refugia in promoting prudent use of biological resources. *Theoretical Population Biology* 40(2): 211–29.

Ludwig, D., Hilborn, R. and Walters, C. 1993. Uncertainty, resource exploitation and conservation: lessons from history. *Science* 260: 17–36.

Malhotra, K.C. 1990. Village supply and safety forest in Mizoram: A traditional practice of protecting ecosystem. In *Abstracts of the Plenary, Symposium Papers and Posters*, p. 439. Presented at the Vth International Congress of Ecology INTECOL, Yokohama, Japan, 23–30 Aug.

Nietschmann, B. 1985. Torres Strait islander sea resource management and sea rights. In *The Traditional Knowledge and Management of Coastal Systems in Asia and the Pacific*, pp.125–54, ed. K. Ruddle and R.E. Johannes. Indonesia: UNESCO.

Nolan, M.L. 1981. Types of contemporary Western European pilgrimage places. In *Conference on Pilgrimages: The Human Quest*. Pittsburgh.

Risley, H. 1908. *The People of India*. Bombay: Thacker, Spink & Co.

Roy Burman, J.J. 1994. Sacred groves and the modern political economy. *Lokayan Bulletin* 10 (5,6): 41–52.

Seeland, K. 1986. Sacred world view and ecology in Nepal. In *Recent Research on Nepal*, pp.187–99, ed. K. Seeland. Kohn: Weltforum Verlag.

Slobodkin, L.B. 1968. How to be a predator. *American Zoologist* 8: 43–51.

Watchel, P.S. 1993. Asia's sacred groves. *International Wildlife*, March/April: 24–7.

White, L. 1967. The historical roots of our ecological crisis. *Science* 155: 1203–5.

3

Learning by fishing: practical engagement and environmental concerns

GÍSLI PÁLSSON

Introduction: engaging with the natural and the social

Much Western thinking assumes a fundamental distinction between ecological and social systems, between nature and society. Hollingshead provided one formulation of such a dualism, speaking of 'the ecological and sociological orders' (1940: 358): 'The former is primarily an extension of the order found everywhere in nature, whereas the latter is exclusively, or at least almost, a distinctly human phenomenon . . . The ecological order is primarily rooted in competition, whereas social organization has evolved out of communication.' Such a theoretical position has been reinforced by a rigid academic divison of labour; the sociological order has remained the subject of anthropologists and sociologists while the ecological one has belonged to professional ecologists. Having established a fundamental dichotomy, Hollingshead qualified his thesis, emphasizing that nature and society were not to be seen as *totally* separate spheres but dialectically inter-linked; each order 'compliments and supplements the other in many ways' (Hollingshead 1940: 359). In recent years, however, the distinction itself between nature and society has increasingly been subject to critical discussion in several fields, including anthropology (Descola and Pálsson, 1996). One kind of evidence against the theoretical dualism of nature and society relates to practical, ecological expertise. Given the perspective of practice theory (Lave, 1988; 1993), such expertise is the result of a simultaneous engagement of the human actor with nature and society – to the extent that the distinction between the two 'environments' may not be very meaningful.

In many societies there is little attempt to draw upon the knowledge that practitioners have achieved in the course of their work, for the purpose of ecological research and decision-making. The current ignorance, if not denigration, of practical knowledge has been supported by a powerful 'modernist' paradigm in bio-economics and resource management which assumes that ecosystems are characterized by linear relationships and that only a market approach, emphasizing private ownership of resources (usually privileging capital rather than labour), will ensure stewardship and responsible resource use. The estimate of total allowable catch (TAC) and the allocation of individual transferable resource quotas (ITQs), it is often argued, are the only feasible and efficient management strategies.

Increasing empirical evidence and a growing body of theoretical scholarship suggest that there are good grounds for questioning orthodox theories of learning and modernist management. According to traditional theories of learning, the novice individual, alienated from the environment, gradually becomes a competent person by internalizing a cultural code or a superorganic script. Recent research indicates, on the other hand, that the current restrictive emphasis on disembedded knowledge needs to be revised (Gergen and Semin, 1990; Fischer *et al.,* 1993). Some scholars focus on the need to maintain or 'recapture' the disappearing knowledge of marginal groups (see, for instance, Chapin, 1994), others have emphasized the potential importance of such knowledge for sustainable resource use (DeWalt, 1994), and still others have emphasized the need to defend the intellectual property rights of the groups they have studied against the hegemonic practices of states and multinational companies, focusing on legal claims about patents and royalties (Brush, 1993). At the same time there has been a growing theoretical interest in 'outdoor' learning (Pálsson, 1994), cognition 'in the wild' (Hutchins, 1995), and the divide between practical and scientific knowledge (Latour, 1994; Agrawal, 1995).

Research emphasizing the uncertain nature of many marine ecosystems suggests that managers modify their hierarchical notion of linearity and expertise. Multi-species ecosystems, it is argued, are highly unpredictable, with constant fluctuations in interactions among species and between species and their habitat (Smith, 1991; Wilson *et al.,* 1994). This does not mean that governance is impossible; it suggests, however, increasing reliance on a finer spatial and temporal scale, a scale that only the skilful practitioner is able to apply. It may be essential, therefore, to pay attention to practical knowledge, allowing for contingency and extreme fluctuations in the ecosystem. Some form of self-governance and co-management may be a practical necessity, strange as it may sound to those accustomed to the theory of the 'tragedy of the commons' which assumes that overfishing is inevitable as long as access is 'free' for everyone. Focusing on Icelandic fishing, this chapter discusses the practical knowledge of fishers and professional scientists and the extent to which public resource management would benefit from their collaboration.

From the time of the settlement of Iceland in the ninth century, subsistence production on the coast always included the exploitation of marine resources; thus, early on, the value of a boat was considered to be equal to the value of a cow. In some cases, where cultivable land was plentiful, fishing was of minor importance, while in others, where good land was scarce, fishing was the mainstay of the economy. Fish formed a staple of the Icelandic diet, together with the milk of cattle and sheep. Soon after the union with Norway in 1262, fish replaced woollens as the island's main export and for centuries dried fish was the most important monetary standard. The importance of fish for foreign powers as well as Icelanders themselves can be seen from the fact that ten 'cod wars' have been fought on the fishing grounds around Iceland, the first one starting in 1415 when the Norwegian king charged English fishers with illegal trading with Icelanders.

With the growth of the market economy, especially after political independence in 1944, fishing effort multiplied. In 1948, the Icelandic Parliament passed laws about the 'scientific protection of the fishing grounds in the coastal zone,' to be able to prevent overfishing of its major fishing stocks, particularly cod; some marine biological research already occurred by the end of the nineteenth century, but full-time research started later, in the 1940s. In 1952, Iceland announced that it would extend its territorial jurisdiction from three to four miles and in 1958 it unilaterally extended its jurisdiction to 12 miles. In 1976, the Icelandic government extended the national fishing limits to 200 miles, which marked the end of the last cod war with Britain and West Germany. However, the domestic fishing fleet continued to grow and catches, relative to effort, continued to decline. The first serious limitations on the fishing effort of Icelandic boats were temporary bans on fishing on particular grounds. Later, more radical measures were developed, including limitations on total allowable catch and a system of individual transferable quotas.

In the expanding Icelandic economy, fishers were central agents. As a result, marine scientists had to carve a space for themselves in the role of collaborators and apprentices. With the birth of the modernist fisheries regime and the grand narrative of marine science, however, the voice of fishers was gradually subdued. The practical knowledge of those engaged in fishing on a daily basis was assumed to be of little value for resource management. Because of the persistent threat of overexploitation, Icelandic fishing was subject to both scientific control and increasingly stringent public regulations. The tone of humility and mutual learning typical for the pioneering biologists during the first half of the century was replaced by claims about scientific certainty and folk 'misunderstanding.' Recently, attempts have been made at bridging the gap between fishers and scientists, at regaining some of the trust that earlier characterized relations between them, an important example being the 'trawling rally' (*togararall*) – a procedure whereby a group of skippers regularly follow the same trawling paths, identified by biologists, in order to supply detailed ecological information. While the trawling rally represents an interesting endeavour, it has its shortcomings. It is important to look for alternative ways of engaging fishers, of using knowledge obtained in the course of production for the purpose of responsible resource use and sustainable management.

Practical knowledge

For several centuries, Western discourse has tended radically to separate scientific understanding and everyday accounts. Scientists, it has often been assumed, are objective explorers of reality, proceeding by rational methods and detached observations, while the lay person is locked up in a particular natural or cultural world, driven by genetic make-up, ecological context, superstitious beliefs, or local concerns. One of the consequences of such a Cartesian scheme is the tendency to reduce local environmental knowledge to mere trivia and to assume that what people have to say about ecological matters and human–environmental interactions is pure ideology, of relevance only as cultural data. Accordingly, sustainable resource-use and sensible management become the privileged business of outsiders formally trained in public institutions.

In recent years, the dualist theory of knowledge on which the modernist production regime is based has been challenged on a number of fronts. For one thing, the notion of the absolute objectivity of science, the idea of some scientific Archimedean standpoint outside nature and history, is frequently subject to critical discussion, with the growing awareness that the modernist perspective fails to account for the actual practice of modern science. As Latour (1994) argues, modern science has never been able to meet its own criteria; paradigms and *épistémès* are inevitably social constructs, the products of a particular time and place. At the same time, practical knowledge has been firmly placed on both the theoretical and management agenda. The community of modellers has been both expanded and redefined, empowering the local voice and relaxing modernist assumptions of privilege and hierarchy (Gudeman and Rivera, 1990). This is evident from current interest in practical knowledge in development agencies on the international scene as well as in academic studies of learning and expertise (Lave and Wenger, 1991; Williams and Baines, 1993).

While it is true that an extensive body of local knowledge has often been set aside, if not eliminated, in the course of Western expansion and domination and that there are good grounds for attempting to recapture and preserve what remains of such knowledge (Chapin, 1994), the reference to the 'indigenous' and 'traditional' in such contexts tends to reproduce and reinforce the boundaries of the colonial world, much like earlier notions of the 'native' and the 'primitive'. How old does a particular skill or body of knowledge have to be to count as 'traditional'? Where does it have to be located to be classified as 'indigenous'? Such terms are not only loaded with the value judgements of colonial discourse, they are fraught with ambiguity: 'on the one hand we find striking differences among the philosophies and knowledges commonly viewed as indigenous, or western. On the other hand we may also discover that elements separated by this artificial divide share substantial similarities' (Agrawal, 1995: 415).

Another contested issue relates to the meaning of knowledge and learning. Orthodox theories tend to present the learning process in highly functional terms, presupposing a natural novice who gradually becomes a member of society by assimilating its cultural heritage. Knowledge becomes analogous to grammar or dictionaries, invested with the structural properties and the stability often attributed to language. Given such a perspective, indigenous knowledge is sometimes presented as a marketable commodity – at times with 'missionary fervor' (DeWalt, 1994: 123). It may be useful and quite legitimate in some contexts to think of practical knowledge as a bounded, tradable object, for instance when encoding indigenous knowledge for the protection of intellectual property; indeed, it signifies a long overdue move to question the assumption of a hierarchy of knowers central to the modernist project. The practitioner's knowledge, however, situated in immediate experience and direct engagement with everyday tasks, is not, as Lave emphasizes (Lave, 1993: 12), 'a collection of real entities, located in heads'. The attempt to conserve practical knowledge and store it *ex situ* in archives and data bases is, therefore, likely to fail. Not only is it an unrealistic task simply because of the fleeting character and ever-changing nature of the knowledge involved, it would probably reinforce the hierarchy of knowledge which it is trying to avoid (Agrawal, 1995), for the benefits of textual experts and management elites.

Focusing on practical knowledge represents one attempt to resolve such conceptual ambiguities and theoretical difficulties. Such a focus does not assume a cultural or temporal boundary, the radical separation of producers and scientists, participants and observers, traditionalists and modernists. There may be different ways of knowing – contextualized constructions, on the one hand, and, on the other, de-contextualized abstractions, generalizations across contexts. On some occasions, however, all of us seek to formulate our tacit knowledge in general terms, by verbal or textual means. Likewise, practical knowledge is not restricted to any particular group of people, for none of us (including practising scientists) would manage to live without it. Scientific knowledge, of course, involves practical knowledge obtained in the course of engagement and experimentation. De-contextualization, too, is a form of local practice, generated and maintained in particular contexts, within a community of scholars.

Learning by fishing

Informed by the notions of situated action and mutual enskilment, the theory of practice emphasizes direct engagement with everyday tasks (Lave, 1988; 1993). Such a perspective resonates with some aspects of the discourse of Icelandic fishers (Pálsson 1994, 1995). For them, 'real' schooling is supposed to take place in actual fishing, not in formal institutions. As one skipper put it: 'Naturally, most of the knowledge one uses on a daily basis is obtained by experience. One learns primarily from the results of personal encounters, that is what stays with you.' The emphasis on 'outdoor' learning is highlighted in frequent derogatory remarks about the 'academic' learning of people who have never 'peed in salty sea' (*migið í saltan sjó*). Even a novice fisherman, skippers say, with minimal experience of fishing, is likely to know more about the practicalities of fishing than the teachers of the Marine Academy. Therefore, there is little connection between school performance and fishing success. Questioned about the role of formal schooling, skippers often say that what takes place in the classroom (during lessons in astronomy, for instance) is more or less futile as far as fishing skills and differential success are concerned, although they readily admit that schooling has some good points, preventing accidents and promoting proper responses in critical circumstances involving the safety of boat and crew.

Such claims seem to be supported by fishing statistics. A recent study by the present author of the fishing fleet of Sandgerði in south-west Iceland has found no significant relationship between the catch for a season and skippers' grades in their final examinations in the Marine Academy (the details of this study have not been published). The data used in this exercise relate to the winter season in 1981, two years before a quota system was

Table 3.1. *Pearson correlations: fishing success*

Variable	No. of trips	Boat size	Average grade	Skipper's age	Years of skipperhood
Catch	**0.682**	**0.714**	−0.001	−0.00 1	0.110
	(39)	(39)	(16)	(37)	(34)
	0.00	0.00	0.99	0.97	0.53
No. of trips		0.185	−0.072	0.244	0.204
		(39)	(16)	(37)	(34)
		0.26	0.79	0.146	0.25
Boat size			0.234	−0.145	−0.021
			(18)	(41)	(34)
			0.35	0.37	0.91
Average grade				0.228	0.142
				(18)	(15)
				0.36	0.61
Skipper's age					**0.665**
					(34)
					0.00

Note: Significant correlations in bold.

introduced. (Under a quota system, the catch of a boat is more or less given in advance in the form of quota holdings and, as a result, catch-data are not very meaningful for the exercise in question.) All vessels using gill nets or long lines were considered. A few trawlers which landed their catches locally were excluded from the calculations since their fishing trips vary in length, from one to two weeks; boats with long lines and gill nets, in contrast, land their catches daily (see Pálsson (1991) for more ethnographic details). Several measures were considered of both fishing success – including total catch (tons), catch per trip, and catch per ton (boat size) – and school performance – grades in mathematics and navigation and average grade. There was no relationship between catch and average grade in the Academy (Table 3.1), the Pearson correlation between total catch and average grade being −0.001 (p= 0.99). Even more importantly, there was no correlation either when controlled for number of trips and/or boat size. Figure 3.1 shows a scattergram for these variables ($n=16$), distinguishing between small and large vessels (<40 tons and >40 tons). From this evidence, success at school has no relation to success in fishing.

'Fishiness,' the ability to catch fish, seems to be fairly consistent over time. Indeed, there are strong correlations between skippers' catches for different years, especially for two consecutive years; the correlations

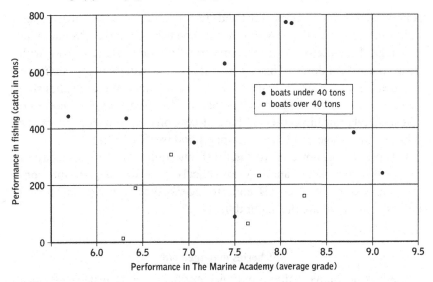

Figure 3.1.   Scattergram: performance in fishing and at school (Sandgerði, 1981).
('□' represents boats under 40 tons, '•' represent boats larger than 40 tons.)

between the 1981 catch and those of earlier years – 1980, 1979 and 1978 –
is 0.92 ($n=28$; $p=0.00$), 0.86 ($n=23$; $p=0.00$), and 0.67 ($n=11$; $p=0.03$),
respectively. The reasons for this consistency are complex. Consistency is
partly explained by the amount of capital involved, in particular boat size:
the correlation between boat size and the catch for the season is 0.71. It is
difficult, however, to separate material and personal factors (see Pálsson,
1994) and, no doubt, one of the reasons for consistency in catches has to
do with the skills of the skipper. It seems unreasonable to assume that the
skipper's skills are given in advance; more likely they are developed in the
course of practice, in the company of tutors (more experienced skippers)
and other crew.

Skipper education recognizes the potential importance of practical
learning, since earlier participation in fishing, as a deck-hand (*háseti*), is a
condition for formal training, built into the teaching programme; this is to
ensure at least some knowledge about the practice of fishing. Once students
in the Marine Academy have finished their formal studies and received their
certificate, they must work temporarily as apprentices – as mates
(*stýrimenn*) – guided by a practising skipper, if they are to receive the full
licence of skipperhood. It is precisely here, in the role of an apprentice at
sea, that the mate learns to attend to the environment *as a skipper*. Working
as a mate under the guidance of an experienced skipper gives the novice the
opportunity to develop attentiveness and self-confidence, and to establish

skills at fishing and directing boat and crew. The attitude to the mate varies, however, from one skipper to another; as one fisher remarked, 'Some skippers regard themselves as teachers trying to advise those who work with them, but others don't'. This suggests that what matters is the *quality* of practical experience (in particular, the apprentice–tutor relationship between novice and skipper) rather than its duration. Indeed, there is no statistical relationship between fishing success, on the one hand, and, on the other, skipper's age and years of skipperhood (see Table 3.1).

It is one thing, however, to catch fish and quite another to manage a fishery. Is it possible to translate the practical expertise of skippers into concrete management proposals and strategies? What are the advantages involved and what are the major difficulties?

Fisheries management

Nowadays, decisions on the scope of fishing operations are usually informed by marine sciences, setting the limit of the total allowable catch for a fishing season on the basis of measurements and estimates of stock sizes and fish recruitment. The science of resource economics, however, has played an even more important role in fisheries management than marine biology, providing the theoretical framework and the political rationale for a quota system and, by extension, private property. In many ways, resource economics has *replaced* marine biology as the hegemonic discourse on Icelandic fishing. While the original, formal demand for the quota system came from within the fishing industry, it would hardly have been instituted if it had not been advocated by influential Icelandic economists. Not only did they play a leading role within the major political parties as well as on a series of important committees that designed and modified the management regime, their writings, in newspapers, specialized magazines and scholarly journals, paved the way for the 'scientific' discourse on efficiency and the 'rational' management which the quota system represents. Some of the economists argued, with reference to the 'tragedy of the commons', that the only realistic alternative – euphemistically defined as 'rights-based' fishing – was a system of individual transferable quotas. Assuming a sense of responsibility among the new 'owners' of the resource (the quota holders) and a free transfer of quotas from less to more efficient producers, economists argued, a quota system would both encourage ecological stewardship and ensure maximum economic efficiency. In Iceland, a quota system was introduced in the cod fishery in 1983 to prevent the 'collapse' of the major stocks and make fishing more economical (Pálsson, 1991: Chapter 6).

The quota system divided access to the resource among those who happened to be boat owners when the system was introduced, largely on the basis of their fishing record during the three years preceding the system. Each fishing vessel over ten tons was allotted a fixed proportion (*aflahlutdeild*) of future total allowable catches of cod and five other demersal fish species. Catch-quotas (*aflamark*) for each species, measured in tons, were allotted annually on the basis of this permanent quota share. And the fortunate quota-holders were the owners of vessels, not crews. This arrangement did not go uncontested, for there have been heated debates about what to allocate and to whom. The issues involved illustrate the discursive contest between different groups of 'producers'. Boat owners argued for '*catch*-quotas', to be allocated to *their* boats. Some fishers, on the other hand, advocated an '*effort*-quota', to be allocated to skippers or crews. After more than a decade of stringent quota management and redistribution of assets, the major Icelandic fishery (the cod fishery) is still in a critical phase. Even worse, there are no signs of ecological recovery. Stock sizes and recruitment rates continue to be far too low, given earlier estimates of maximum sustainable yield.

Some evidence suggests that quota management results in the erosion of ecological responsibility. Discarding of small and immature fish during fishing operations – 'high-grading' in everyday language – and the dumping of species of relatively low economic value – 'bycatch' of low-value species that are 'accidentally' caught and discarded on the spot – seem to be major problems in many fisheries, including the Icelandic one. A recent report provides an estimate of the discards of bycatch in commercial fisheries of 27.0 million metric tons every year (Alverson *et al.*, 1995). While much of this happens in the absence of quotas and a general pricing mechanism, general commoditization may significantly *contribute* to the waste of living resources, increasing discards. Since quotas are fixed and excessive catch is often treated as a crime, a quota holder tends to land only the portion of the catch which generates the highest income. This usually entails discarding small and immature fish.

Wilson and associates suggest that the 'numerical' approach of current resource economics and marine biology, emphasizing linear relationships and states of equilibrium, fails to account for the chaotic aspects of many fisheries (c.f. Wilson *et al.*, 1994; Fogarty, 1995). Their work indicates that while fisheries are deterministic systems, because of their extreme sensitivity to initial conditions, even simple fish communities have no equilibrium tendency. As a result, management faces forbidding problems when trying to explain the noise in ecological relationships, for example the relationship between recruitment and stock size, often a key issue for managers: 'the degree of accuracy and the completeness of knowledge required for prediction are far beyond any capabilities we might expect to achieve in a fisheries environment' (Wilson *et al.*, 1994: 296). Therefore, it becomes difficult, if not impossible, to know the outcomes of management actions such as quotas. Given the level of uncertainty and the limits of scientific ecological knowledge, it is reasonable to try to draw upon the knowledge of those who are directly involved in resource use on a daily basis. After all, they are the ones who are likely to have the most reliable information as to what goes on in the system at any particular point in time.

An important issue in current discussions of practical knowledge is the extent to which technical and economic changes influence levels of practical skill, leading to deskilling or upskilling (Gallie, 1994). Somewhat paradoxically, with the crisis in the world's fisheries, fishing technology has been revolutionized. While in some fisheries technological changes seem to have resulted in rapid deskilling, there is little reason to believe that this has generally been the case. Some skills are inevitably lost; in Iceland, old and retired skippers sometimes point out that fishing has been radically transformed by electronic technology (including the computer) and artificial intelligence, emphasizing that 'natural signs' (birds in the air, clouds in the sky, etc.) are increasingly redundant. However, attentiveness continues to be one of the central assets of the good skipper and, just as before, it demands lengthy training. The skipper's universe is very different from that of his colleagues of earlier decades, but what shows on the screens of the radar, the computer, and the fish-finder is no less a natural sign, directly sensed, than birds in the air or natural landmarks.

For the skilled skipper, fishing technology – the boat, electronic equipment, and fishing gear – is not to be regarded as an 'external' mediator between the person and the environment but rather as a bodily extension in quite a literal sense. Experienced skippers often speak of knowing the details and the patterns of the 'landscape' of the sea bottom 'as well as their hands'. Thanks to technological extensions, the experienced skipper is able to 'see' the fish and 'probe' the landscape of the sea bed. For many landlubbers, no doubt, the sea is primarily fishing *space*. For skippers, in contrast, it is a three-dimensional world, with variable bottom-features, migrating fish, and stratified masses of water.

The continuous engagement of the skipper with the three-dimensional world of the sea is potentially important for ecological monitoring. Often skippers alert marine biologists and resource managers to particular conditions in the sea – for instance, the relative amount of immature fish in catches on particular grounds, the exceptional abundance of a given species, and unexpected ecological conditions. They may voice their opinions and observations in the mass media or in conversations with biologists and state officials. Skippers' monitoring, however, is a relatively untapped resource. Their reporting is minimal and, in any case, on a personal basis, granted in the absence of a supportive institutional framework, a framework that would allow for mutual trust and routine collaboration between skippers, on the one hand, and, on the other, scientists and government official. Nevertheless, there are interesting attempts at collaborative monitoring on both a local and a national level. One example of the former pertains to the recent negotiation and demarcation of fishing limits around the Vestman Islands. In this case, small-scale fishers, biologists and local politicians united in an attempt to reduce effort as well as to exclude destructive fishing technology on local grounds, within three miles, in response to fishers' concern about drastic reductions in catches. Such collaboration might well be extended to other communities. Bringing management decisions to the local level is likely to encourage responsible resource use and sustainable management.

Bridging the gap: the 'trawling rally' and beyond

In an attempt to encourage a more general co-operation between scientists and fishers and to involve the latter in the collection of detailed ecological data on the state of the seas and the fishing stocks, the Marine Research Institute hires a group of skippers to fish regularly along the same pre-given trawling paths on their commercial vessels (see Pálsson *et al.*, 1989); this is the so-called 'groundfish project' – in everyday language, the 'trawling rally' – initiated in 1985:

The cooperation with fishermen is based on the main objective of the project; to increase precision and reliability of stock size estimates of relevant fish stocks, especially cod, through the integration of fishermen's knowledge of fish behavior and migrations, as well as the topography of the fishing grounds (Pálsson *et al.*, 1989: 54).

Because mature cod sometimes migrate from Greenland to Icelandic fishing grounds, thereby making traditional stock assessment on the basis of fishing statistics relatively unreliable, a systematic fishery-independent survey was seen to be essential. Such a large-scale project was beyond the capabilities of Icelandic research vessels and, therefore, commercial vessels were hired for the task. For two weeks in March every year, five vessels survey the same research stations (595 in the beginning, then 600), jointly

60   *Gísli Pálsson*

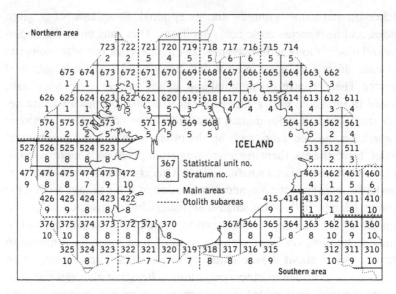

Figure 3.2.   The survey areas of the 'trawling rally' within the 500-m depth contour.
(From Pálsson *et al.* (1989), reproduced with the authors' permission.)

selected by skippers and scientists through a semi-randomly stratified
process (Figure 3.2). The vessels (stern-trawlers) are identical in overall
equipment and design, in terms of size, fishing gear, engine power, etc. This
is seen to be important to ensure comparable data-sets, allowing for reliable
estimates of changes in the ecosystem from one year to another.

The trawling rally is an interesting experiment. While it is partly a diplo-
matic endeavour on behalf of the Marine Research Institute, in order to
reduce the tension that has developed between biologists and fishers in
recent years as a result of stringent scientific management, and to improve
the image of the institute among the general public, there is obviously more
to the story. No doubt, the trawling rally yields extensive comparative
information that the biologists could not possibly gain otherwise, given
their limited funding. The catches of the trawlers in question are analysed
in detail and the results are used, along with data gathered by other means,
to develop predictive models about stock sizes and fish migrations.

Nevertheless, as an attempt to cultivate effective interactions between
fishers and biologists for management purposes, it has significant limitations. To begin with, while the design of controlled surveying has an
obvious comparative rationale, it is also a straitjacket, preventing a more
flexible and dynamic sensing of ecological interactions in the sea.
Skippers fail to be impressed with the scientific design, criticizing the

biologists for 'isolating themselves temporarily on particular ships, pretending to practice great science', to quote one of the skippers. Many skippers pointed out during interviews that, fixed to the same paths year after year, the rally fails to respond to fluctuations in the ecosystem, thus providing poor estimates. Also, the reliance on trawling, skippers say, is likely to produce biased results; often, those fishing with gill nets on nearby grounds offer a very different picture. From the skippers' point of view, a more intuitive and holistic approach, allowing for different kinds of fishing gear and greater flexibility in time and space, would make more sense. Indeed, skippers discuss their normal fishing strategies in such terms, emphasizing constant experimentation, the role of 'perpetual engagement' (*að vera í stanslausu sambandi*), and the importance of 'hunches' (*stuð*) and tacit knowledge.

Despite the occasional lip service in the reference in public statements to 'collaborating' (*hafa samráð*) with fishers, there is little real dialogue. Significantly, a recent two-volume survey of the history of marine research on Icelandic fishing grounds by Jónsson (1988, 1990), former director of the Marine Research Institute, only has one page on the trawling rally (1990: 122–3). The negligible attention paid to the trawling rally in Jónsson's detailed analysis need not be that surprising, however, given the discursive tone and framework of his general account, focusing on the triumph of modern science. Jónsson does mention that research stations were defined 'equally' by skippers and marine biologists and that the former were guided by 'experience', but his comment on the reasons for resorting to skippers' experience is blunt, emphasizing the sheer amount of data required and the practical constraints of their collection – the fact that 'the Marine Research Institute did not have at its disposal the fleet required for such a data collection' (Jónsson, 1990: 122).

One of the reasons for the lack of collaboration between fishers and scientists is the difference between the two groups in terms of the knowledge they seek. Like their colleagues in other parts of the world, Icelandic biologists have often focused on one species at a time, modelling recruitment, growth rates, and stock sizes, although recently they have paid increasing attention to analyses of interactions in 'multi-species' fisheries. 'Erecting an ivory tower around themselves', one skipper argued, 'biologists are somewhat removed from the field of action; they are too dependent on the book'. While such comments have to be seen in the light of cultural and economic tension between social classes and between centre and periphery, they should not be rejected on that ground alone as they also have some grain of truth. Biological estimates and fisheries policy are often literal and rigid in form, unable to deal with variability and to respond to changes in the ecosystem. While the knowledge of scientists is largely normative and textual, preoccupied with theoretical ways of knowing, that of fishers, often tacit, is tuned to practical realities in the ever-changing sea, to the flux and momentum of fishing. Fishers question some of the basic assumptions of scientists, arguing that understanding of fish migrations and stock sizes is too imperfect to make reliable forecasts. Scientists and practitioners, after all, have somewhat different methods and motives.

The fact that skippers often fail to express what they know by verbal means, since much of what they have learnt is tacit and intuitive, presents a formidable 'translation' problem. One Icelandic skipper explained his approach in the following terms: 'It's so strange, when I get there it's as if everything becomes clear. I may not be able to tell you exactly the location, but once I'm there it's as if everything opens up'. Indeed, a frequent comment in interviews with skippers on their fishing tactics is simply 'I cannot quite explain'. How can one elicit and reformulate such tacit knowledge in general terms, in order to incorporate it into biological models and decision-making? If much of the relevant knowledge that skippers obtain in the course of their work is the result of first-hand experience, embedded in the practical world of fishing, its mediation to landlubbers is obviously a difficult task. One of the problems entailed by the collaboration between skippers and scientists, to draw upon J. Kloppenburg and B.R. DeWalt, is to transform practical knowledge which produces 'mutable immobiles' – that is, relatively flexible knowledge geared to the details of a given task of a particular locality – into 'mutable mobiles' (see DeWalt, 1994) – general, holistic knowledge that can be applied to similar phenomena in other contexts. 'An art which cannot be specified in detail', Polanyi points out (Polanyi, 1958: 53), 'cannot be transmitted by prescription, since no prescription for it exists. It can be passed on only be example from master to apprentice'.

The difference between fishers and scientists, however, should not be exaggerated. While much of fishers' knowledge is tacit and non-verbal, one should not forget that they often discuss their observations and theories in fairly clear terms, verbalizing their personal knowledge and their decision-making; as we have seen, fishers often voice their opinions on what goes on in the sea and how to manage the fisheries. Likewise, to repeat, scientific knowledge does involve practical knowledge, obtained in the process of data collection and experimentation. The theoretical de-contextualization of the marine biologists, much like the contextualization of the practising skipper, is necessarily generated and maintained within a community of practice.

One possible avenue for drawing upon skippers' knowledge for the purpose of sustainable fisheries management is to have biologists observe practising skippers on different kinds of commercial vessels during actual fishing trips, using a variety of fishing gear at different times of the year. This would allow them to learn by fishing, much like a novice mate learns from his skipper in the course of production. Biologists would thereby periodically become apprentices, guided by experienced skippers. Knowledge obtained in such practical encounters may later become important for ecological assessment, for the estimation of stock sizes, recruitment, migrations, and carrying capacity. Such an approach seems to have been advocated early on in the writings of Bjarni Sæmundsson, a pioneering Icelandic ichthyologist. In the 1890s, he spent much time learning from practising fishers:

I had the opportunity to observe various kinds of newly-caught fish, to look at fishing gear and boats and to listen to the views of fishermen on various matters relating to fishing and the . . . behaviour of fish   *(Ægir, 1921, 14: 115).*

Sæmundsson seems to have thought of himself as a 'mediator' (*milliliður*) between foreign scientists and Icelandic fishers (*Ægir,* 1921, 14: 116), eager to learn from both groups. Scientific knowledge, along with the 'practical knowledge' (*reynslupekking*) of fishers, he suggested, was 'the best foundation for . . . the future marine biology of Iceland' (*Ægir,* 1928, 21: 102). Sæmundsson did not only regard himself as an apprentice, he was moderately optimistic about the immediate achievements of the scientific enterprise, the prospects of dealing with 'the old mystery, the migration of fish' (*Ægir,* 1924, 17: 144).

Conclusions

Policy makers in fisheries often remain firmly committed to a modernist stance, presenting themselves as detached observers, independent of the 'partial' viewpoints and the trivial, practical knowledge of the actors. A major problem with current bio-economic theory relates to its tendency to separate experts and practitioners. Management is often presented and practised as a hierarchical exercise, the business of privileged professionals. However, the view which presents the pursuit of environmental knowledge as a relatively straightforward accumulation of 'facts' and radically separates knowledge of nature and the social context in which it is produced, has come increasingly under attack in several fields of scholarship, including anthropology, economics and environmental history. We may well be

advised to search for alternative epistemologies and alternative management schemes, democratizing and decentralizing the policy-making
process. It may simply be more effective. As has been argued in this chapter,
learning by doing entails the development of detailed practical knowledge,
the accumulation of personal wisdom which is potentially of crucial importance for any project of resource management that seeks to ensure
resilience and sustainability. Science and practical knowledge should be
seen as complimentary and interactive sources of wisdom, not mutually
exclusive.

The relative failure of the quota system and scientific management of
recent years to deliver the goods they promised, and the severe social and
ethical problems of inequality they have raised (Pálsson and Helgason,
1995), suggest that it may be wise to look for alternative management
schemes emphasizing the practical knowledge of the fishing industry, in
particular the people who are directly engaged with the ecosystem. An
important task on the management agenda is to look for ways in which
practical knowledge can be employed for the purpose of responsible
resource management, bridging the modernist gap between scientists and
practitioners. While the Icelandic trawling rally represents an interesting
endeavour in this respect, it has its shortcomings. It is important to look for
ways of collectively engaging fishers and scientists, of using knowledge
obtained in the course of production and research for the purpose of
responsible resource use. It may be tempting either to submit to the pop-
ulist notion that privileges the indigenous or to contribute to the opposite
enterprise, the reproduction of the master narrative of science. A more realistic and democratic approach, however, would be to search for an egalitarian discursive framework akin to the ethics of the 'ideal speech situation'
identified by Habermas (1990: 85), a communicative strategy for recognizing differences and solving conflicts in the absence of repression and
inequality.

Such an institutional context is particularly important nowadays, given
the mutual integration of many fishing societies and the importance of co-
ordinated fisheries policy. Several studies of European fishing communities
(see, for instance, LiPuma and Keene Melzoff, 1994), informed by theor-
izing on the process of 'globalization', have drawn attention to the ways in
which local concerns are articulated within a larger regional, national and
international context, emphasizing that in order to understand recent
developments and to act responsibly it is necessary to move beyond the
study of either 'local' or 'external' influences to the wider encompassment
of the local community within larger contexts.

Acknowledgements

Parts of this chapter are also presented in 'Learning by fishing: practical science and scientific practice,' published in *Property Rights in a Social and Ecological Context: Case Studies and Design Applications*, edited by Susan Hanna and Mohan Munasinghe, The Beijer Institute of Ecological Economics and the World Bank, Washington DC, 1995. The study on which the article is based relates to a collaborative research project, 'Common Property and Environmental Policy in Comparative Perspective', initiated by the Nordic Environmental Research Programme. It has also received financial support from the Nordic Committee for Social Science Research, the Beijer Institute of the Royal Swedish Academy of Sciences, the Research Center of the Vestman Islands and the University of Iceland, and the Icelandic Science Foundation. I thank Agnar Helgason (University of Cambridge) and Jónas G. Allansson (University of Iceland) for their help with practical logistics in the Vestman Islands as well as interviews and data collection.

References

Ægir 1921–28. Reykjavik. (The journal of the Icelandic Fisheries Association).

Agrawal, A. 1995. Dismantling the divide between indigenous and scientific knowledge. *Development and Change 26*: 413–39.

Alverson, D.L., Freeberg, M.H., Murawski, S.A. and Pope, J.G. 1995. *A Global Assessment of Fisheries Bycatch and Discard*. FAO Fisheries Technical Paper, 339. Rome: FAO.

Brush, S. 1993. Indigenous knowledge of biological resources as intellectual property rights: the role of anthropology. *American Anthropologist 95*: 653–86.

Chapin, M. 1994. Recapturing the old ways: traditional knowledge and Western science among the Kuna Indians of Panama. In *Cultural Expression and Grassroots Development: Cases from Latin America*, pp.83–101, ed. C.D. Kleymeyer. Boulder and London: Lynne Rienner Publishers.

Descola P. and Pálsson, G., eds. 1996. Introduction. In *Nature and Society: Anthropological Perspectives*, pp.1–21. London and New York: Routledge.

DeWalt, B.R. 1994. Using indigenous knowledge to improve agriculture and natural resource management. *Human Organization 53*: 123–31.

Fischer, K.W., Bullock D.H., Rotenberg E.J. and Raya, P. 1993. The dynamics of competence: how context contributes directly to skill. In *Development in Context: Acting and Thinking in Specific Environments*, pp.93–117, ed. R.H. Wozniak and K.W. Fischer. Hillsdale, NJ: Lawrence Erlbaum.

Fogarty, M. 1995. Chaos, complexity and community management of fisheries: an appraisal. *Marine Policy 19*: 437–44.

Gallie, D. 1994. Patterns of skill change: upskilling, deskilling, or polarization? In *Skill and Occupational Change*, pp.41–75, ed. R. Penn M. Rose and J. Rubery. Oxford: Oxford University Press.

Gergen, K.J. and Semin, G.R. 1990. Everyday understanding in science and daily life. In *Everyday Understanding: Social and Scientific Implications*, pp.1–18, ed. K.J. Gergen. London: Sage Publications.

Gudeman, S. and Rivera, A. 1990. *Conversations in Colombia: the Domestic Economy in Life and Text*. Cambridge: Cambridge University Press.

Habermas, J. 1990. Discourse ethics: notes on a program of philosophical justification. In *The Communicative Ethics Controversy*, pp.60–110, ed. S. Benhabib and F. Dallmar. Cambridge, Mass.: MIT Press.

Hollingshead, A.B. 1940. Human ecology and human society, *Ecological Monographs* 10: 354–66.

Hutchins, E. 1995. *Cognition in the Wild*. Cambridge, Mass.: MIT Press.

Jónsson, J. 1988. *Hafrannsóknir við Ísland I: Frá öndverðu til 1937*. Reykjavik: Bókaútgáfa menningarsjóðs.

Jónsson, J. 1990. *Hafrannsóknir við Ísland II: Eftir 1937*. Reykjavik: Bókaútgáfa menningarsjóðs.

Latour, B. 1994. *We Have Never Been Modern*. Cambridge, Mass.: Harvard University Press.

Lave, J. 1988. *Cognition in Practice: Mind, Mathematics and Culture in Everyday Life*, Cambridge: Cambridge University Press.

Lave, J. 1993. The practice of learning. In *Understanding Practice: Perspectives on Activity and Context*, pp.3–32, ed. S. Chaiklin and J. Lave. Cambridge: Cambridge University Press.

Lave, J. and Wenger, E. 1991. *Situated Learning: Legitimate Peripheral Participation*, Cambridge: Cambridge University Press.

LiPuma, E. and Keene Melzoff, S. 1994. Economic mediation and the power of associations: toward a concept of encompassment. *American Anthropologist* 96: 31–51.

Pálsson, G. 1991. *Coastal Economies, Cultural Accounts: Human Ecology and Icelandic Discourse*, Manchester: Manchester University Press.

Pálsson, G. 1994. Enskilment at sea. *Man* 29: 901–28.

Pálsson, G. 1995. Learning by fishing: practical science and scientific practice. In *Property Rights in a Social and Ecological Context: Case Studies and Design Applications*, pp.85–97, ed. S. Hanna and M. Munasinghe. Washington DC: The Beijer Institute of Ecological Economics and the World Bank.

Pálsson, G. and Helgason, A. 1995. Figuring fish and measuring men: the quota system in the Icelandic cod fishery. *Ocean and Coastal Management* 28(1–3): 117–46.

Pálsson, Ó.K., Jonsson, E., Schopka, S.A., Stefánsson, G. and Steinarsson, B.A.E. 1989. Icelandic groundfish survey data used to improve precision in stock assessments. *Journal of Northwest Atlantic Fisheries Science* 9: 53–72.

Polanyi, M. 1958. *Personal Knowledge: Towards a Post-Critical Philosophy*, Chicago: University of Chicago Press.

Smith, M.E. 1991. Chaos in fisheries management. *Maritime Anthropological Studies* 3: 1–13.

Williams, N.M. and Baines, G., eds. 1993. *Traditional Ecological Knowledge: Wisdom for Sustainable Development*. Canberra: Centre for Resource and Environmental Studies.

Wilson, J.A., Acheson, J.M., Metcalfe, M. and Kleban P. 1994. Chaos, complexity and community management of fisheries. *Marine Policy* 18: 291–305.

4

Dalecarlia in central Sweden before 1800: a society of social stability and ecological resilience

ULF SPORRONG

Introduction

The landscape around Western and Eastern Dalälv is quite different in character from the large alluvial plains that are typical of Southern Dalecarlia in Sweden. Settlements are extensive and close together: here are the largest villages in the country, while the area devoted to arable is generally small. The forest dominates, but the attentive observer sees that parcels are considerably fragmented both in forest lands and in areas under cultivation, which is particularly noticeable in forests where felling is in progress.

Fragmentation is often regarded as something negative, as indeed it has been for centuries. Whether or not this has been the case in Upper Dalecarlia will be the focus of this chapter, which describes how the land tenure system was embedded in the local community, in particular in relation to inheritance and acquisition of land, and analyses how the social mechanisms of this unique institution contributed to social resilience and sustainable use of local resources. Following a brief description of the area of study, the chapter continues with a review of the partial inheritance system, the social code that governed land owner-ship in Upper Dalecarlia. Then follows an analysis and discussion of the partial inheritance system of Upper Dalecarlia. The chapter ends with some conclusions about the role of partial inheritance in socio-ecolog-ical linkages. In contrast to the conventional view, this case-study shows that the partial inheritance system was a very resilient institution which contributed to sustainable land use without fragmenting the economic basis of the local community. The period under investigation starts at the beginning of the eighteenth century and ends 100 years later (c. 1820).

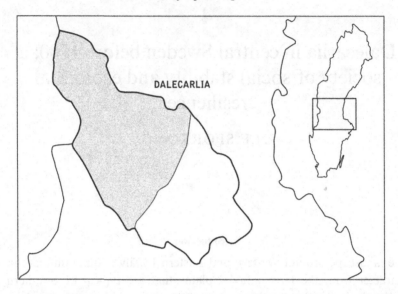

Figure 4.1.   The province of Dalecarlia in central Sweden. The area practising
partial inheritance before 1962 is shaded.

Upper Dalecarlia

The name of the region under investigation requires some explanation. In
the judicial sense, Dalecarlia was divided into Upper and Lower Dalecarlia
as well as the Dalecarlian mining region (*Dalabergslagen*). Upper
Dalecarlia may be said to have consisted of the judicial districts of Ovan-
Siljan, Nedan-Siljan, Nås and Malung. We often differentiate between
Eastern and Western Dalecarlia, depending upon which branch of the
River Dalälv is referred to. Parishes situated on the eastern branch of the
river thus belong to Eastern Dalecarlia. From now on, we shall speak of
Leksand parish in Eastern Dalecarlia which, together with Western
Dalecarlia, forms what we call Upper Dalecarlia or Dalecarlia Proper. The
actual investigation is focused mainly on two central villages in Leksand
parish, Tibble and Ullvi. At the beginning of the nineteenth century these
villages consisted of about 100 farms (Figure 4.1)

Partial inheritance: the social code governing ownership of land

The term 'partial inheritance' (*Sw realarv*) refers to the dividing up of real
estate between all the siblings in one family. The actual partitioning might
be done in slightly different ways, depending on traditions. The opposite of

partial inheritance is primogeniture. The word *realarv* has apparently never been used by the people of Dalecarlia, and should therefore be regarded as a scientific term in this connection. It is derived from the German Realteilung. In recent years, the term *likarättsprincipen* (the principle of equal rights) has been current in Dalecarlia for this form of inheritance (Nilsson, 1929).

Sometimes literature on the subject gives the impression that partial inheritance is the original form of passing on assets from generation to generation. It is mentioned in Jewish law and is founded on the pleasing thought that a good father divides up his assets equally between all his children (Thirsk, 1978). This view may seem quite natural in many ways. The details of how the dividing up is done may vary, but a son generally inherited twice as much as a daughter. The assets were usually transferred when the parents died, but it might be done earlier, as was common in Dalecarlia. In certain cases, vital parts of the inheritance, for instance the main farm on a large property, fell to the lot of the eldest son. Sometimes, this son might also receive a larger inheritance than the other siblings.

Particularly where Upper Dalecarlia is concerned, it is noticeable that the authorities held a negative view of the effects of partial inheritance on the ownership of land and consequently on the condition of agriculture in the region. On his travels in Dalecarlia (in Orsa), Hylphers wrote in 1757: 'When parents who owned a *bolbyställe* (a small part of a mansus unit) and between 3 and 6 shealings (*Scand* seter) die, none of the children will allow any of their land to be redeemed, but each hold on to their part of each of these places, where they build houses, bring new land under cultivation and share the taxes between them' (Hylphers, 1757). However, the fact remains that my investigation showed that in Upper Dalecarlia people often tried to divide up the estate *between* heirs so that sons inherited as large a share in central areas as possible. This meant that sons, and especially the eldest son, often received their shares in toft and arable while daughters received theirs in meadowland, outfields and shielings, that is to say in places where the geographical precision was less clear and land perhaps less productive – boundaries in outfields were not fixed until enclosure. The main reason for this, I think, was that young women were expected to be married to a land owner and, preferably, to a head of a certain family. In such a way the most fertile land could be kept under the control of the tribe. Here there are parallels with other parts of Sweden, where one of the sons generally inherited the farm while the other siblings could inherit land in more peripheral areas, outfields etc., or received their inheritance in the form of assets other than land. The difference between partial inheritance and primogeniture was thus not always as hard and fast as the principles would indicate (Sporrong and Wennersten, 1995).

Primogeniture is certainly also of ancient origin. Its main function was probably to prevent the fragmentation of property, especially among the upper classes, where land ownership was connected with political power. There are signs in older literature that primogeniture in Western Europe is medieval (Thirsk, 1978), but it may well be even older. On the other hand, we know that the practical application of the principle varied from place to place.

Some people, however, are of the opinion that partial inheritance was retained in many areas to the detriment of the political power structure. Typical of areas practising partial inheritance is that they also have lower birth rates (Knodel, 1988: 122–3; 291, Sporrong and Wennersten, 1995: 147). The fragmental effect on ownership could thus be ameliorated.

The geographical spread of partial inheritance

The principle of partial inheritance is dominant in the south of Europe and in the north-western part of the continent. However, it is also to be found in Eastern Europe as well as on the Atlantic islands, and it is probably in the context of these islands that we should see the existence of partial inheritance in Dalecarlia. In general it may be said that partial inheritance belongs to the south of Europe while primogeniture dominates in the northern part (Schwarz, 1966: 51; Thirsk, 1978).

Partial inheritance is typical of regions where the means of sustaining life are poor, and it is tempting to ascribe its effects and the consequent fragmentation of land to this fact. In districts practising partial inheritance, other occupations besides agriculture are often to be found, not infrequently early forms of industry, proto-industries or fishing. The geographical connection with mountainous regions is common in all parts of Europe where partial inheritance is to be found. As has been pointed out, these regions have a multitude of ecological zones that can be utilized on a small scale. A diversity of occupations is also typical of mountainous regions – the different trades complement each other, often in different seasons. All these conditions seem to be fulfilled in Upper Dalecarlia.

On the social level, regions where primogeniture reigns are characterized by class distinctions among the population, whereas what is typical of regions with partial inheritance is social equality, not least between older and younger generations of landowners. The right to belong to a legitimate class of landowners despite a small holding of land often leads to great density in settlements – 'Vermehrung durch Veringerung'. Several families might share a house, but it was quite common to build new dwellings – settlements were filled with houses (Schwarz, 1966: 162). The same is not always true of the amount of arable belonging to farms (Bentley 1987), a matter discussed later in connection with Upper Dalecarlia – apparently one of the few places in Europe where the source material enables a close study to be made of how the inheritance of real estate in its entirety is transformed generation after generation. The problem is that there are few 'ecological' data from the actual period, but there are some estimations made by travellers, local politicians and land surveyors.

As already mentioned, partial inheritance is not an exclusively European phenomenon. It is to be found in the Far East in connection with artificial irrigation. Immigrants brought the principle with them to some parts of North America, while in other parts primogeniture was the rule. In actual fact, partial inheritance has been widely practised all over the world at different times. If we look at relatively modern conditions, it is to be found in large parts of Asia, Africa and Central America (Bentley, 1987).

Common to many of these areas is the fact that authorities and agronomic experts have maintained that, on the whole, partial inheritance has had a detrimental effect on the development of farming. This has also been the prevailing view in Upper Dalecarlia ever since the reign of the Vasa dynasty (sixteenth century). It has often been thought that partial inheritance leads to the fragmentation of land at the same time as it results in over-population in relation to the meagre resources of the region. To prevent fragmentation going too far, a minimum size for farms and a minimum total area of arable land were introduced as threshold values. In Dalecarlia, a minimum size for what might reasonably be regarded as a mansus unit was discussed, and varied from district to district.

It should be pointed out, however, that scientific literature often contains information showing that, in practice, partial inheritance did not result in such enormous fragmentation of properties as might have been expected. It has obviously been possible to deal with the question – in somewhat different ways as it turns out. Precisely this question is the central point in the present studies of Upper Dalecarlia, an account of the findings of which are given later in this chapter. What should be emphasized here, however, is the information indicating that no change was made to the physical shape of fields, arable or meadowland, by dividing them up and sharing the land in smaller parcels among the heirs: instead, these assets were handed over intact. Several examples from the Central European Alps indicate that the practice there was similar (Bentley, 1987). Furthermore,

farmers commonly supplemented their ownership of land in every way, for
instance by marriage, or by taking over land from unmarried relatives
without issue, or from kinsmen whose assets were not sufficient for them to
be reckoned as owning a mansus unit. It is characteristic of regions practising partial inheritance that kinsmen have great influence on what is to be
done with land, particularly in matters concerning its sale to persons who
are not kinsmen (prior option, pre-emption).

Partial inheritance in Upper Dalecarlia: the case study
The rural landscape in Leksand between 1734 and 1819

My investigation confines itself to the period between the two parish land-
measurement programmes of 1734 and 1819. This does not prevent me
from looking further backwards or forwards in time. Unfortunately, no
proper maps were made before enclosure. Even though the minutes of the
land-measurement programmes contain a great deal of detailed information, they do not show how the rural landscape was organized in all par-
ticulars; in other words, we do not know what it actually looked like.

In most cases, however, we do know the names of the parcels and we also
know what they were used for and how large they were. It has been possible
to identify nearly 90% of the names in the cadastral rolls, either in later
maps or by means of verbal information, and therefore to obtain a good
idea of the fragmentation of holdings and the geographical distribution of
parcels belonging to one and the same farm. So, in brief, comprehensive
information is available about how the landscape was used in the parish of
Leksand, but not about the exact geometrical formation of the parcel
system. This means that unfortunately a great deal of the information
about the genesis of the rural landscape has been lost.

Enclosure maps show the situation that existed after enclosure had
been carried through, and on them the boundaries are quite normal for
those days – relatively large blocks with straight sides. If one studies these
maps more closely, however, one can see the rough outline of how the land
was partitioned before enclosure. The surveyor had mapped the areas
under cultivation, and one can frequently see that the parcels had been
laid out in rectangular blocks, which is what one might expect in view of
the fact that the arable landscape had developed out of what had originally been a one-field system of compact holdings. The names used in
older estate inventories also indicate that the parcels differed from each
other in shape. The pieces of land were small, generally not larger than 10
snesland (about 1300 m^2). They were often much smaller, only a few

hundred square metres in size. The rolls even include land that is only about ten square metres.

A distinction was made between measured land and so-called reduced land, which means that the land quality was graded on a scale running from one to ten (where four to five was the normal figure), and the measured area was multiplied by one tenth of the value of the land. An area that measured 6 *snesland* and had a land-value of 4 was given a reduced value of 2.4 *snesland*, or 2 *snesland* and 4 *bandland*. The relationship between measured land and reduced land differed in other parts of Upper Dalecarlia. In Floda, for example, the relationship was 1 to 6 (Pallin, 1977: 18, 154).

The minimum acceptable size for a farm was set at 1/16 of a mansus unit or 4 *spannland* of reduced land. The method of reducing land was probably introduced in conjunction with the Crown land-measurement programme in the 1670s.

A parcel was usually a cultivated area as well, although there were exceptions. In certain cases, a parcel had a quality rating that was not the same for the whole area, something that was carefully noted in the cadastral rolls. Furthermore, different parts of the land in the same parcel might be used for different purposes, for example some of it might be arable while other parts were meadow. Data pertaining to those areas in the same parcel which gave a different yield, or were used for different purposes, were given in brackets in the cadastral rolls. These brackets thus indicated that the figures within them differed in some respect from the others. All figures in one pair of brackets therefore referred to the same parcel. This is an important piece of information, since it tells us that parcels in a certain field belonging to one farm were probably not contiguous, but separated from each other, unless the figures were included in the same pair of brackets. This gives us a clear indication that a system of fragmented holdings existed.

Arable land was divided into fields, and the fields were bounded by wooden fences 'supported by stones'. During the seventeenth century the villages investigated used a four-course rotation, but in the more outlying parts of the landscape a one-field system was also to be found. There are also examples of two-course rotation, and when we approach the time of enclosure, Ullvi was using three-course rotation. The somewhat unsystematic way in which arable land is divided up is typical of relatively young agricultural districts where the village organization was weak in the sense that the effects on the landscape of inheritance and landownership were continually changing from generation to generation. The ownership situation was in a state of constant flux.

The size of the fields also varied; in the central parts of the village area, they were relatively large with a well-developed open-field system. On the outskirts, they were often small, adapted to the terrain and not infrequently privately owned. Meadowland was generally a fenced-in part of the field but there was no permanent, clearly marked boundary between the different parts of the parcel. Meadow was delimited with boundary stakes of stones. There was also meadowland on the shielings, but it was on poorer land, gave a poorer yield and was often called *hackslog*. Outfields were unfenced.

However, it is possible to reproduce to some degree the geography and structure of properties in a cartographic sense. Since such a large number of the old parcels had names that give an indication as to their geographical situation, it has been possible to produce property maps showing where each farm had its land. Figure 4.2 shows how the land belonging to farm number 2 in Tibble village was distributed over the village territory. The farm was one of the larger properties right in the core of the village, and we can see that it owned parcels of land practically everywhere in the village domain, at least in the centrally situated parts. For that matter, the same is generally true of all large, centrally situated farms. On the other hand, the land belonging to farms more on the outskirts of the village, for instance Lilla Tibble to the south, was distributed on a different pattern. These farms had their land concentrated in the southern part of the village (Figure 4.3). This implies that the land belonging to new and relatively small farms was less fragmented, which seems quite natural if one considers the effects of partial inheritance on older farms.

In both the eighteenth and the nineteenth centuries, the size of farms varied markedly. In the eighteenth century, the smallest farm in Tibble had five shares of the arable and meadowland while the largest had 123. Around 1800, the corresponding figures were 26 and 284. We can, of course, break down the statistics for each farm quite a long way. The land owned by individual farms can be studied down to the last detail, and many interesting differences between them come to light. Here, however, I wish merely to point to one similarity between properties. However one structures the ownership of land at the farm level, the mean value of the land for taxation purposes is about the same for all farms. This must be interpreted as meaning that all landowners have shares in all kinds of land. There is nothing to show that newer farms had inferior land in an outlying site or vice versa. This phenomenon, too, is probably a result of partial inheritance – heirs inherited land both in central and outlying situations.

One further comment concerning the physical shape of landowning: even though the mean value of land was approximately the same for all

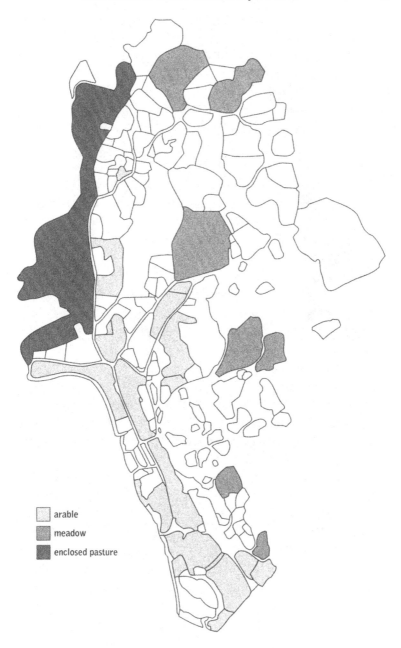

Figure 4.2.   Land belonging to a farm in the central part of Tibble village, Leksand.

Ulf Sporrong

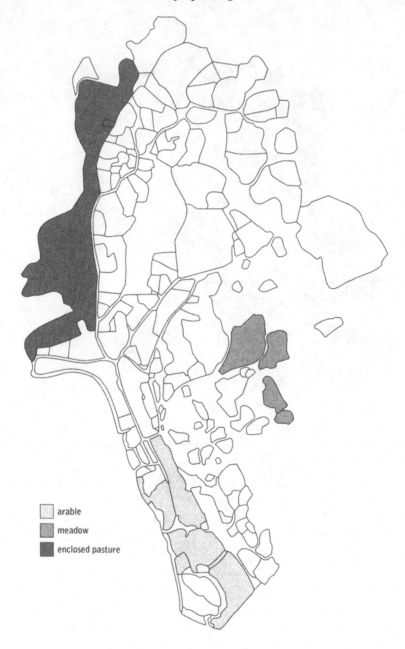

Figure 4.3.   Land belonging to the farm 'Lilla Tibble' in the southern outskirts of
Tibble village, Leksand.

farms, different parcels within the same field could give very different yields. To sum up: both the number of plots a farm had in a certain field and its share in the best parts of that field varied considerably from case to case.

The original crops were mostly rye and root vegetables, but during the eighteenth century barley, peas and mixed grain were introduced. Cattle farming was said not to be particularly successful and though there were many shielings, they gave very little return. Arable was probably farmed individually, each man doing as he wished, which indicates that the inner village organization of natural resources was weak. However, co-operation did occur. One example was when several farmers joined forces to fence in arable land that was to lie fallow, which in principle was done every four years. The relationship between arable and meadow is not very clear but, judging by the cadastral rolls for 1734, both were regarded as having the same value (the same scale for evaluating them was used). At various times it was therefore possible to till meadowland and cultivate it temporarily when the arable land was lying fallow – the land use was very flexible.

Finally, if we imagine we are looking out over the rural landscape round Leksand just before enclosure began, we could describe what we see as follows. Settlements lie concentrated in tribe-related clusters, i.e. around several centres in the same village domain. Arable land is divided into parcels. These parcels are small in area, on average about 300 m². The name of each parcel tells us where it is situated. Where the arable land is extensive, arable land enclosed within the same fence is given a name ending in -*åker*, for instance Våtåkern ('the wet field'), Krokåkern ('the hook field') etc. In outlying areas the word *gärde* is used, for example Trangärdet ('the crane field'), Råggärdet ('the rye field') etc. Smaller fenced-in fields have their own names such as Erkers *åkern* ('Erker's field').

An arable landscape looked something like a patchwork quilt, with the smallest fields centrally placed in the landscape. Areas of meadowland were often larger. The outfields nearest to the village acted as grazing ground; the land here was cropped bare but offered a broad view. It is typical of this landscape that everything is on a small scale and open to view. One more thing that should be emphasized is that, although there was great mobility in ownership, the boundaries between parcels of land changed little. This was due to the effects of partial inheritance – farmers did not willingly divide up such small parcels of land as these; they were part of the system, so to speak. These parcels were seldom partitioned when they were passed on to the next generation; heirs were generally allotted whole parcels. And, according to contemporary observations, the agricultural production in Upper Dalecarlia functioned very well. In his report in 1822, the province governor, Hans Järta, certified a yield above mean compared with Sweden as a whole (Bortas, 1992).

Land measurement and results

It is obvious that the unstable system of land ownership of Upper Dalecarlia might often be thrown out of gear. A landowner could certainly keep account of his parcels, but for sheriffs and supervisors it was a different matter. This instability was probably also the reason why the parish, not the individual farm, was the taxation unit in Dalecarlia for long periods. Land-measurement programmes of various kinds were carried out from time to time to try to eliminate the uncertainty about who owned what. These measurements might be at the farm, village or parish level, this last certainly being the basis on which taxes were calculated or, as in 1820, the basis for enclosure.

The parish land-measurement programmes of 1670–72, 1734 and 1819–20 are well known. The measurements themselves were made by farmers who were skilled in the art (*revkarlar*), under the supervision of trustees from the neighbouring parishes. Parcels were divided up into geometrical figures, measured, and the area of each calculated, although no regard was paid to any slope in the ground. The area was then expressed in terms of *bandland* (12.7 m^2) or *snesland* (equal to ten *bandland*).

Let us describe the routines around the measuring process. For each parcel, the landowner prepared two wooden tallies in advance. On these tallies a note was made of how the land was used as well as of its quality rating. These tallies were then countersigned by the *revkarl* . One tally was kept by the owner of the land and the other was handed over to the recorder in the parish council building, who wrote down the information in the cadastral ledger. Later, the landowner dropped in to check the information, the two tallies were compared and, if they agreed, each tally was ticked off with a red cross and the landowner took his tally home with him again. Tallies were threaded together on a piece of string, and could act as title deeds to the land. During the enclosure process, these tallies were regarded as part of the documentation (Archives of the Museum of National Culture, EU 3576 and 4136).

In some places, a set of tallies was accepted as full proof of ownership. The second set was kept in the village tally store and looked after by a special tallyman. If a landowner wished to change his holdings, he added new tallies to his stock or handed over some of his old tallies, according to whether he acquired or relinquished land, and corresponding changes had to be made to the second set in the tally store. Elsewhere, old debts could be settled and goods could be paid for with tallies at the local shop. Indeed, at times they were treated purely as commodities. When the title to land was to be registered, the claimant might authenticate his right of ownership in court by producing a bag full of tallies which he emptied on to the magistrate's table (SOU, 1931: 103).

Tallies thus represented the actual ownership of land, and this way of recording the fact has parallels in other parts of the world. England also used a system of tallies – long, thin pieces of wood with a hole in the middle. When a transaction had been concluded, the piece of wood was broken in two, each party retaining his half as proof of what had been decided. If a dispute about the matter arose later, the two pieces could be produced and compared to see if the broken edges fitted together. The system is ancient, going back at least to the early Middle Ages when the King of England borrowed money from his noblemen. Each tally represented 100 pounds– 'the stock'. The two broken halves of the tally were exchanged when the loan was handed over – the stock exchange (The Universal Dictionary, 1936).

The parish land-measurement programme of 1734

As far as we know, the1734 land-measurement programme gives the oldest detailed account of the holdings of arable in the whole of Leksand parish at one time. Comparisons between different villages can be made.; the following is an account of conditions in just two of them: Tibble and Ullvi.

In 1734, Tibble and Ullvi were approximately the same size: Tibble had 37 properties or farms and Ullvi had 33. The fiscal term 'mansus unit', has already been used in this chapter; units thus classified in the cadastral rolls were later subject to taxation. Here, the modern term 'property' is equivalent to 'mansus unit'. The word farm is also used in the following account but is applied first and foremost to family or household units.

The mean number of parcels registered in the cadastral rolls per property was 42 in Tibble and 45 in Ullvi, so here, too, there is a fair amount of agreement. As the countryside is much the same in the two neighbouring villages, it is not surprising to find that the structure of the properties is also similar. The mean value of the parcels on the scale of 1 to 10 was 4.25 in Tibble and 4.4 in Ullvi. The total number of parcels in Tibble was 1600 and 1500 in Ullvi. The mean size of the measured land was 2.5 *snesland* in Tibble and 2.7 *snesland* in Ullvi. The corresponding reduced areas were 1.05 and 1.16 *snesland,* respectively.

It is therefore highly probable that the quantitative figures in the cadastral rolls of 1734 can be compared with each other. The picture this source gives us of the rural landscape conveys an impression of landownership that was on a very small scale and highly fragmented. In both cases, about 1500 parcels of land with an average size of only about 300 m^2 were scattered over central parts of the village domain. Thus the question we must ask is whether these small parcels were the result of a partitioning of farms that had gone a long way by 1734, or whether the parcels were older. Might they have been in use for a very long time? Perhaps for several hundred years?

To penetrate deeper into the statistical picture we have also distributed the parcels according to their quality rating. Figure 4.4 highlights some interesting aspects which, in part, are common to both villages.

Ullvi used the entire scale of evaluation: eight parcels have a value lower than one while no less than 29 were awarded the highest value (10), which made their owners very proud according to oral traditions that are still alive – fancy owning a field worth a 'ten'! In Tibble, there were no parcels with the lowest value, but as many as 23 had the highest. There is thus no great difference between the villages in this respect, either.

The statistical values for both Tibble and Ullvi are peculiar insofar as they are bimodal. In Tibble, the material centres around the values 2 and 6; in Ullvi, 2 and 5. These small differences might indicate that, in spite of the mean values given above, conditions in Tibble were in fact slightly more favourable than in Ullvi. What, then, is the explanation of the bimodal distribution of land classification? We will return to this question later when we have compared the development of the two villages up to the time of enclosure, but it is probably connected with the fact that land here, as in other parts of Upper Dalecarlia, was classified either as arable or as meadow when enclosure was begun (Pallin, 1977: 18). In actual fact, the dividing line between arable and meadowland was not always so hard and fast as one might assume. This is confirmed by a map of Tibble from 1692, which mentions that certain enclosed arable lands sometimes lie fallow and are mowed, but when the map was made they were marked as arable.

The measurement in preparation for enclosure in the 1820s

Compared to 1734, the number of parcels registered had risen dramatically by the 1820s. It is difficult to determine to what this increase was due. Tibble records slightly more than 6800 parcels, distributed among 68 properties, while the figures for Ullvi are 7600 and 50, respectively. More properties were partitioned in Tibble than in Ullvi, with the result that the mean

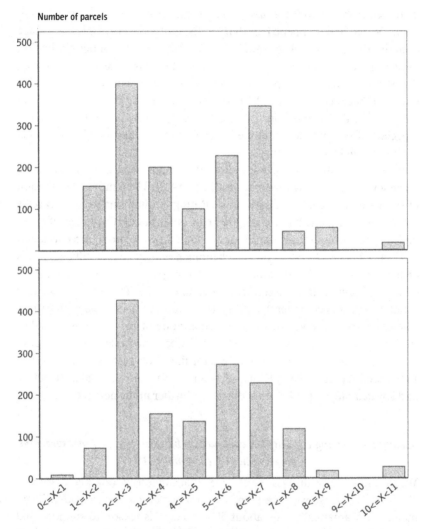

Figure 4.4. The parcels belonging to Tibble (above), and to Ullvi (below)
according to their quality rating in 1734.

number of parcels per property was 100 in Tibble, but as high as 152 in
Ullvi.

The mean value of the parcels had fallen drastically during the period –
from 4.25 to 2.75 in Tibble and from 4.4 to 3.0 in Ullvi. The reason for this
drop in mean value is almost certainly that the large number of newly cultivated parcels gave a lower yield than did the old udal land. The supposi-
tion is supported by the fact that the mean size of the measured plots

increased from 2.5 to 6 *snesland* in Tibble and from 2.75 to 3.5 *snesland* in Ullvi. What had happened is that in the outlying parts of the village domain, the new parcels of land that had been brought under cultivation were larger per property than were the old fields. It is remarkable, however, that the mean area of the pieces of land expressed in reduced land had also increased between 1734 and 1819. The mean value of the parcels rose from 1.0 to 1.3 in Tibble and from 1.2 to 2.1 in Ullvi. This may seem surprising, especially if one imagines that the fragmentation of land steadily increased during the eighteenth century.

If we take another look at the statistical distribution of quality ratings, we make several interesting observations (Figure 4.5). The bimodal distribution of the material has almost disappeared. Ratings around 1 on the quality scale are the most frequent in both villages. There is admittedly the vestige of a secondary peak around the ratings 3 and 4, but it is not so pronounced as it was in 1734. As the explanation of this phenomenon may be that different kinds of land are given different ratings, we listed the parcels in the two villages as either meadow or arable (see Figure 4.6a and 4.6b). The result shows that meadowland was consistently given a lower rating – about 1 in both cases – whereas arable land was mostly assigned a rating of between 3 and 4. Also, the land in the two lowest ratings (1 and 2) was almost exclusively meadow. This means, firstly, that the type values for the whole material were lower in 1819 than they had been in 1734 and, secondly, that the bimodal distribution had levelled out by 1819. This is discussed further in the next section.

Changing carrying capacity? A comparison between the land-measurement programmes of 1734 and 1819

We should perhaps discuss the changes that obviously took place between 1734 and 1819. As we have seen, the number of parcels in both villages increased dramatically – by about 500%. There is reason to suppose that the details in the land measurements on which this observation is based are not entirely comparable. However, they undoubtedly indicate that a great deal of new land had been brought under cultivation, especially if we consider the changes in the size of parcels that have taken place: they did not decrease in size, i.e. they were not partitioned between the two measurements. So, the number of properties doubled while the number of parcels per property increased by 300%. But, while the number of farms more or less doubled, the relative proportion of small to medium-sized and large farms remained surprisingly constant.

This indicates that there has been no fragmentation in the properties. On

Number of parcels

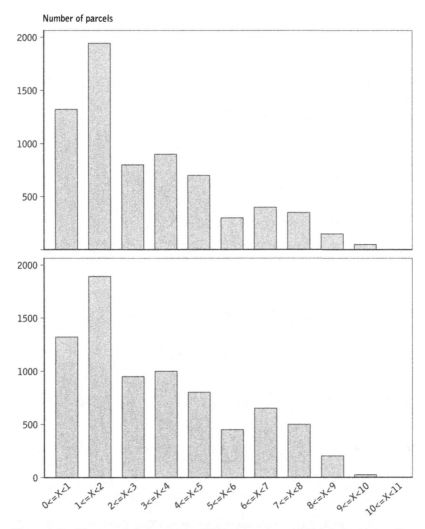

Figure 4.5.   The parcels belonging to Tibble (above), and Ullvi (below) according
to their qualitative rating around 1820.

the contrary, farms have on average, kept or even increased their holdings
of arable. New ground was broken to compensate for the increase in
population. On the other hand, there is reason to doubt the data showing
a dramatic increase in the number of pieces of land, and to query the proportion of arable to meadow in the 1734 cadastral rolls. May one of them
be under-represented? We can assume that a farm's economy rested on the
same type of land use in 1734 as it did in 1819. The 1734 cadastral rolls

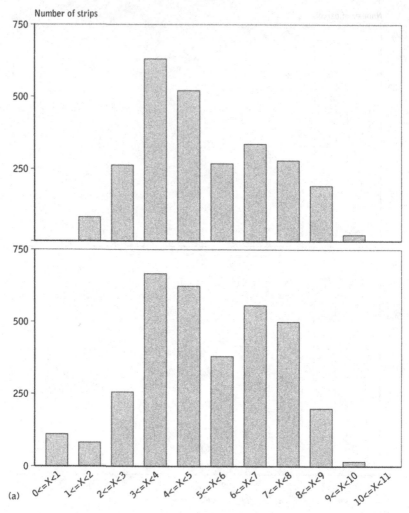

Figure 4.6. Strips of (a) arable and (b) meadow according to quality rating in Tibble (above) and Ullvi around 1820 (below).

showed data bimodally distributed: according to the present interpretation, this is due to the fact that arable and meadow are recorded in the rolls without any distinction being made between them. It can be seen that in 1734 the proportion of arable to meadow was 1:1 whereas in 1819 it was 3:4. The percentage of meadowland had thus increased. Furthermore, land with a very low quality rating was included in the rolls in 1819, but not in 1734. There is reason to believe that the 1734 rolls contain a certain misrepresentation; at any rate, no one seems to have bothered to record the

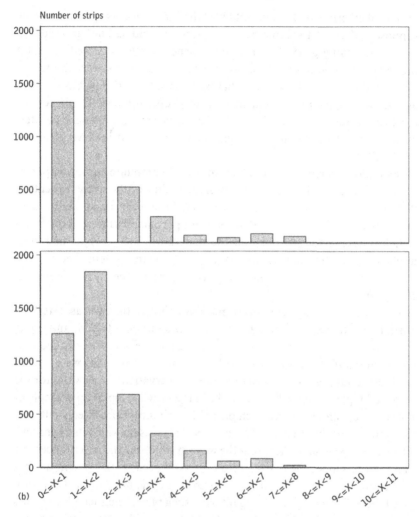

Figure 4.6. (*cont.*)

poorest areas of land. The alternative is that meadow lying outside the centrally situated fields (*Sw hackslog*) was not included in the rolls.

The observations made concerning properties and crops are among the
most important results of the investigation. They show that the 'pernicious
fragmentation of land' that central authorities referred to whenever conditions in Dalecarlia were discussed, and which had to be stopped, had in fact
never taken place. On the contrary, the size of farms and pieces of land had
actually increased during the period! The explanation is that the authorities never understood that the system of land ownership was the product

of partial inheritance. They had allowed themselves to be hypnotized by the supposed effects of the principle of equal rights, and had not grasped that in the succeeding generation the consequences might very well be that a property increased in size, for example as the result of a suitable marriage. It is interesting to note, however, that far from sharing the negative view of agriculture in Upper Dalecarlia held by their superiors, civil servants on the local level often held a positive view of conditions in the region – an interesting reflection on the lack of insight shown by the central powers (Bortas, 1992: 92).

It should be mentioned here that an investigation into the development of settlements in Leksand as early as in the middle of the sixteenth century does not show any clear signs that farms had been fragmented. The number of farms was quite modest at that time, but partial inheritance should nevertheless have made its mark on the material if the principle had been applied. The matter must clearly be examined in greater detail before there can be any certainty about the antiquity of partial inheritance (Svensson, 1987: 48).

Two assertion which can now be made are that, in the two cases studied, partial inheritance did not lead to the fragmentation of land, and estate inventories have shown that the parcels themselves were not divided into smaller units when they were handed down; in most cases they were distributed intact among the heirs. Should these observations prove valid for the whole of Upper Dalecarlia, we can hold the province up as an example of how a European region in which partial inheritance is practised dealt with the matter without fragmenting the economic basis on which its farms rest. There is a great deal to show that the ways in which partial inheritance were practised in Upper Dalecarlia were preserved because the system did not lead either to social deprivation or to an economic or ecological catastrophy. Upper Dalecarlia was an egalitarian society with neither a lower class nor a real upper class. The growing population could continue to be a class of freehold farmers right up to the time of enclosure. As far as we can see, the people of Upper Dalecarlia had created a sustainable society founded on multiple use of the natural resources built on different ecological zones in the landscape.

Enclosure in Leksand and its effects on a sustainable society

Enclosure in Leksand was, as in other parts of Upper Dalecarlia, an extensive and important process – a tour de force on the part of surveyors that was not to be repeated until present times. The course of events is described partly in the records of various surveyors, which are split up among several archives, and partly in the minutes of parish council meetings. The original reports, which contain the rules and the reasoning behind enclosure, are now kept in the Archives of Local History in Leksand. The minutes of the parish council in 1817 show that the farmers in Leksand wished the plans for enclosure to proceed provided that the state paid compensation as it had promised.

However, such a positive attitude to enclosure was rather unique for the region. Circumstances surrounding the implementation of enclosure in Upper Dalecarlia are described in several places, including in parish monographs, and it is clear that it was opposed by many people. In Leksand, things were somewhat different. For further comments on enclosure in this parish, see Bortas (1992), Skommar (1941) and Sporrong (1987). However, one or two things of importance to this monograph should be mentioned here.

The actual process of surveying and the local land measurement programme were kept quite separate. A large number of *revkarlar* worked in the vanguard using the traditional method of measuring land. Their task was first and foremost to determine the size and quality rating of properties on a well-tried local scale, according to which the poorest land was given the rating 0·25 and the best 10. A tallyman was appointed to take care of the tallies which, on this occasion, probably numbered between 100000 and 200000 and were kept in a storehouse near the parish church specially acquired for the purpose.

For their part, the surveyors measured the area and mapped the rural landscape without paying any attention to how the parcels were partitioned. This is why maps of the enclosure lack information about the actual appearance of the strip field system before enclosure. All that the surveyors took from the *revkarlar* were the quality ratings, which they used to adjust the size of properties and then made proposals for new boundaries between properties. In most cases these were probably very different from the old ones. This is the conclusion one arrives at after studying details on the maps, for instance sloping ground where terraces and other obstacles to cultivation are mapped beside the new boundaries, which were often marked in red on the copies. When the arable land had been mapped, work was begun on the forests and shealings.

It was many years before enclosure was completed. Time and again adjustments and alterations had to be made to this unstable system, which could never be mapped entirely correctly – in some place or other, land was always being sold, pledged or transferred.

After enclosure, however, it was the map one turned to in matters concerning the ownership of land in the village – the time for tallies was over. But the moment enclosure was enforced, the ground was cut away from beneath the well-functioning system of inheritance that Upper Dalecarlia had developed over the centuries. It was based on verbal agreement on the partition of land between the parties concerned. There were no legally binding boundaries, and the authorities were seldom involved. It was a way of evading insight, which was certainly part of the reason why the authorities regarded the situation as unsatisfactory.

The new boundaries between the properties in the village thus included old infields, outfields and shealings. The boundaries between properties ran like long ribbons across the terrain, as they still do today in some places. Boundaries were often drawn right across the fields, up towards high ground. These strips were not infrequently several kilometres long but only 20–30 m broad. Certain exceptions were made for old farms and new land brought under cultivation in forests, which were often allowed to retain their irregular boundaries, as can clearly be seen on modern economic maps.

What those people who pressed for enclosure thought they were achieving was a way of doing something in a positive sense about the fragmented state of land ownership and gathering the small fields together into larger units. Farm land was joined to farm land in this process, which was modelled on what had been done in other parts of the country, even though enclosure there had already become out of date and had been replaced by more radical forms of enclosure which later served as models for *enskifte* (1803) and *laga skifte* (1827). It is not too much to say that when enclosure was begun in Dalecarlia, that phase in the structural transformation of agriculture was already over in the rest of the country.

What the enthusiasts had not reckoned with was that landowners would continue to practise *sämjedelning*. Enclosure froze the ownership of land since the boundaries of all land had been fixed and the land assigned to some definite property. The flexibility was gone. However, the old principles of partitioning were nevertheless still applied, generally without authorized confirmation. This state of affairs did not finally cease until *sämjedelning* was prohibited by law in 1962. Between enclosure and the 1960s, the 'wrong' kind of partitioning was applied to the new system of dividing up land – a form of partitioning that the new system had not been designed for. Here we come across a new sort of fragmentation – one that could not be neutralized by the buffer effect of the old traditional means whereby one could compensate losses by marriage, purchasing small pieces of land or other systematic measures to avoid fragmentation. The old system was independent of 'economic boundaries'. The consequence of the change was that village land, both in forest and arable, was split up into ever smaller parcels within the framework of this 'new' enclosure. The effects have proved lasting and are, in the opinion of many people, detrimental to modern economic management, particularly in the forests.

Furthermore, this fixing of fields led to the gradual extinction of a social system that had been conspicuous for mobility between families and farms. General changes in society also allowed commercial interests to gain in strength. A landless class emerged, and the picture we have of impoverished Dalecarlia and the resultant emigration to America is, in this author's opinion, largely a nineteenth-century phenomenon and one of the consequences of enclosure. Without exaggerating too much, one might say that enclosure, which was meant to solve the problem of fragmentation, successively killed off the old customs connected with land rights and brought about a new and more negative form of fragmentation – which is exactly what enclosure was not intended to do. A well-functioning society was knocked out because official thinking was not in tune with local traditions, and the consequences were the opposite of what had been intended. It is these effects that various plans for rationalization hope to cure today. Bortas (1994) gives a most illuminating and detailed example of what happened to the development of properties in the village of Hagen. The principles of partial inheritance and property-building were applied in the old way, but using the 'new' fixed boundaries (Figure 4.7).

Partial inheritance as applied in practice – some conclusions for analysing the link between social and ecological systems

Circumstances that are apparently unique enabled a detailed study to be made into how partial inheritance was applied to land ownership and the partition of land in Upper Dalecarlia. The small parcels involved were functional in the system described. On the death of their parents, all the children inherited land in central areas, girls half as much as boys. The aim was to make it possible for sons to hand on the farm intact. The custom of leaving land to girls lived on up to the 1950s.

When an estate was distributed, each heir received a number of whole parcels of land so that the total corresponded to his share of the inheritance. A parcel was not normally divided among the heirs. To cope with the registration of these small areas, a special system of measurement was introduced and measuring techniques were developed which, although they

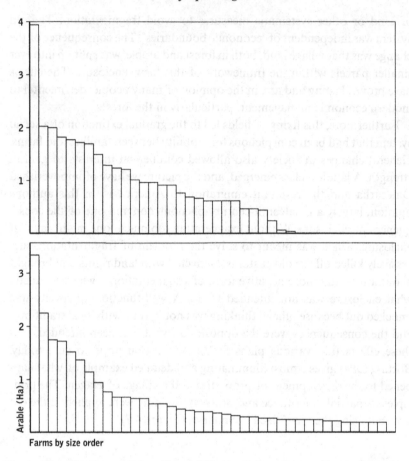

Figure 4.7.   The farms of the village of Hagen, Leksand, according to farm size 1906 (above) and 1970 (below.). (Source: Bortas, 1994.)

might have been used in other parts of the country, may have been unique to a region which practised partial inheritance, since the two phenomena belong together. There is, at any rate, no evidence to show that units such as *bandland* and *snesland* have even been used in Sweden other than in the region with partial inheritance.

It seems as if heirs began the reconstruction of their new property immediately, first and foremost by making a suitable marriage, but also by buying and selling land. One property was partitioned and at the same time a new one was built up. Practically everyone in the village took part in the process. Few people were landless. On the other hand, fairly large estates were sometimes built up, thereby creating a certain social differentiation, a situation that at least the well-to-do tried to preserve. In all these transactions, the origin of the land was always borne in mind. A wife's land was often marked in a special way in cadastral rolls. In the same way, land that had been redistributed to take care of the rights of children born in a previous marriage when a widow/widower remarried (*Sw avvittring*) was marked as belonging to the children etc.

The factors described above indicate that the ecological, physical and social structure of the rural landscape was fairly stable. Few boundaries were changed. New farms or groups of farms were constantly springing up, but at the same time other farms were dismembered and the buildings pulled down. The ownership of any one parcel of land, on the other hand, was in a constant state of flux. In the abstract sense, the system was extremely fluid.

When there was a scarcity of land, new ground could be broken in outlying areas. If the owner had sufficient land to give him the right to build on the newly cultivated land, the farm might be situated there. It seems apparent that partial inheritance was a system that worked excellently in Upper Dalecarlia, both ecologically and socially. In some respects, it worked considerably better than the corresponding system in other parts of the county where the struggle for social status and ecological sustainability was much tougher, or at any rate had more drastic consequences for those who were obliged to give up their position in society. The rural community in Upper Dalecarlia was egalitarian and in good ecological balance.

There are few remaining reliable 'ecological data' from Dalecarlia regarding the situation as early as in the eighteenth century, but civil servants on the local level held a positive view of the conditions, for instance when they reported figures to their superiors on yield, cattle breeding, etc. However, there are sources available from which some conclusions can be drawn. There was a great expansion of people and arable between 1734 and 1819, but the mean size of the farms seemed to be constant over time. No landless people appeared and every family was, by tradition, using a great variety of ecological 'zones' in the landscape. So, in spite of increasing pressure on the resources, the local system of production maintained its resilience. The main reason for this seems to be the tight social control over people and land that was apparent in this society. Partial inheritance is an essential part of that system. A high degree of flexibility seems to be another.

The land was owned individually within the frame of a family-based tribal system where the village elders held command over the social and ecological reproduction of resources, i.e. all the complex links between land use, property rights, marriages etc. There is an interesting amalgamation of individual rights and the local society's social control, building upon a diversified use of the resources in the forest and mountain district. There are parallels in the rest of Europe, for instance in the Alps and the Pyrenees. The most striking thing in the Swedish example, however, is the social homogeneity of the population: everyone was allowed to take part in all socio-economic decisions of the community. This feature of Dalecarlia is as well known in Sweden's history as it is unfamiliar to the rest of the country – something that often caused tensions between the central government and the remote rural district around Lake Siljan in Upper Dalecarlia.

This was the reason why enclosure was introduced fairly late in the district. It had been quite a success in the rest of the country where, in its way, it announced what we today look upon as Western resource management. In Dalecarlia, however, where production aimed at self subsistence was still going on, very few asked for the reform. To the traditional social and ecological system of Dalecarlia, the land reform was a catastrophy. Since then there has been an ever-decreasing farm size, starting at the beginning of the nineteenth century leading to poverty, fragmentaton of land and emigration from the area. However, partial inheritance was still practised within the frame of new, fixed economic boundaries. The consequence was that farms got smaller and smaller, and this was the real and gradual fragmentation of land! Of course, there were other driving forces involved as well, such as an increasing population and a growing commercial interest in forest land for industrial use.

The curious thing is, however, that today people seem to be adapted to the 'new' situation derived from the enclosure movement in the nineteenth century. Now, when the Land Survey of Sweden wants to reorganize the ownership pattern (especially in the forest lands) for economic and strategic reasons, there is once again – in some parts of the province at least – a strong resistance to a change of the traditional, fragmented ownership pattern. This could be looked upon as resisting progress, but the use of this fragmented landscape with parcels spread out over vast areas has become something of a life-style of the region. 'Those small plots belong to me and I want to go on with my hunting and fishing in different places. No one has the right to interfere, not even the forest companies.' This attitude is founded on a tradition going far back in history.

Of course, this is an interesting phenomenon and there are some general conclusions that can be drawn. Before new ideas and new techniques are introduced, local people must be informed about the conditions under which change is going to take place. The authorities must have some kind of platform for communication with local people, and those who want to introduce something new must have respect for specifically regional characteristics, traditions and knowledge. At least in a short time perspective, it is often a question of optimizing the use of the resources of a certain area rather than maximizing the economic output. What we can learn from the story of Ullvi and Tibble villages in Leksand parish is that, in the long term, little can be changed without the acceptance of people involved in the actual transformation. Thus, anchoring ideas among ordinary people is a necessity – this is the lesson to be learned from the struggle between small landowners and 'external' interests which has been going on for more than 150 years in the area investigated.

At present, the Land Survey of Sweden and other authorities involved in the transformation of the cultural landscape and its ownership pattern in Upper Dalecarlia, are going to lose substantial financial support from the Swedish government. So, the question is whether there is any interest in going on with the reconstruction of the fragmented ownership pattern in this remote part of Sweden. The landowners themselves cannot afford it. We do not know what will come out of this, but so far the fragmented holdings and the small-scale landscape seem to be attractive to modern urban people. It is in a way exotic. For the first time since 1920, the 'rural' population of the area is increasing as its urban population decreases.

References

Bentley, J.W. 1987. Economic and ecological approaches to land fragmentation. *Annual Review of Anthropology* 16: 31–67.

Bortas, M. 1992. *Storskiftet i Dalarna 1803–04 med exempel från Leksands och Åls socknar.* Leksand: Leksands kulturnämnd.

Bortas, M. 1994. *Från storskifte till ägaderättsutredning. Förändringar i markägoförhållandena 1827–1971 i Hagen, Leksand,* Leksand: Leksands kulturnämnd.

Hylphers, A. 1757. *Dagbog öfwer en resa genom de, under Stora Kopparbergs lydande Lähn och Dalarne.* Västerås.

Knodel, J.E. 1988. *Demographic Behavior in the Past. A Study of Fourteen German Village Populations in the Eighteenth and Nineteenth Centuries.* Cambridge Studies in Population, Economy and Society in the Past Time 6, Cambridge: Cambridge University Press.

Nilsson, S. 1929. En dalabys utveckling och dess historia under 1800-talet. Dalarnas hembygdsförbunds tidskrift. *Årgång* 9: 41–58

Pallin, B. 1977. *Bälg och Bondelag.* Meddelande från Kulturgeografiska Institutionen i Stockholm, B 34.

SOU 1931 The Governmental Dala Report from 1931.

Schwarz, G. 1966. *Allgemeine Siedlungsgeographie.* Berlin: Walter de Gruyter.

Skommar, E. 1941. *Jordförhållanden i Leksands socken.* Examensarbete i jorddelningsrätt vid Kungliga Tekniska Höskolan i Stockholm. Manuskript.

94   *Ulf Sporrong*

Sporrong, U. 1987. Om storskiftet och det föregående äldre tegskiftet i Öster- och Västerdalarna, *Bebyggelsehistorisk tidskrift* 13: 157–76.
Sporrong, U. and Wennersten, E. 1995. *Marken, gården, släkten och arvet. Om jordägandet och dess konsekvenser för människor, landskap och bebyggelse i Tibble och Ullvi byar, Leksands socken 1734–1820.* Stockholm: Leksands sockenbeskrivning X.
Svensson, J-E. 1987. Bebyggelseutvecklingen i Leksands socken 1539–1573. Paper produced at Stockholm University, Department of History.
Thirsk, J. 1978. The European Debate on customs of inheritance 1500–1700. In *Family and Inheritance. Rural Society in Western Europe 1200 – 1800*, ed. J. Goody, J. Thirsk and E.P. Thompson. Cambridge: Cambridge University Press.
The Universal Dictionary 1936. London: Herbert Joseph Limited.

Part II

Emergence of resource management adaptations

Introduction

The chapters in this section use an evolutionary perspective and illustrate some of the long-term changes and adjustments through which resource management systems emerge in response to their particular social, economic and ecological contexts. The four chapters in the section provide a wide spectrum of cases from subarctic Canada, rural Brazil, Nigerian tropical agro-ecosystems, and coastal resources of the State of Maine, USA, and cover both locally used subsistence resources and commercial resources, with various levels of government-level management.

Is the evolution of resource management a continuous or a discontinuous process? Chapter 5, by Berkes, explores the idea that resource management crises may trigger learning and re-design of management systems, based on a study of current and historical use patterns of three key resources (caribou, beaver and fish) by Cree Amerindians of James Bay in subarctic Canada. Evidence from oral history and contemporary practice of caribou hunting suggests that management may have evolved only in the last century, following a resource crisis, and was encoded in ethical and cultural beliefs. In the case of beaver, contemporary management may have evolved as a consequence of commercialization and periodic depletions since the eighteenth century, by combining a changing indigenous land tenure system with European notions of resource management. Both cases support the crisis-and-subsequent-learning model. By contrast, there is no evidence for resource collapse and learning in the case of fishing. Cree fishing practices appear to have been influenced neither by commercialization nor by European fishery management. They depend on a set of practices that includes the thinning of fish populations by the use of a mix of gill net mesh sizes, a practice that appears to be particularly significant for maintaining the resilience of fish populations.

Chapter 6, by Begossi, covers aspects of the environment in which two groups of mixed-culture rural Brazilians, *caiçaras* and *caboclos,* live, the Atlantic Forest coast and the Amazon, respectively. The chapter also deals with their knowledge concerning the use of natural resources, and resource rights and institutions of the two groups. The concept of resilience, defined as the magnitude of disturbance that can be absorbed before a system changes, is used by the author to explore the idea that neo-traditional populations may have higher cultural flexibility to help them deal with environmental changes. Begossi argues that cultural flexibility, which helped the two groups to change and survive, is the function of a mix of traditional and new behaviours; the capacity of *caiçaras* and *caboclos* to cope with change increases the resilience of the social – ecological system. The two groups have followed different paths. *Caiçaras* have rules and property rights to deal with the environment based mostly on kinship ties, but little in the way of political organization. By contrast, some *caboclos* (especially *seringueiros* or rubber-tappers) have organized political movements that created the Extractive Reserves in the Amazon, an institutionalized and legally established form of local common-property management.

Chapter 7, by Warren and Pinkston, describes the interactions between social and ecological systems in the Nigerian rainforest at Ara, Osun State, as reflected in changing indigenous agricultural and natural resource knowledge and decision-making systems. The study compares and contrasts these systems in the pre-colonial, colonial, and post-colonial periods. Although the forest cover has diminished due to cash cropping and population pressure, the authors argue that the introduction of exotic tree and crop species, as well as numerous new varieties of arable crops, has increased useable biodiversity during the past half-century. These changes have also resulted in the emergence of new indigenous knowledge. For example, a major lesson about the relationship between tree cover and soil fertility has been learned by farmers who moved from an agroforestry system to a bush-fallow farming system in which the tree cover is entirely removed. The negative impacts of decreased soil fertility due to the bush-fallow system and its mitigation are now part of indigenous knowledge. The authors conclude that the social–ecological system of the Yoruba of Ara is still resilient. But it is in a transition period, during which a number of negative social and ecological factors could result in non-sustainable outcomes.

Adaptive systems with local management are not confined to the Third World or tribal societies. Chapter 8, by Hanna, deals with the soft shell clam fishery in Maine in the northeastern United States. Hanna shows how the present management system has evolved over a period of some two centuries. Resource use rights of coastal residents were formally established by law in 1821, followed by the rights of coastal towns to make rules of harvest; rule enforcement responsibility of local communities was established in the 1960s. Maine's soft shell clam fishery is characterized by the integration of informal local knowledge and locally generated formal scientific information, and a co-management process for the sharing of management rights and responsibilities between the State of Maine and the local community. The author traces the role of a number of important factors that have shaped management, including political ideology (private property versus equal access by all residents), market forces (soft shell clams have become a major seafood industry), and transactions costs (their distribution and the problem of keeping them low). Hanna helps make the point, building on the chapters by Gadgil *et al.* and Sporrong in Part I and on the three other chapters in Part II, that management should not be considered an end-point but a process of change; indeed, all management systems are 'in transition'.

5

Indigenous knowledge and resource management systems in the Canadian subarctic

FIKRET BERKES

Introduction

Historically, resources have been used under prescientific, traditional systems, and in some cases they have persisted for long periods of time without degradation, although not all traditional societies have lived harmoniously with their environment. Indigenous systems have come under scholarly inquiry in recent years for a number of reasons, including their potential survival value and the adaptations they represent, and for the design of sustainable ecosystem management strategies. The last two items are of particular interest in this chapter; they are about indigenous systems which appear to conserve biological diversity (Berkes, Folke and Gadgil, 1995) and ecosystem processes (Alcorn, 1989), and to maintain ecological resilience.

The chapter will pursue the above questions by examining indigenous resource management systems of Cree Amerindians of the Canadian eastern subarctic. The Cree are the aboriginal hunting–gathering people who occupied the eastern and central subarctic zone of North America. Many Cree groups still maintain resource use practices that seem to be unique, along with social institutions that are related to land and resource use. Hunting and human–animal relationships continue to be a spiritual and religious matter, even though the Cree have been Christians for a century or more (Preston, 1975; Tanner, 1979; Brightman, 1993). The Cree worldview, still dominated by respect for nature and a belief in the equality of all beings, is distinctly different from that of the dominant society that surrounds them.

There is an ongoing scholarly debate among anthropologists and others on the nature and origin of conservation practice among aboriginal hunters of North America. Many scholars are sceptical of the image of 'the indigenous conservationist', even though this image has become part of the identity and political posture of native peoples of the Canadian North. Researchers such as Brightman (1993: 287–9, but also see 258–9) have pointed out that there is much ethnohistorical evidence, mainly from the eighteenth century, that there was neither conservation ideology nor conservation practice among many subarctic Cree groups.

This evidence is contradicted, however, by literature documenting both conservation ideology and practice among contemporary Cree groups (e.g. Speck, 1915; Feit, 1973, 1987). There is evidence for sustainable resource use in the study area, inhabited by the Chisasibi First Nation of the Cree: (a) none of the species used by the Cree have become locally extinct since the glaciers departed the eastern subarctic some 4000 to 5000 years ago; (b) all the major species in the harvest are presently found in healthy populations; and (c) there is no evidence of damage to ecosystem structure and function from resource use practices of the Cree. These conditions satisfy the operational definition of conservation in the World Conservation Strategy (IUCN/WWF/UNEP, 1980).

The above analysis applies to the Chisasibi Cree area (see Figure 5.1), and not necessarily to the whole Cree area which covers a large part of the Canadian subarctic from Labrador to Alberta. The continued abundance of wildlife populations is generally true for the east coast of James Bay (Berkes, George and Preston, 1991) as well as the west coast (Berkes *et al.*, 1994), areas in which the Cree exercise control over their traditional property rights over wildlife. These rights are particularly strong in the case of the east coast (Quebec). Legal rights of the Cree to land and resources were established by the *James Bay and Northern Quebec Agreement,* signed in 1975.

To advance the analysis of Cree indigenous knowledge and management systems, the chapter addresses two issues: (a) what are the mechanisms which lead to sustainable resource use? and (b) how did those mechanisms evolve? An attempt is made to answer these questions by comparing the circumstances surrounding the use of three different kinds of resources (caribou, beaver, fish) over time, and by comparing the evolution of the management systems of each. The premise is that we can learn as much about Cree responses to resource abundance or scarcity by what has changed in some situations as by what has not changed in other situations.

Regarding the issue of mechanisms, it is known, for example, that in some indigenous cultures, there appears to be a knowledge base and 'rules of thumb' that provide guidance on how to deal with feedbacks and to respond to environmental change (Lewis and Ferguson, 1988; Gadgil and Berkes, 1991). This capability is of interest because human intervention in

100   *Fikret Berkes*

conventional Western scientific resource management systems typically results in reduced resilience; the very success of management may 'freeze' the ecosystem at a certain stage of dynamic change, making it more fragile and inviting unpredictable feedbacks from the environment (Holling, 1986). The identification of any features of Cree traditional resource use systems that may be relevant to the question of maintenance of ecological resilience and hence sustainability is of interest.

Regarding the origin of the observed resource-use practices and their possible evolution, ethnohistorical evidence is contradicted by the evidence from current practice, as noted by Martin (1978) and Brightman (1993). Yet the ethnohistorical material offers no explanation for this contradiction. If both sets of evidence are valid, the interesting possibility that emerges is whether conservation practice may have evolved among the Cree in historical times. Finding direct evidence for such cultural evolution is perhaps as difficult as finding direct evidence for biological evolution. But the hypothesis is plausible theoretically, as both social and ecological systems have an evolutionary character (see Chapter 13). A particularly promising model of cultural evolution, provided by Gunderson, Holling and Light (1995), consists of sudden 'lurches' of institutional learning and redesign that follow pathological exploitation and resource management crises. What kinds of experiences precipitate learning, and how does a society redesign management? The *process* of evolution of indigenous resource management practice may be of even greater interest than the *nature* of the system itself, in that it may provide insights into the adaptive processes by which human societies might respond to crises in the future.

Thus, the chapter has two interrelated objectives. The first is to evaluate the evidence of adaptation in Cree resource management practice which is related to the maintenance of ecological resilience. The second is to seek evidence for learning and evolution in Cree resource management systems, and the mechanisms by which adaptation may have occurred.

Following some social and ecological background for the study area, the chapter will explore the use of caribou, beaver, and fish resources, together with indigenous knowledge, property rights systems, and common-property institutions that pertain to each. The caribou hunting story illustrates how traditional worldviews may be used to interpret new observations and how this information may be transmitted. The case shows the use of newly created local knowledge simultaneously for validating cultural values and for elaborating new resource management applications. The beaver case illustrates the close linkages between local knowledge and the design of an ecologically elegant management system. As with caribou hunting, the beaver case provides some indirect evidence of the redesign of an exploitation system following crisis and subsequent learning. The fishery case concentrates on the identification of practices consistent with the maintenance of ecological resilience in the subarctic ecosystem, and proposes a resource management principle that may be applicable widely.

Background: the James Bay area

Eastern subarctic Canada has been occupied for thousands of years by Algonquian-speaking native peoples, of which the Cree are the major group. Historically, the Cree of the James Bay area lived in small, scattered bands and subsisted by hunting, trapping, fishing and gathering. They followed the seasons, travelling through the boreal environment forest by canoe in summer and snowshoes and toboggan (sled with no runners) in winter. Their pursuits were well adapted to what their environment produced; they developed technology, skills, customs, institutions and ethics appropriate for a hunting way of life.

The land of the Chisasibi Cree (Figure 5.1) is located near the northern edge of the Northern Coniferous Forest Biome which stretches as a broad belt across North America and Eurasia. The eastern and northern part of the Chisasibi area is a lichen–woodland transition zone. These ecosystems are characterized by relatively low biological productivity; relatively few species; pronounced seasonal periodicity; large year-to-year variability in both the physical environment and in animal populations; and general unpredictability of environmental conditions and animal distributions.

The James Bay region supports a relatively small number of species; not many can adapt to the extremes of environmental conditions in the area: mean annual temperatures of about −3° C; mean daily maximum temperature of −17° C in January and +20° C in July; 80 frost-free days; 65 cm mean annual precipitation; and 300 cm mean annual snowfall. There are some 530 species of vascular plants in the area, 139 species of birds, 37 species of land mammals, and 23 species of freshwater fish (Berkes, 1979; Hydro-Quebec 1993).

The number of species may be small, but some are found in large numbers at certain times and places. For example, only seven species account for 80% of the waterfowl population and, during certain seasons, these populations represent some of the largest aggregations of waterfowl in the world. The animals were the real 'wealth of the land', as the fur traders discovered in the seventeenth century (Francis and Morantz, 1983). But this wealth was available only to those who knew how to find

Figure 5.1. The James Bay area and the hunting territory of the Cree Amerindian community of Chisasibi. Hydroelectric development sites are also indicated along La Grande River.

the animals, which were distributed at low population densities most of the time over a vast landscape. The Europeans never did master the art of finding the beaver and the other valuable animals, relying instead on indigenous hunters who already possessed the appropriate knowledge and skills.

The James Bay area is a low-lying land dominated by rivers, streams, marshes and fens. Waterways provide transportation corridors for the people of the area; they also provide habitat for the most important species and govern productive processes. For example, the annual spring flood drives some of the most important food chains in the local ecosystem. The characteristic vegetation in the Northern Coniferous Forest Biome are evergreen trees, including various spruce, fir and pine species (Rowe, 1972). In the Chisasibi area, the typical vegetation is black spruce, sphagnum moss and ground lichen in the higher ground. In fens, typical vegetation is sedge, birch and tamarack. Stands of white spruce, balsam fir, trembling aspen, balsam poplar and white birch occur in the better-drained areas. Deciduous tree species such as alder, poplar and willow are found mainly along waterways, and mark a more productive zone than the conifer-dominated uplands.

The caribou of the James Bay area are usually found in small groups. They feed on grasses and sedges in low areas and near the coast in summer, and move back inland in the autumn to take shelter and to feed under the trees where the snow remains powdery and where lichen, the main winter food, is accessible. Many of the herbivorous mammals, such as beaver, moose, muskrat and snowshoe hare, are found in the deciduous vegetation zone, and many of them are dependent on adjoining wetlands as well. Rivers and creeks support populations of beaver which live in lodges in family groups and feed on deciduous vegetation. The rivershore or riparian environment is also home and source of food to muskrats and to snowshoe hare.

Wetlands and the coast are important for waterfowl. Two major species of goose and some 20 major species of ducks, as well as loons, swans and other waterfowl migrate twice yearly along the James Bay coast, which acts as a funnel concentrating bird populations from a large span of the eastern arctic and subarctic North America. When spring comes, the landscape becomes alive with migrating waterfowl. Birds from three of the major North American flyways use the James Bay coast. Of the two most important species, Canada geese nest in the wetland interior, and the lesser snow geese nest in colonies further north.

The most abundant fish species of the region is whitefish, found in lakes, larger rivers and in estuaries along James Bay. Cisco, a smaller relative of whitefish, occurs in large aggregations in some estuaries. Northern pike and walleye are the two most common predatory fish. Lake trout are restricted to a few lakes, and sturgeon to a few large rivers. Brook trout occur in some large sea-run populations. Fish populations are unusual among the animal resources used by the Cree in that fishing is productive and predictable from year to year; hence, the Cree refer to fish as a 'staple' resource.

This contrasts with small game (snowshoe hare and several species of grouse and ptarmigan) which show large year-to-year variability. Some of these species exhibit 8–10-year population cycles throughout the subarctic (Keith and Windberg, 1978). In addition to the cyclic species, first recognized in ecology by Elton (1942), a number of others, including caribou, marten, muskrat and porcupine, have been known to fluctuate in numbers over time. But there is little scientific agreement, even in the case of caribou, as to why these fluctuations occur. There is a general consensus, however, that subarctic ecosystems are characterized by uncertainty and variability, as reflected, for example, in the title of Ingstad's (1931) book, *The Land of Feast and Famine*.

Winterhalder (1983), working with the Cree and Ojibwa of Northern Ontario, drew attention to the dynamic and unpredictable nature of subarctic environments. Using optimal foraging models to help identify decision rules that the Cree seem to follow, Winterhalder studied how hunters coped with day-to-day challenges. He found that they used their powers of observation and bush skills to respond to situations with adaptive flexibility, and that their foraging strategies took into account the unexpected, thus responding in a manner consistent with adaptive management (Winterhalder, 1983).

Many Cree groups continued to follow a traditional way of life, with a seasonal cycle of migration, until the 1950s and the 1960s. Chisasibi (formerly Fort George) changed from a summer gathering place to a year-round settlement in the 1960s. Although about one-third of the families still spend several months on the land, the majority of the Cree in the 1990s live year round in permanent settlements. The Cree participate in a mixed economy, relying on wage income and transfer payments, and hunting-related activities still continue to be important in the overall regional economy. In western James Bay, for example, the replacement value of the fish and wildlife harvest is one-third as large as the total cash economy, and this harvest is still large enough to fulfil the protein requirements of the local population (Berkes *et al.*, 1994).

In eastern James Bay, and the Chisasibi area in particular, the local land-based economy has been impacted by a large (12000 MW) hydroelectric development, the James Bay project (Berkes, 1988a; Hydro-Quebec, 1993). This project has altered the environment in a major way, resulted in the closing down of the fishery in the La Grande system after the mid-1980s because of high mercury levels (Berkes, 1988a), and generally affected the relationship between the Cree and their environment, perhaps permanently. But the development also led to a 1975 treaty under which the rights of the Cree to manage their own resources was recognized by the government.

A caribou story

A subspecies of the Eurasian reindeer, the North American caribou has not been domesticated but roams free in the tundra and the lichen–woodland transition zone. As the most abundant large mammal of arctic and subarctic North America, it has a special place in the traditional economy of the aboriginal peoples of these environments. For Charles Elton (1942), one of the fathers of modern ecology, caribou has a special place in ecological theory in demonstrating population fluctuations in subarctic ecosystems. But to the ecologist, caribou-population increases and declines are a scientific problem yet unresolved. The conventional biological view of caribou does not include population cycles.

By contrast, the Inuit, who live to the north of the Cree above the tree line, seem to believe that there is a 80-year or so natural population cycle in caribou (M. Freeman, University of Alberta, personal communication). To the Cree, caribou population fluctuations are cyclical but these are not predictable, periodic cycles. The Cree wisdom predicts the return of the caribou but not its timing; as one Amerindian saying goes, 'no one knows the way of the winds and the caribou' (Munsterhjelm 1953: 97). To the Cree, caribou declines are mysterious only in part. They are partly explainable in terms of hunter–animal relationships and have to do with the ethical transgressions of hunters. Whereas Elton's data came from biological science and the records of missionaries and traders, the 'data' of the Cree hunter come from stories told by elders and from the hunter's own day-to-day observations. The caribou are part of the living landscape shared by the Cree.

The hunting grounds of the Chisasibi Cree Indians are rich with caribou-related place names. Examples include Point Attiquane (Caribou Point), where caribou antlers from ancient hunts may still be found, and Maanikin Lake ('maanikin': caribou-aggregating device, a corral). Chisasibi hunting lore is likewise rich with caribou natural history. Examples include, 'How do you tell the sex of animals in the herd you are following?' ('From the shape of digging marks in snow'); 'How do you tell if there is a really big bull in the group, so that the hunters should be wary?' ('His tracks in the snow would go wide around trees' – that is, the big bull takes care not to entangle his large antlers).

The notion of 'Cree traditional knowledge of caribou' is at odds with the fact that most Chisasibi Cree had never seen a caribou until the 1980s. Caribou were last seen in the area in the 1910s (Speck, 1935: 81; Elton, 1942), but records of the Hudson Bay Company from the 1600s to the 1800s indicate that caribou were at least periodically abundant for most of that period. They were one of the major food resources of the James Bay Cree in the area north of Eastmain, and a source of irritation to HBC traders because Cree hunters would periodically take off after the caribou instead of concentrating on trapping furs for the HBC (Francis and Morantz, 1983: 7).

Chisasibi hunters saw their first large caribou hunts of this century in the winter of 1982/83. According to information from hunters, most of the kills occurred in the area between Caniapiscau and the LG 4 dam, and amounted to some 100 animals. The following winter, large numbers of caribou appeared further west, in an area accessible by road; in fact, many were right on the road serving the LG 4 power station. 'Large numbers' were taken (the actual kill was unknown), even though the caribou stayed in the area only for a month or so. Chisasibi hunters used the road, bringing back truckloads of caribou. There was so much meat that, according to one hunter, 'people overdosed on caribou'.

However, some community leaders were unhappy, not because of the large numbers killed, but because some hunters had been shooting wildly, killing more than they could carry, and not disposing of wastes properly. Chisasibi Cree hunters' code speaks strongly against wastage and calls for burning or burying of animal waste. The leaders were worried that the hunters' attitude and behaviour signalled a lack of respect for the caribou, a serious infraction of the traditional code in which ritual respect ensures that animals will continue to make themselves available. It is a system of mutual obligations: 'show no respect and the game will retaliate', Cree hunting leader Robbie Matthew explained later. In the following winter, 1984/85, there were almost no caribou on the road. Hunters in trucks waited and waited and many left empty handed. Those who had the skills to go into the bush and hunt without causing disturbance nevertheless came back with reasonable kills. According to information from hunters, about 300 caribou were taken, only a fraction of the hunt in the previous year. Back in town, many people worried: had the caribou decided not to come to the Chisasibi hunting grounds after all?

Meetings were called. Two of the most respected elders stepped forward and retold the story of the disappearance of the caribou shortly after the turn of the century. Caribou had been declining on the coast in the 1880s and the 1890s but continued to be plentiful in the Caniapiscau area, near the centre of the Labrador Peninsula where the Cree of Chisasibi, Mistassini and Great Whale, the Naskapi and Montagnais (Innu) of Labrador and the Inuit of Ungava Bay seasonally got together to hunt the great migrating herds of caribou as they crossed the Caniapiscau River.

In the 1910s a disaster occurred, the elders told. Hungry for caribou and equipped with repeating rifles which had just become available (Banfield and Tener 1958), previously respectful hunters lost all self-control and slaughtered the caribou at the crossing points on the Caniapiscau. Instead of 'taking care of the caribou', the hunters killed too many and wasted so much food that the river was polluted with rotten carcasses. The following year, the hunters waited and waited, and there were no caribou. None at all. The caribou disappeared for generations.

The slaughter and the subsequent disappearance of the caribou were etched in the collective memory of Chisasibi hunters, and became part of their oral history. But for the Cree, all changes occur in cycles, and all was not lost. The caribou would once again be plentiful, the wise men predicted at the time. The caribou would return one day, but the hunters had to take good care of them if the caribou were to stay. This was the story that the elders were now retelling in Chisasibi. The elders' words had a profound effect on the younger hunters. The caribou had indeed come back, true to the old peoples' prediction. Oral history was validated. However, by violating traditional ethics, were they about to lose the caribou once again?

In the winter of 1985/86, the hunt was carried out very differently and it was productive. Some 867 caribou were taken, about two per household, according to the survey done by the Chisasibi Cree Trappers Association (CTA) which had now taken on the role of monitoring the hunt. Overseen by the elders, hunting leaders and other hunters who make up the membership of the CTA, the hunt was conducted in a controlled and responsible manner, in accordance with traditional standards. There was little wastage, no wild shooting. The harvest was transported efficiently and butchering wastes were cleaned up promptly. The Cree exercised their self-management rights under the *James Bay Agreement*; government resource managers were never involved in the solutions that the Cree hunters devised themselves (Drolet *et al.,* 1987).

The caribou kept coming. It was almost as if they were responding to the restoration of proper hunting ethics, and they were moving much deeper into the Chisasibi area. Some of the largest numbers were seen halfway between the coast and the LG 4 area. Hunters were ecstatic. In spring 1986, caribou were seen right on the James Bay coast for the first time in recent memory. In the next two to three years, many hunters passed up the chance to hunt the smaller, scattered groups of caribou near the coast, until those caribou re-established permanent populations in these areas; instead the hunters concentrated on the larger aggregations of caribou to the east. By 1990, hunters' observations of tracks showed that caribou had reached the sea all along the James Bay coast, re-establishing the former range of the 1900s. Their observations were consistent with the results of biological surveys (Figure 5.2).

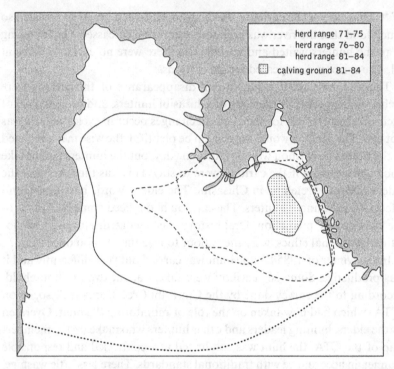

Figure 5.2.   Range expansion of the George River caribou herd, 1971–84 (Messier *et al.* 1988). The dotted line indicates additional range expansion after 1984, according to Couturier *et al.* (1990).

Beaver stewardship: evolution of a system

The history of beaver resources in James Bay has some notable similarities to, as well as differences from, the caribou story. The beaver was the main species in the fur trade that opened up Canada, with exclusive rights granted by the British Crown to the Hudson's Bay Company (HBC) in 1670 (Francis and Morantz, 1983). In contrast to caribou, which remained as a subsistence resource, beaver was a major trade item. Also in contrast to caribou, which were hunted over the communal territory (and inter-communally in some areas such as the Caniapiscau), beaver were harvested on the basis of family territories. Like the caribou, beaver was (and still is) a respected species, and respect was symbolized by rituals such as hanging beaver bones from trees near bush camps. Like the caribou, the beaver of the eastern and central Canadian subarctic went through major changes in abundance during the last 200–300 years, with lows in the early 1900s, the 1800s and the 1700s.

Perhaps the most notable feature of present-day beaver management in the James Bay area is the territorial system (Feit 1991). The community hunting area of Chisasibi (see Figure 5.1) is divided into 40 'traplines' or hunting territories, representing traditional family areas and formalized by the government into a resource management system. Each territory is occupied seasonally by a hunting group, usually consisting of two or three nuclear families, and led by a 'beaver boss' or a steward, as Feit (1986) prefers to call them. The steward is a senior hunter who acts as leader for the group, and who is part of the collective leadership provided by the CTA. He (and it is usually a he) oversees that all the codes and rules for proper hunting, not just for beaver but for all resources, are carried out. The steward acts as a 'gatekeeper' for the area and controls access; any persons who wish to hunt or fish in an area are expected to obtain permission from him.

He also keeps a mental inventory of resources and harvests in his area. It is not unusual for a steward to have a mental map of all the beaver colonies in an area of several hundred square kilometres, and a good idea about the numbers and age composition of the beaver in each. By integrating the knowledge from past hunts with trends in abundance, a hunter can set his objectives for the next season (Feit, 1973; 1986). However, these objectives are flexible. If there are no animals where there should have been plenty, the hunter quickly adjusts his expectations. These adjustments are articulated in terms of respecting the wishes of the animal that it does not want to be caught yet, the principle being that it is the animals (and not the hunter) which are in control of the hunt (Berkes, 1988b). The Chisasibi Cree ideal of beaver management involves maintaining a proper, respectful relationship with the beaver. Hunting areas are 'rested' so that the animals can replenish themselves. As explained by one steward, this involves dividing up a territory and rotating the parts:

A trapper paces himself, killing only what he needs, and what can be prepared by the women (that is, skin preparation), so that there is no wastage of meat and fur, and respect for the animals is maintained. He should also make sure that the area is rested. He divides up his area and concentrates on one part at a time. . . The remainder of his area is rested. Normally, a trapper should rest parts of his trapline for two or three years but no longer than four years. If he leaves it, say, six or ten years, he is not properly using his area, and the beaver will not be plentiful.

Figure 5.3 illustrates the actual practice of rotation over a four-year cycle in family territory in the north-central portion of Chisasibi lands by a family group of three adult couples. The leader and his family occupied this land over the winter months, and hunted and fished a variety of animals,

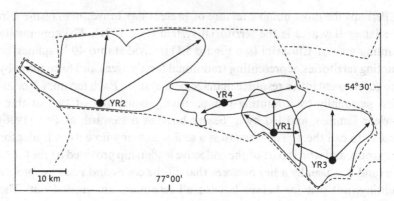

Figure 5.3.   A four-year cycle of land and resource use in a Chisasibi Cree family hunting territory. Much of the territory is rested in any given year, as the group of hunters concentrates on one part of the area.

including beaver (although international markets for beaver fur declined sharply after 1982–3).

The concept of rotation is rooted in the local knowledge of the relationship between beaver and beaver food. There will be many empty beaver lodges in an area which has not been trapped for a long time; this may be due to disease from overcrowding or due to food depletion. Scarcity of beaver food such as aspen in the vicinity of the lodge signals overgrazing. The Cree believe that part of their obligation to animals is to keep them productive, and that continued harvesting is not only good for the hunter but also good for the beaver (Berkes, 1988b). The trapper monitors the health of the beaver–vegetation system by observing vegetation changes, teeth marks on cut wood to estimate the age composition of beaver in lodges, and other evidence of overcrowding, such as fighting among the beaver. The Cree see the interaction between beaver and vegetation as a discontinuous relationship, a system that exhibits a kind of catastrophic behaviour. The Cree practice of resting an area followed by heavy harvesting of beaver keeps the system from reaching the critical threshold at which food would be depleted and the system would flip or show catastrophic change (Figure 5.4).

Cree traditional ecological knowledge and understanding of the natural world are of course not restricted to beaver. Hunters are experts on the natural history of a number of species, and on food chain and habitat relationships. Figure 5.5 is taken from a study of the environmental impact of the damming of La Grande River, conducted from the point of view of the Cree. The stewards emphasized the importance of the wetland habitat

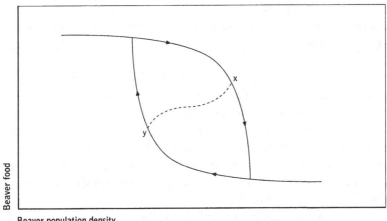

Beaver population density

Figure 5.4.   Cree trappers believe that as the beaver population increases, beaver food such as willow and aspen decreases until, at a certain point (shown as X in the figure along the upper attractor), the area can no longer support beaver. When this happens, the area will be without beaver for many years until the overgrazed vegetation recovers (shown here as point Y along the lower attractor). The Cree management recipe for keeping the system from reaching the critical point X, at which the system would flip, is to trap beaver at least every four years.

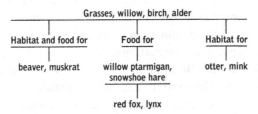

Figure 5.5.   The main habitat and feeding relationships in the wetlands of the lower La Grande River, according to Cree hunt leaders (stewards) (Berkes, 1988a).

along La Grande for biological productivity and the hunting economy, displaying a coherent ecological view of the wetland system (Berkes 1988a). Such understanding of interactions makes it tempting to speculate whether Cree hunters are capable of achieving sustainable management not only of individual populations but of the whole ecosystem.

In the Cree view of the natural world, the hunter is part of the system, and key Cree values (such as respect, sharing, reciprocity, generosity, taking care) apply to relationships between animals and people as well as to those among people (Preston, 1975). The hunt leader acts as a steward of the resources on behalf of the community (Feit, 1986), and he also acts as a social leader, ensuring, for example that no one goes hungry. No doubt the leader in the old days was also a spiritual leader, as hunting was (and to some extent still is) a religious and spiritual pursuit (Preston, 1975; Tanner, 1979).

As Brightman (1993: 287) put it, 'in the 1700s, as in the present, the dominant (Cree) ideology was that of the animal benefactor who 'loves' the hunter and voluntarily surrenders its body. In return, the animal exacts diverse expressions of ritual deference and respect. . .'. Hence there is reciprocity between the hunter and the animal. As seen in the caribou and beaver cases, limiting the harvest and avoidance of waste are two of the main ways, along with various rituals, in which respect is shown. The ethnohistorical record, however, says something very different in this regard: 'there is no evidence at all that either limited killing or total utilization was included among these' [expressions of respect]. 'The Crees in the 1700s believed – or told the traders they believed – that the more animals they killed, the more they would kill in the future' (Brightman, 1993: 287).

Brightman goes on to present a great deal of evidence from the eastern and central subarctic of the 1700s that (1) there was much wasteful killing, and (2) hunting was not sex or age selective but indiscriminate and unlimited, although there were also a few examples of selective harvesting (Brightman, 1993: 258). He argues that, among the Cree and other subarctic peoples, there was a conception of animal resources as infinitely renewable: 'The numbers of animals available to hunters in the future could be influenced by ceremonial regeneration; the numbers killed or the parts utilized were irrelevant' (Brightman, 1993: 289). Even allowing for the likely differences in the notion of 'waste' between European and Amerindians, and for the limitations of Europeans to recognize Amerindian resource management, Brightman's evidence still holds. This leads to a paradox.

If both ethnohistorical evidence, such as that provided by Martin (1978) and Brightman (1993), and evidence from the present-day Cree, such as that provided by Feit (1973; 1986), Tanner (1979) and others are valid, the hypothesis that emerges is that practices of limiting the kill and avoiding waste must have evolved among the Cree in historical times. How, then, did beaver conservation come about?

Although the subarctic is a vast area, there had been a number of crises as far back as the 1700s, corresponding to periods of heavy competition between the HBC and its rivals, and resulting in the depletion of beaver. One major period of competition and resource crisis started with the competition between the HBC and the Northwest Company in 1763

(Brightman 1993: 260). Waves of depletion spread north and west, using steel trap–castoreum technology, castoreum being a chemical attractant. Ojibwa and Iroquois, whose lands had already been depleted, moved into Cree areas, reaching the Churchill River area in the central subarctic by 1790 (Brightman, 1993: 265).

The rivalry resulted in resource depletions, and beaver-fur returns gradually declined until the HBC absorbed its rival in 1821, establishing monopoly once more. The HBC instituted a series of conservation measures, including area and season closures; age and breeding season management (1823); and quota management (1826–9) (Brightman, 1993: 305). In the Rupert's River District (James Bay), returns were at a low in 1825–9, improved in 1830–34, declined again with the incursions of outsiders through the 1830s and the 1840s, and recovered again after 1860 when intruders were finally expelled from Cree areas (Francis and Morantz, 1983: 130).

The next major beaver-depletion cycle occurred in the 1920s following the building of railways which opened up the southern James Bay area to non-native trappers from the south, at a time when fur prices were high (Feit, 1986). By their own accounts, the Cree themselves contributed to the depletion of beaver. Unable to control access, the Cree management system broke down. As Feit (1986) surmises, Cree hunters must have suspended conservation ethic and proper practice to join in the race to overhunt the beaver, rather than let the outsiders take them all.

By 1930, beaver were depleted and the outsiders left, leaving the Cree to re-establish their resource management institutions, this time backed by government regulation. Starting with one beaver reserve established in southern James Bay in 1930, non-native trappers were excluded from the communal territories of indigenous groups, family hunting territories were mapped, and beaver bosses/stewards were officially put in charge of their own territories. Beaver populations recovered by the mid-1950 in the Chisasibi area, and harvests gradually increased until the 1970s, as measured by fur statistics and government records. Productive (and presumably sustainable) harvests continued until the international fur markets collapsed in the 1980s. Since then, trapping diminished in economic importance but has continued mainly for the meat, as in western James Bay (Berkes *et al.,* 1994).

Keeping fish populations resilient

For the Cree, fish are a staple resource; one can count on fishing even when other resources fail or become unavailable, as in the beaver case. Unlike many other animal resources, the Cree take their fish almost for granted; in Chisasibi, there are no rituals and ceremonies that involve fish. As with all animals, however, there is respect. One does not waste fish; one does not abuse fish by swearing at them or by 'playing' with them. (The Cree are horrified by catch-and-release sport fisheries.) Nor does a fisherman boast. Boasting brings quick retaliation from fish – they stop making themselves available to the fisherman. The principle that animals are in control of the hunt holds for fish as well.

In the past, the Mistassini Cree had rituals for the largest lake trout (Speck, 1935) and the Waswanipi Cree had special respect for large, old sturgeon (Feit, personal communication), but it is probably fair to say that there is no overt spirituality in evidence as rituals and ceremonies in present-day Cree fishing. This is in contrast to beaver and about a dozen or so other highly respected species, including black bear, for which rituals have continued into recent years (Tanner, 1979).

Most of the Chisasibi Cree fishery takes place in medium and large-sized lakes, on the James Bay coast, and in estuaries of rivers. Fishing success is predictable partly because habits of fish at different times and places are predictable, as any fisherman knows. For example, fishermen know that in spring the best catches of whitefish are obtained following the melting ice edge in bays (Berkes, 1977). In August, fishermen know where the pre-spawning aggregations are, and in September they know that whitefish is best harvested over a sand–gravel substrate at certain depths of water.

Fishing seasons are part of the seasonal cycle of harvesting activities, and they are signalled by biophysical events in the landscape such as the spring ice break-up in the river and change of colour of the vegetation in September. In the past, major harvesting activities were marked by ceremonies; few of these now remain. However, fishermen know how to recognize and respond to a variety of environmental feedbacks that signal what can be fished where and when. Fishing in any given location ceases in response to one important indicator, a drop in the catch per unit of effort, at which point the fishing operation relocates. Under the guidance of master fishermen, often the older family heads, fishing effort is spread over a large area traditionally fished by a family group, as in beaver trapping (Berkes, 1977; 1979). Unlike beaver trapping, however, stewards make little attempt to limit access. On the James Bay coast, for example, it is usually one's knowledge, or lack thereof, of the coast and navigational ability that indirectly limits one's access to the resource.

Fishing is productive because of the great store of culturally transmitted knowledge about the habits of fish. But there is another reason as well. As noted by Johnson (1976) and others, species such as lake trout and white-fish in Northern Canadian lakes are found in populations that represent the accumulation of as many as 50 year-classes of fish, analogous to old trees in mature tropical forests, and characterized by a high ratio of biomass to biological production.

Slow growth and long life in species such as whitefish give rise to a characteristic bimodal length–frequency distribution (Johnson, 1976; Power, 1978). How such a peculiar distribution comes about can be shown mathematically through the summation of overlapping size-classes of older fish, for any long-lived species that has low growth rates and low mortality rates after first maturity (Figure 5.6). The presence of many year-classes of large and slow-growing fish presumably represents a life-cycle adaptation to fluctuations in the ecosystem. Multiple-spawning in fish populations elsewhere has been shown to be of adaptive value in dampening the effects of environmental variability, especially those leading to poor reproductive success for two or more years in a row (Murphy 1968).

What this means in practical terms for northern indigenous peoples is that the fish are available as easily harvestable, large units, not because the populations are highly productive (they are not), but because they consist of many years of accumulated production. An appropriate ecological adaptation to such an environment is to 'bank' one's food by not fishing any one lake year after year but 'pulse fishing' instead, that is, returning to a given lake every few years and fishing intensively over short periods of time. This is, in fact, exactly what Cree fishermen do in Chisasibi's subsistence fisheries (Berkes, 1977; 1979).

In contrast to subsistence fisheries, commercial fisheries come under government regulation, and are conducted differently. (However, in Chisasibi there has never been a commercial fishery; it is too far from markets.) The conventional scientific management systems for subarctic commercial fisheries in Canada have employed some combinations of the following tools: restrictions on gill net mesh size and fishing gear used, minimum fish size, season closures, and prohibition of fishing at times and places when fish are spawning. Catch quotas and maximum sustainable yield calculations based on population dynamics of the stock have also been used in larger-scale fisheries.

Yet many of these commercial fisheries based on whitefish and other species in the Canadian North have collapsed after a few years. One possible explanation for these observed stock failures is the reduction of repro-ductive resilience by the selective removal of larger fish. The mechanism can be modelled. Figure 5.7 shows the effect of single mesh sizes on a model

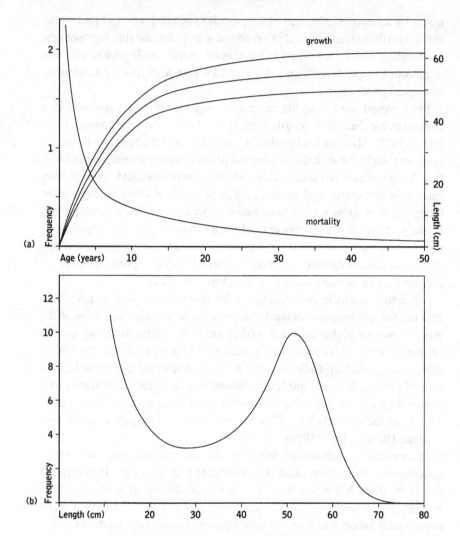

Figure 5.6. Mortality and growth curves of a model whitefish population (a), and the resultant length–frequency structure (b) (Berkes and Gonenc, 1982).

whitefish population (Berkes and Gonenc, 1982). The use of a single large mesh size is efficient in maximizing short-term yields; a large biomass is initially available to 139.7 and 127.0 mm (5.5 and 5 inch) nets, mesh sizes used in commercial fisheries. However, the catchable population represents the accumulated growth of many years, and the reproductive resilience of the population is diminished as the multiple age-classes of the spawners are depleted. For example, if all but two reproductive year-classes are depleted,

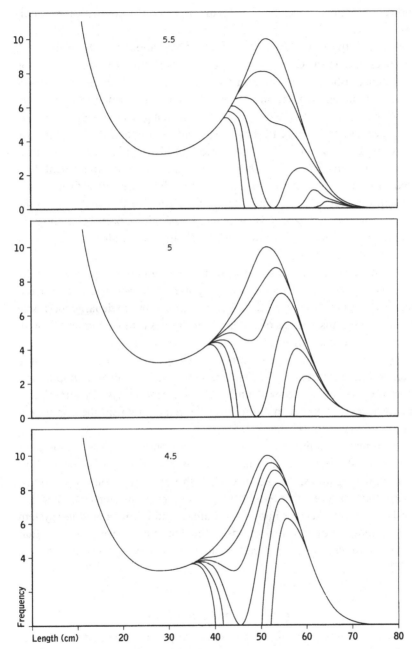

Figure 5.7. The change in length–frequency structure of a model whitefish population when fished with single gill-net mesh sizes. The contour lines represent different fishing intensities (Berkes and Gonenc, 1982).

and reproduction is poor for three years in a row, the population may collapse.

How can the reproductive resilience of the population be conserved? Practices used in the Cree fishery suggest one solution. As with indigenous subsistence fisheries elsewhere in Canada, Chisasibi fisheries are exempt from government regulations. Cree subsistence fishermen use the most effective gear available, the mix of mesh sizes that gives the highest possible catch per unit of effort (CPUE) by area and season, and they concentrate on aggregations of most efficiently exploitable fish (Berkes, 1977). In short, Cree fishermen violate just about every conservation/management tool used elsewhere by government managers. The subsistence fishery is a conventional resource manager's nightmare. Yet records going back to the 1930s show that *Coregonus* fisheries in Chisasibi have been sustainable; in fact, the age-class composition of whitefish appears similar over a span of 50 years (Berkes, 1979).

The management secret seems to be that the thinning of populations by the use of a mix of mesh sizes conserves population resilience, as compared to the wholesale removal of the older age groups by single large mesh size (Figure 5.8). Thus, the use of a mix of mesh sizes is more compatible with the natural population structure than the use of a single large mesh size. Many reproductive year-classes remain in the population even after fishing (Berkes and Gonenc, 1982). The reduction of the overall population density brings about renewal and increased productivity by stimulating growth rates and earlier maturation in the remaining fish, the population compensatory effect (Healey, 1975).

To summarize, a number of practices contribute to the sustainability of Chisasibi fisheries: switching fishing areas according to falling CPUE; rotating fishing areas; using a mix of mesh sizes to thin out populations; keying harvest levels to needs; having a system of master fishermen/stewards who control access into harvest areas; and having a land use system with common property institutions at the community level to ensure that resources are used under principles and ethics agreed upon by all (Berkes, 1989).

Discussion and conclusions

These three cases (caribou, beaver and fish) provide rich material for the study of the dynamics of resource management regimes, and of principles that have the potential for application to other areas. James Bay is a 'laboratory' with several long-term human ecology case studies, good archival

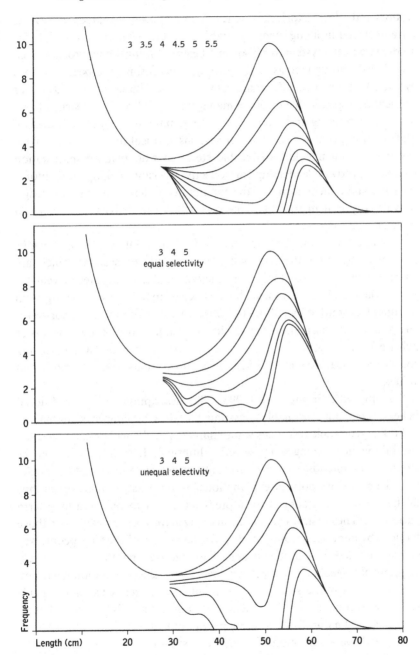

Figure 5.8. The change in length–frequency structure of a model whitefish population when fished with a mix of mesh sizes (Berkes and Gonenc, 1982).

records and ethnohistorical evidence. Also, some key resource crisis events have occurred in living memory, making cultural evolution accessible for study. Two of the systems (caribou and beaver hunting) appear to have been redesigned through crisis and learning; one (fishing) appears to have evolved without a specific management crisis. The essential features of resilient and sustainable resource management, as Cree have learned to do it, includes rotating areas, pulse harvesting, maintaining multi-age classes, and renewing over-mature natural systems. Critical features of the Cree system also include knowledgeable stewards who manage information feedbacks, provide leadership for collective decision making, and enforce the rules and ethical norms of the community. Elders transmit knowledge, provide corporate memory, and interpret ritual knowledge to help redesign management systems.

Cree practices for the management of fish and wildlife, taken as a whole, are consistent with the harvesting of subarctic ecosystems which are characterized by low biological productivity and a high degree of year-to-year variability. Feedbacks used by the Cree seem to be based on long-term learning transmitted by culture (as in the case of caribou), and short-term environmental observation that builds on detailed local knowledge (as in fisheries). Cree wildlife management can be best understood as an evolving social–ecological system in which local institutions reinforce rules and values.

Starting with Brightman's (1993: 280) description of the traditional belief system, it is reasonable to assume that the Cree conceived animals as 'infinitely renewable resources whose numbers could neither be reduced by overkilling nor managed by selective hunting'. It stands to reason, as Brightman contends, that the Cree did not associate hunting with depletion. There are two possible explanations for this. First, animal populations in the subarctic are subject to unpredictable year-to-year and long-term variations. These natural fluctuations may have been greater than those induced by hunting. Second, the population density of the Cree people was so low, and they were mobile over such a large area that they were unlikely to get reliable feedback on the consequences of their exploitation strategies. Either way, the hunters in their small numbers over the vast subarctic landscape were probably not able to read the signs and signals from their lightly harvested and naturally fluctuating prey populations.

If the above assumption of underharvesting is correct, it would make sense for hunters to perturb the prey population (as in adaptive management) to try to obtain some signals. The ethnohistorical finding that 'the Crees of the 1700s believed . . . that the more animals they killed, the more they would kill in the future' (Brightman, 1993; 287) can thus be reinterpreted as an attempt at adaptive management and a means to stimulate population compensatory responses (e.g. Healey, 1975). It is notable that scientific resource managers believe now, as the Cree of the 1700s apparently believed then, that increasing the exploitation level increases the productivity of a lightly harvested population. The ancient Cree hunters had experience with underexploitation. What they lacked, however, was experience with overexploitation. It was the latter experience with caribou and beaver that apparently triggered learning.

In the case of caribou, the Chisasibi Cree and their neighbours experienced overhunting perhaps for the first time around 1910, with the last big caribou hunt in Caniapiscau. Several factors may have been at work. Availability of the repeating rifle may have made the hunters dizzy with new-found power over animals. It is likely from Elton's (1942) account that the caribou may already have been declining at that time for other reasons, including fires in the caribou range, climate trends, and lichen depletion by the previous caribou population high (Banfield and Tener 1958). They may have been declining as part of a natural population cycle, even though the conventional biological view of caribou does not include population cycles.

In any case, as far as the Crees were concerned, the disappearance of the caribou was unambiguously linked to the last, big, wasteful hunt. The lesson of the presumed transgression, once learned, survived for 75 years in oral history, and it was revived precisely in time to redesign the hunting system when the caribou returned. Had there been government intervention to regulate the Chisasibi caribou hunt, it could not possibly have had as much impact on the hunters as did the teachings of the elders (Drolet *et al.*, 1987). The lesson delivered (not to kill too many and not to waste) came right at the heels of the validation of the elders' prediction that the caribou would return one day, and it was too powerful to take lightly, even for the most sceptical young hunter.

The beaver case provides an even clearer instance of learning by crisis. When the beaver is abundant, selective harvesting and prevention of waste make little sense. In fact, some extra predation on a superabundant herbivore makes for resilient ecosystem management (see Figure 5.5). Local depletions would not have mattered as long as there were other beaver nearby to repopulate the area. Assuming that the Cree had a knowledge base at that time to provide guidance in interpreting environmental information such as change in vegetation in beaver habitat, it is possible that the Cree were actively modifying the environment by managing feedbacks. The textbook wisdom on hunter–gatherers precludes habitat management. But there is, in fact, evidence that some of the Northern hunter groups practised a variety of environmental management measures, for example by the use of fire (Lewis and Ferguson, 1988).

We know little of the reaction of the Cree to the beaver depletion of the 1700s. By the 1820s, however, the Cree must have learned enough about overhunting that they were willing to practise European-style harvest limitation. (If the measures had made no sense to the Cree, the HBC would not have been able to enforce them.) However, in each beaver crisis, recovery follows the revival of communal property rights and the re-establishment of land tenure, not merely harvest limitation. This is true also in the most recent beaver crisis in the period 1920–30. An interesting twist there is that the Cree (or at least some of them) must have known from oral history exactly what was happening when the new incursions started. By that time, they must have had an understanding of European conservation as well as their own common-property land-use system. Unable to control the access of outsiders and caught in a 'tragedy of the commons', all they were able to do was to contribute to the depletion of the beaver. They were only able to regain control over their area about one decade later. At that point the stage was set for a revival of the old and the incorporation of new management measures to redesign a sustainable system.

There is controversy over the nature of traditional hunting territories of the Cree and neighbouring groups. The conventional anthropological interpretation of beaver-trapping territories is that it is a response to commercialization. That is, territoriality was not an aboriginal institution but came into being as a result of the shift from a subsistence economy to a commercial one (Bishop and Morantz, 1986). Evidence from the James Bay area property rights systems does not support this hypothesis (Berkes, 1989). It supports the idea that communal and family-based territories form a continuum. Intensification of resource use, as caused, for example, by population increase and trade, tends to create pressures to shift the property-rights system from a looser form (e.g. with communal territories) to a tighter form (e.g. with family territories). Tighter property rights through territoriality improves the incentives for conservation. These considerations apply to species for which territoriality makes ecological and economic sense, such as beaver, but not to species such as caribou which range over such large areas that no family group could possibly try to enforce its property rights over them (Berkes, 1989).

According to the above model, then, fur trade would have stimulated the tightening of the management system. More care had to be taken with the harvest, and more of the responsibility must have shifted to hunting leaders who knew their areas particularly well. Over time, their responsibility came to include rule making and monitoring, and their collective leadership (presently manifested as the Cree Trappers Association) became the key common-property resource management institution. The powers of the stewards, however, are at the pleasure of the community. Stewards talk about 'their land' and can pass on their land to their sons, but the land in question is far from being private property. Stewards can regulate access but cannot exclude other community members from using that land; they are expected to share the products of that land (including beaver meat but not the fur); as high-status hunters, they are expected to be generous; and they certainly cannot sell that land. There are cases in which the community has stripped a steward of his powers for violating community norms in the conduct of trapping.

The changes in both the caribou and the beaver management systems provide a good fit with the Gunderson *et al.* (1995) model: pathological cases of exploitation and resource management crises may trigger sudden lurches in understanding, and a redesign of management systems for flexibility and innovation towards sustainability. The caribou and beaver cases, however, bring out another important aspect of redesign not touched upon by Gunderson *et al.* (1995). It is not only the management system that is redesigned; worldview and ethics which provide the background to hunting institutions and practices are modified as well. Now that the hunters know that it is possible to deplete animals by overhunting and that wastage does matter, their value systems change accordingly.

The fishery case does not fit the Gunderson *et al.* (1995) model. There is no evidence of learning and redesign by crisis, and the existing system shows no evidence of European resource management thinking. The Cree system is as different as it can be from the biological management system applied to subarctic commercial fisheries. There are no clues as to the origin of practices such as the use of a mix of gill net mesh sizes that help conserve the reproductive resilience of the fish populations. Nor is there an easy explanation of how such practices may have evolved. The general 'rule of thumb' is that Chisasibi Cree fishermen will use that mesh size which gives the highest CPUE for a given location and time of year; this has been checked and verified by field experiments (Berkes, 1977).

There is no easy explanation either of how the practice of rotation may have evolved. We do know that rotation is not specific to fisheries or to the subarctic but occurs in a great diversity of indigenous resource management systems (Berkes *et al.*, 1995). In the subarctic, rotation is not specific

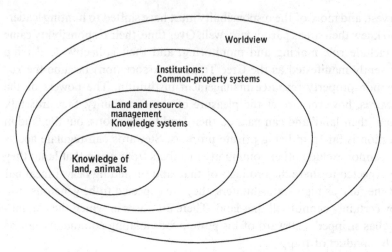

Figure 5.9.   Levels of analysis in traditional knowledge and management systems.

to fishing and applies to other species such as beaver. Territories which are rotated and rested for two or more years produce significantly higher yields of beaver as compared to non-rotated territories, based on Feit's (1986) data for the Waswanipi Cree. Rotation provides higher yields in fisheries also, and the practice of 'pulse fishing' borrowed from native fisheries has been recommended for northern Canadian commercial fisheries in lakes and char (*Salvelinus alpinus*) streams as well (Johnson, 1976).

The two practices – rotation/pulse fishing and population thinning by the use of a mix of mesh sizes – may be mutually supporting. Their combination may help operationalize a management principle with potential applicability to any population in which multiple reproductive year-classes are adaptive: *harvest more year-classes at a lower rate by the use of nets of different mesh sizes (as opposed to selective harvest of older year-classes at a higher rate).* Rotation, in turn, allows the fishermen to maximize the CPUE, and also serves to protect the stock in case of possible miscalculation of the rate of harvest. That is, the CPUE provides the feedback to the fisherman regarding the state of the stock; rotation is the design principle that brings the stock back to high values of CPUE and, at the same time, protects the stock in case the environmental signal (the level of CPUE) monitored by the fishermen has been misread.

The three cases of traditional resource management in James Bay illustrate the idea that traditional knowledge and management systems may be considered at several, nested levels (Figure 5.9). The first level is about local knowledge of animals and the landscape, such as habits of caribou, the behaviour of beaver and the seasons of fishing. All local knowledge has obvious survival value but would not be sufficient by itself for resource management. The second level of analysis is about the management system, that is, a set of tools, techniques and practices. For example, for each of the three resources (caribou, beaver and fish), the management tools and practices are different, although some of the principles used may be the same (such as rotation with both beaver and fish, but not with caribou).

All systems of management require appropriate institutions. That is, for a group of interdependent hunters to function effectively, there has to be a social organization for co-ordination, co-operation, rule making and rule enforcement. Thus, the third level of analysis is about institutions. Common-property institutions that apply to each of the three resources are different. For example, caribou hunters are not territorial and caribou is hunted on a communal-territory basis, and even the communal territories may overlap. By contrast, beaver is hunted on a family-territory basis, and access is managed by stewards. Fishing is carried out both on a communal-territory and a family-territory basis; stewards have the capability to control access but usually do not, given that fish are abundant relative to needs. Thus, each resource differs in the way access is controlled, but the same key institution governs all three. This key institution is the collective leadership of the senior hunters or stewards, who make and enforce hunting rules on behalf of the community, and seem to have the capability to self-regulate, as in the case of the steward who violated an important trapping rule and lost his status.

The fourth level of analysis is the worldview or cosmology which shapes human–nature relations and gives meaning to social interactions. The Cree consider that human–animal relations basically follow the same rules as social relations; for example, concepts of reciprocity, respect and generosity apply equally well to both sets of relationships. Customs and rituals help people remember the rules and interpret signals from the environment appropriately. Rituals with the bones of beaver convey respect and help hunters remember the importance of this animal. Not all rituals and taboos are explainable in resource management terms, nor should they be. But many are ecologically adaptive, although the value of some of them may be obscure. Rituals of traditional hunters–fishers may not be very different in kind from the many rituals observed by formal institutions of resource management, such as government agencies (Hilborn, 1992). But there is one large difference between traditional systems and Western scientific systems. In traditional systems, morality and ethics are explicitly a part of

126   *Fikret Berkes*

the management system; in Western scientific systems they are merely implicit.

Acknowledgements

The author would like to thank Rob Brightman, Harvey Feit, Milton Freeman, Carl Folke, C.S. (Buzz) Holling and Lyn Pinkerton for comments and inspiration. Chisasibi hunters and leaders, George Bobbish, James Bobbish, George Lameboy, Robbie Matthew and many others contributed to my understanding of their system. The research work in the James Bay area was supported over many years by the Social Sciences and Humanities Research Council of Canada (SSHRC).

References

Alcorn, J.B. 1989. Process as resource. *Advances in Economic Botany* 7: 63–77.
Banfield, A.W.F. and Tener, J.S. 1958. A preliminary study of the Ungava caribou. *Journal of Mammalogy* 39: 560–573.
Berkes, F. 1977. Fishery resource use in a subarctic Indian community. *Human Ecology* 5: 289–307.
Berkes, F. 1979. An investigation of Cree Indian domestic fisheries in northern Quebec. *Arctic* 32: 46–70.
Berkes, F. 1988a. The intrinsic difficulty of predicting impacts: Lessons from the James Bay hydro project, *Environmental Impact Assessment Review* 8: 201–20.
Berkes, F. 1988b. Environmental philosophy of the Cree people of James Bay. In *Traditional Knowledge and Renewable Resource Management in Northern Regions*, pp.7–21, ed. M.M.R. Freeman and L. Carbyn. Edmonton: Boreal Institute, University of Alberta.
Berkes, F. 1989. Cooperation from the perspective of human ecology. In *Common Property Resources*, pp.70–80, ed. F. Berkes. London: Belhaven.
Berkes, F., Folke, C. and Gadgil, M. 1995. Traditional ecological knowledge, biodiversity, resilience and sustainability. In *Biodiversity Conservation*, pp.281–99, ed. C.A. Perrings, K-G. Mäler, C. Folke, C.S. Holling and B-O. Jansson. Dordrecht: Kluwer.
Berkes, F., George, P.J. and Preston, R.J. 1991. Co-management. *Alternatives* 18 (2): 12–18.
Berkes, F., George, P.J., Preston, R.J., Hughes, A., Turner, J. and Cummins, B.D. 1994. Wildlife harvesting and sustainable regional native economy in the Hudson and James Bay Lowland, Ontario. *Arctic* 47: 350–60.
Berkes, F. and Gonenc, T. 1982. A mathematical model on the exploitation of northern lake whitefish with gillnets. *North American Journal of Fisheries Management 2*: 176–83.
Bishop, C.A. and Morantz, T., eds. 1986. Who owns the beaver? Northern Algonquian land tenure reconsidered. *Anthropologica* 28 (1/2). 219 p.
Brightman, R.A. 1993. *Grateful Prey. Rock Cree Human–Animal Relationships.* Berkeley: University of California Press.

Couturier, S., Brunelle, J., Vandal, D. and St.-Martin, G. 1990. Changes in the population dynamics of the George River caribou herd, 1976–87. *Arctic* 43: 9–20.
Drolet, C.A., Reed, A., Breton, M. and Berkes, F. 1987. Sharing wildlife management responsibilities with native groups: case histories in Northern Quebec. *Transactions of the 52nd North American Wildlife and Natural Resources Conference*, pp.389–98.
Elton, C. 1942. *Voles, Mice and Lemmings: Problems in Population Dynamics*, Oxford: Oxford University Press.
Feit, H.A. 1973. Ethno-ecology of the Waswanipi Cree; or how hunters can manage their resources. In *Cultural Ecology*, pp.115–25, ed. B. Cox Toronto: McClelland and Stewart.
Feit, H.A. 1986. James Bay Cree Indian management and moral considerations of fur-bearers. In *Native People and Resource Management*, pp.49–65. Edmonton: Alberta Society of Professional Zoologists.
Feit, H.A. 1987. North American native hunting and management of moose populations. *Swedish Wildlife Research Vitlrevy* Suppl. 1: 25–42.
Feit, H.A. 1991. Gifts of the land: Hunting territories, guaranteed incomes and the construction of social relations in James Bay Cree society. *Senri Ethnological Studies* 30: 223–68.
Francis, D. and Morantz, T. 1983. *Partners in Furs. A History of the Fur Trade in Eastern James Bay 1600–1870*. Montreal: McGill-Queen's University Press.
Gadgil, M. and Berkes, F. 1991. Traditional resource management systems. *Resource Management and Optimization* 8: 127–41.
Gunderson, L.H., Holling, C.S. and Light, S.S., eds. 1995. *Barriers and Bridges to the Renewal of Ecosystems and Institutions*. New York: Columbia University Press.
Healey, M.C. 1975. Dynamics of exploited whitefish populations and their management with special reference to the Northwest Territories. *Journal of the Fisheries Research Board of Canada* 32: 427–48.
Hilborn, R. 1992. Can fisheries agencies learn from experience? *Fisheries 17 (4)*: 6–14.
Holling, C.S. 1986. The resilience of terrestrial ecosystems: Local surprise and global change. In *Sustainable Development of the Biosphere*, pp.292–317, ed. W.C. Clarke and R.E. Munn Cambridge: Cambridge University Press.
Hydro-Quebec 1993. *Grande-Baleine Complex. Summary. Feasibility Study.* Montreal; Hydro-Quebec.
Ingstad, H. 1931. *The Land of Feast and Famine*, 1992 edition, Montreal: McGill-Queens University Press.
IUCN/WWF/UNEP 1980. *The World Conservation Strategy*, Gland, Switzerland: International Union for Conservation of Nature and Natural Resources.
Johnson, L. 1976. Ecology of arctic populations of lake trout, *Salvelinus namaycush*, lake whitefish, *Coregonus clupeaformis*, arctic char, *S. alpinus*, and associated species in unexploited lakes of the Canadian Northwest Territories. *Journal of the Fisheries Research Board of Canada* 33: 2459–88.
Keith, L.B. and Windberg L.A. 1978. A demographic analysis of the snowshoe hare cycle. *Wildlife Monographs* 58: 1–70.
Lewis, H.T. and Ferguson, T.A. 1988. Yards, corridors and mosaics: How to burn a boreal forest. *Human Ecology* 16: 57–77.
Martin, C. 1978. *Keepers of the Game*. Los Angeles: University of California Press.

128 *Fikret Berkes*

Messier, F., Huot, J., Le Henaff, D. and Luttich, S. 1988. Demography of the George River caribou herd: Evidence of population regulation by forage exploitation and range expansion *Arctic* 41: 279–87.
Munsterhjelm, E. 1953. *The Wind and the Caribou.* New York: Macmillan.
Murphy, G.I. 1968. Pattern of life history and the environment. *American Naturalist* 102: 391–403.
Power, G. 1978. Fish population structure in arctic lakes. *Journal of the Fisheries Research Board of Canada* 35: 53–9.
Preston, R.J. 1975. *Cree Narrative: Expressing the Personal Meanings of Events.* Canadian Ethnology Service Papers, Mercury Series No. 30.
Rowe, J.S. 1972. *Forest Regions of Canada.* Canadian Forestry Service Publication No. 1300.
Speck, F.G. 1915. The family hunting band as the basis of Algonkian social organization. *American Anthropologist* 17: 289–305.
Speck, F.G. 1935. *Naskapi: Savage Hunters of the Labrador Peninsula.* Norman: University of Oklahoma Press.
Tanner, A. 1979. *Bringing Home Animals.* London: Hurst.
Winterhalder, B. 1983. The boreal forest, Cree-Ojibwa foraging and adaptive management. In *Resources and Dynamics of the Boreal Zone,* pp.331–45, ed. R.W. Wein, R.R. Riewe and I.R. Methven Ottawa: Association of Canadian Universities for Northern Studies.

6

Resilience and neo-traditional populations: the *caiçaras* (Atlantic Forest) and *caboclos* (Amazon, Brazil)

ALPINA BEGOSSI

Introduction

The aim of this study is to review ecological–cultural aspects of Brazilian native populations that descend from Indian and Portuguese, from the Atlantic Forest coast (*caiçaras*) and from the Amazon (*caboclos*). Following the definition by Berkes and Folke (1992), the objective is to analyse their *cultural capital* or the factors related to *caiçara* and *caboclo* adaptations to the environment. These factors are analysed in the light of ecological concepts, such as *resilience*. Examples from the Atlantic Forest coast will be concentrated on the southern coast of Brazil, and from the Amazon, from the States of Acre, Amazon and Tocantins (Figure 6.1).

The term resilience has widely different meanings in the ecological literature, sometimes related to the concept of stability (Putman and Wratten, 1984; Toft, 1986). Holling (1992) defined cycles as being part of structuring processes in which organisms participate. These cycles are classified in terms of four functions: exploitation, conservation, release and organization. In the cases dealt with here, resilience is determined by a release and re-organization sequence. As used in this book (see Chapter 1), resilience is the magnitude of disturbance that can be absorbed before a system changes.

Cultural behaviours may have an effect on ecological resilience and they have interesting attributes. On the one hand, it is the flexibility of human behaviour that made humans able to adapt to different environments, that helped them to overcome climatic changes, such as those in the Pleistocene, and to respond quickly to a variable environment (Boyd and Richerson, 1981). On the other hand, human behaviour may be very conservative and hard to change (such as traditions) as the result of a phenomenon sometimes called *cultural inertia* (Boyd and Richerson, 1985).

Figure 6.1.   Maps including the major areas mentioned in this study related to *caiçaras* (south-east Brazil) and *caboclos* (northern Brazil).

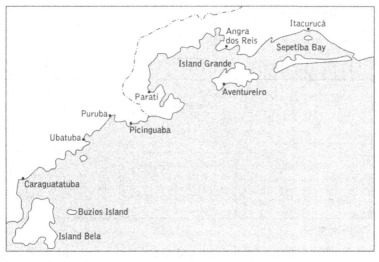

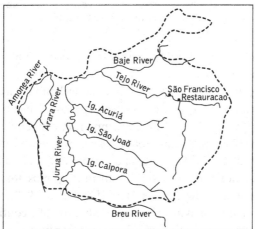

Figure 6.1. (*cont.*)

In ecological terms, we can observe different consequences of cultural inertia. For example, the environment may change and a particular cultural behaviour might no longer be useful in the new set of conditions. Such a cultural behaviour might be considered a 'load' (as in genetic load, deleterious traits or maladapted variants). In other cases, traditional cultural behaviours might improve ecological resilience by preventing overexploitation of systems or by helping them to recover. Or, a cultural behaviour which appears useless might be useful only in certain specific situations.

132　　　　　　　　　　　　*Alpina Begossi*

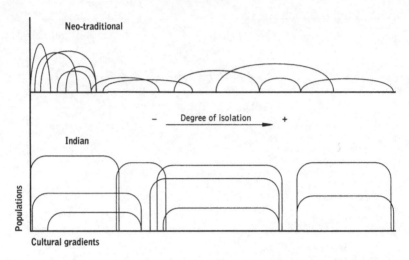

Figure 6.2.　The position of neo-traditional societies associated with the concept of ecological communities by Gleason, contrasted to Indian communities, which follow the concept by Clements.

There are examples of cultural behaviour from *caiçaras* and *caboclos* which fit into all three categories.

In exploring the relationships between cultural and ecological systems, the views of community organization expressed by Clements and Gleason (Whittaker, 1975) may be relevant. Contemporary community ecology recognizes that the Gleasonian concept of *open communities* better represents natural communities. The Gleasonian concept (Whittaker, 1975), suggests that species do not form the sort of distinct groups that characterize bounded types of communities. As stressed by Levin (1992), what we call a community or ecosystem is an arbitrary subdivision of a continuous gradation of local species assemblages. Thus, communities are not well integrated; they include species that respond individualistically to temporal and spatial variations. This approach includes the idea of *edge effects,* effects such as competition with invaders, increased predation and parasitism as well as higher extinction probabilities (Aizen and Feinsinger, 1994). Studies of forest fragments are well suited for the study of edge mosaics. According to Malcom (1994), edge effects give rise to peculiar community structures, because some species increase in abundance and others decrease close to the edge. These concepts from community ecology may have interesting implications for cultural ecology, when comparing native Indians (*indios* or *indígenas* in Brazil) with neo-traditional communities (Figure 6.2).

Neo-traditional resource management systems are defined as including elements from traditional and newly emergent systems (see Chapter 1). In such a context, neo-traditional populations are those with both traditional knowledge and a body of new knowledge coming from outside the population. All populations have new variants of knowledge coming in, but there may be differences or, better, a gradient among the proportion of what is old and new. Naturally, many behaviours learned from the external society are not ecologically adaptive, and not all traditions can be considered as ecologically 'sustainable'. The key point here are the possibilities of acquiring new variants; in other words, of maintaining variability to cope with change.

The flexible cultural boundary of neo-traditional communities might diminish their cultural inertia, and make them more open to new cultural values, which may lead to cultural adaptations and practices that may help increase ecological resilience. Examples are shown in this study, after a description of the Atlantic Forest coastal *caiçaras* and their communities and rules, followed by a description of the Amazon *caboclos*.

The Atlantic Forest coast

Tropical rainforests cover only 7% of the earth's surface, but contain more than half of the world's biodiversity (Wilson, 1988). The Atlantic and Amazon forests are among the most species-rich forests in the world. In general, Brazilian forests are divided into two major groups: those of Amazon and those of Atlantic origins. In the south/southeast of Brazil, the coastal plain forest, slope forest, and high-altitude forest are part of the Atlantic Forest (Joly, Leitão-Filho and Silva, 1991).

In the five centuries of occupation, Atlantic Forest exploitation was concentrated in *pau brasil* (*Caesalpina echinata*), followed later by many other timber woods, extraction of the heart of the palm (*palmito* – *Euterpe edulis*) and *xaxim* (*Cyathea* spp.), and the cultivation of sugar cane, coffee and cocoa (in Bahia State) (Joly *et al.*, 1991). The Atlantic Forest is represented today by sparse rainforest remnants along the Brazilian coast, from the south to the northeastern coast. As part of a 'hot spot' area, which includes habitats with many species in great danger of extinction (Wilson, 1992), the last remnants of the Atlantic Forest – about 5% of the initial vegetation cover (Myers, 1988) – are included in the Biosphere Reserve Program (MAB/UNESCO) (Lino, 1992). Nevertheless, between 1985 and 1990, about 536000 ha were deforested (Capobianco, 1994).

The Atlantic rainforest is a tropical forest with high endemism and diversity. For example, about 55% of the arboreal species are endemic. In the coastal plain forest, where most *caiçaras* live, important families are Myrtaceae, Melastomataceae, Lauraceae, Celastraceae, Fabaceae, Mimosaceae, Anacardiaceae and Compositae. In deforested areas, shrubs and colonizing plants predominate, especially *embaúbas* (*Cecropia*) and *quaresmas* (*Tibouchina*). Soils are shallow, sandy and usually have low fertility (Joly *et al.*, 1991; Leitão-Filho, 1982; 1987).

Demographic reduction or extinction of many animal species resulted from centuries of exploitation of the Atlantic Forest. Atlantic Forest vertebrates are small in size (with the exception of the tapir, *Tapirus terrestris*). The fauna has high endemism and diversity; batrachians are known to show marked endemism. The Atlantic Forest is rich in simians, in spite of having fewer monkey and marmoset species than the Amazon forest (Coimbra, 1991).

The *caiçara* communities studied are located on the northern coast of São Paulo State and the southern coast of Rio de Janeiro State, for example Puruba and Picinguaba (Ubatuba District), Búzios and Vitória Islands (Ilhabela District) and Sepetiba Bay (Itacuruçá District) (see Figure 6.1). All communities have small populations, ranging from 26 families (in Gamboa, Itacuruçá) to 100 (in Picinguaba and Jaguanum Island).

The caiçaras of the Atlantic coast

Caiçaras are descendants of Indians and Portuguese, and have culture and technologies that derive from these influences. Some African influences are also found, especially in religious feasts. African slaves arrived in the eighteenth and nineteenth centuries, representing 39% of the population of Ubatuba in 1836 (Marcílio, 1986). During this century, Japanese immigrants influenced the *caiçaras*, particularly with technological innovations for fishing, such as the *cêrco* (floating chambers of nets) (Mussolini, 1980).

The first inhabitants of the southern Brazilian coast were the Tupinambá Indians. After the arrival of the Portuguese, the people of this coast participated in the economy, such as production of sugar cane (including the production of sugarcane rum) and coffee (França, 1954; Marcílio, 1986). The *caiçaras* are native inhabitants of this coast, whose livelihood is based on the natural resources from the forest and sea, and who practise small-scale agriculture and fishing.

Agriculture is usually based on manioc (the main crop), but it includes potatoes, yam and beans (Begossi, Leitão-Filho and Richerson 1993).

Manioc, being a crop with a low input and a low risk, has spread in South America (McKey and Beckerman, 1993). In the southern coast of Rio de Janeiro State, banana plantations are very important. The processing of manioc, to produce flour, still has strong Indian influences, and includes methods of extracting the cyanidric acid by using round baskets called *tipiti*. Agriculture played an important economic role for the *caiçaras* until around 1950, when it began to be superseded by fishing as a source of cash. The low price paid for manioc compared to fish explains this economic shift (Diegues, 1983; Begossi *et al.*, 1993) and fishing is now the main source of cash for many coastal *caiçaras*. It is done both at sea and in the rivers of the forest, usually close to the river mouths.

Plants are also used by the *caiçaras* for a variety of purposes, such as for food, medicine, handicrafts and construction. For example, at Búzios Island, about 61 species are used for food, 53 for medicine and 32 for house and canoe construction and handicrafts; at Sepetiba Bay, about 100 plants are used for these purposes, and at Puruba and Picinguaba more than 200 species were mentioned as useful (Begossi *et al.*, 1993; Figueiredo, Leitão-Filho and Begossi, 1993; Rossato, 1996). The intense use of fish and plants by the *caiçaras* shows that in spite of their proximity to large Brazilian cities, such as Rio de Janeiro and São Paulo (see Figure 6.1), they still maintain a very close interaction with nature.

Local knowledge

Caiçaras have a deep knowledge of the rainforest coast where they live. Specific knowledge about natural resources includes forms of land cultivation, especially for manioc cultivation; knowledge about animals and plants, such as their uses and avoidances (taboos); knowledge about the classification of nature (ethnosystematics and ethnotaxonomy); and about appropriate technology.

Land cultivation

In spite of a relatively large number of studies on land cultivation by Amazonians (both *caboclos* and Indian populations), there are few studies dealing with land management by the *caiçaras*. In common with the *caboclos*, the *caiçaras* practise slash-and-burn and manipulate species diversity.

Roças (small plots where, amongst other things, manioc, beans and potatoes are cultivated) and *hortas* (small gardens next to houses for green vegetables) are generally included in *caiçara* agriculture. The dry season (July–October, November) is the time to prepare plantations (Begossi *et al.*,

1993). Manioc is the basic crop, especially the many varieties of *Manihot esculenta*. There are varieties of manioc with a high content of cyanogenic glucosides (HCN) in the edible part of the root cortex (bitter manioc or *mandioca brava*) which are used for preparing flour, and varieties with low cyanidric acid content, called *mandioca mansa* or *aipim (*sweet manioc), which are eaten after cooking (McKey and Beckerman, 1993).

Cury (1993) studied the *caiçara* cultivation of manioc in populations of the Vale do Ribeira and of the southern coast of São Paulo State. According to Cury's study, plantations (*roças*) occur in open fields inside the natural vegetation, helping to increase diversity. Cultivated and wild species are planted close together (in *sympatry*) maintaining the gene flow among species through hybridization. The planting of manioc through vegetative propagation, the practice of obtaining old *roças* material for transferral to new *roças*, and the planting of different varieties in a random distribution favour intraspecific hybridization, increasing diversity and allowing natural selection to act on new combinations. McKey and Beckerman (1993) stressed the importance of sexual reproduction and hybridization with wild relatives contributing to variation in the quantity of HCN among species and varieties of manioc.

Fallow periods vary; land used to be left fallow for over 15 years but this has declined due to shortage of land available for small-scale agriculture (Cury 1993). In fact, most areas where the *caiçaras* live are technically federal or state conservation lands, and their traditional methods for cultivation conflict with government environmental policy. Conflicts with environmental agencies about the slash-and-burn technique were mentioned by *caiçaras* in interviews in a variety of localities along the coast, such as Puruba, Picinguaba, Ponta do Almada, Serraria (Ilhabela) and Vitória Island. Hanazaki *et al.* 1996 observed that at Ponta do Almada, 7 families (out of 14) decided to stop their agricultural activities.

In spite of the changes that led *caiçaras* to focus more on fishing than on cultivation, many families still maintain houses, called *casa de farinha* (flour houses), to process manioc. One *casa de farinha* usually provides for an extended family, which includes more than one nuclear family. As residence is usually virilocal (the sons live, after marriage, close to their fathers' house), such as at Búzios Island (Begossi 1989) and Aventureiro (Vilaça and Maia, 1989), the flour houses usually provide for nuclear families that are located close to their fathers' houses.

The traditional management of the *roças* is an example of attitudes that increase the ecological resilience of the *caiçara* community. Oliveira *et al.*

(1994) studied the methods of slash and burn of the *caiçaras* of the community of Vila do Aventureiro (Grande Island, coast of Rio de Janeiro State), which includes about 90 individuals (see Figure 6.1). Manioc is planted in a typical polyculture, with beans, yam, maize, rice, pumpkin, papaya, sweet potatoes and watermelon. The forest is then cleaned and burned; calcium and magnesium increase in the soil, whereas aluminum decreases (increasing soil fertility). In the fallow period, other species are planted, such as the *cobi* (*Anadenanthera colubrina*), an important legume that helps to fix nitrogen. As stressed by the authors, through this method nutrients are replaced, erosion is minimized, and there is no need for chemical pest control. A low population density is a necessary condition for the system to work because land must be available for fallow periods.

Animals and plants: uses and classification

Many animals and plants are used by the *caiçaras*, whose knowledge and practices include animals and plant uses, taboos and folk biosystematics.

Common marine animals used for food and sale are, at Búzios and Vitória Islands, bluefish (*Pomatomus saltatrix*), squid (*Loligo sanpaulensis*) and halfbeak (*Hemiramphus balao*); at Puruba and Picinguaba, snook (*Centropomus parallelus*), mullets (*Mugil* spp.), cutlass fish (*Trichiurus lepturus*), sand drum (*Micropogonias furnieri*), kingfish (*Menticirrhus americanus*), freshwater catfish from the Puruba or Quiririm rivers; and, at Sepetiba Bay, shrimp (*Pennaeus schmitti*), sand drum, mullets, weakfish (many Sciaenidae) and kingfish. Fish is the main source of animal protein used by the *caiçaras*. For example, at Búzios Island, about 68% of the animals consumed are fish (Begossi and Richerson 1993); at Puruba Beach, 52%; and at Gamboa, Itacuruçá Island, 67% (Paz and Begossi, 1996).

There are many fish food taboos among the *caiçaras*, and there seems to be an association of these taboos with different ecological factors: tabooed fish tend to be carnivorous, toxic, or medicinal fish. A study on Búzios Island by Begossi (1992b), showed that carnivorous fish, such as bonito (*Auxis* sp. and *Euthynnus alleteratus*) and bluefish, are usually avoided as food when people are ill; other fish are avoided because they are toxic, such as pufferfish (*Sphoeroides spengleri*) or because they might be toxic (*Gymnothorax funebris*). An example of the latter is the avoidance by *Buzianos* of fish that have been known to cause *ciguatera* in other regions. Ciguatera is an intoxication caused by the ingestion of fish with ciguatoxin. About 300 fish species belonging to 12 families have been responsible for ciguatera fish poisoning, and fish not valued as food may be ciguatoxic but have not yet identified as such scientifically (Habermehl, 1981; Lewis, 1984). *Gymnothorax* species have been found with ciguatoxin in some areas. Such a food taboo might represent an adaptive behaviour conserved by *cultural inertia* if in the past there were cases of ciguatera in the region; or it might be an adaptive behaviour if there are cases of ciguatera in the region (no research or data are available).

Examples of food taboos related to medicinal animals include the avoidance of ray (the eggs are used for haemorrhages) and especially of lizard (81% of *Buzianos* use lizard fat for snake bites, among other things). These taboos on eating may preserve the supply for medicinal use.

Similar fish food taboos were observed in other *caiçara* communities (Gamboa, Puruba and Picinguaba), including, for example, those on pufferfish, catfishes, rays, sand drum and sharks. Some taboos only occur in specific situations, such as avoidances of ray, bonito, catfishes, mullets and sharks. Examples mentioned by men involve diseases and by women, menstruation and after childbirth.

Plant uses reflect both Indian and Portuguese practices. Besides manioc growing and processing, many medicinal plants are common non-native herbs, such as mint (*Mentha* spp.), spearmint (*Laurus nobilis*), wormwood (*Arthemisia absinthium*), cress (*Lepidium virginicum*) and pennyroyal (*Cunila spicata*). The medicinal plants most quoted in the *caiçara* communities studied (see Figure 6.1) were avocado (*Persea americana*), orange (*Citrus sinensis*), boldo (*Coleus barbatus* and *Vernonia condensata*), balm (*Lippia citriodora*), fennel (*Foeniculum vulgare*), and wormseed (*Chenopodium ambrosioides*). On the other hand, most plants used for house and canoe construction are native trees, such as anjelywood (*Jacaranda* sp.), aracurana (*Alchornea iricurana*), guapurubú (*Schyzolobium parahyba*), species of *Tabebuia* and of *Aspidosperma* (Begossi *et al.*, 1993; Figueiredo *et al.*, 1993; Rossato 1996). A relative loss of knowledge about medicinal plants was observed among the young generation of *caiçaras* of Búzios Island and Sepetiba Bay (see Figure 6.1). The loss of local knowledge decreases local cultural variability, which may leave the community less able to cope with changes, affecting the resilience of the local system. Also, a higher diversity of plant uses is observed in areas which, on the basis of island biogeography theory are supposed to have higher plant diversity, (Figueiredo *et al.*, 1993). In Mexico, Benz *et al.* (1995) concluded that the use of plant resources is probably a function of the relative taxonomic abundance of the area flora.

The knowledge of the *caiçaras* about natural resources includes more than just that related to the use of animals and plants. *Caiçaras* classify and name organisms; probably their ethnotaxonomy, as other features of their culture, is a mixture of old and new traits, which may be acquired through contacts with other fishers that come from outside, in a typical neo-traditional sense (see Chapter 1). Begossi and Figueiredo (1995) found 115 folk fish species (corresponding to 105 species) and 73 folk species at Sepetiba Bay (corresponding to 66 species). Binomial names, that is, a name and a modifying adjective, are an important part of the ethnotaxon-omy of the *caiçaras*, ranging from 20% at Sepetiba to 35% at Búzios island. This frequency corresponds to that usually found for small-scale cultiva-tors, which tend to have more detailed taxonomies than foragers (Brown, 1985).

As other neo-traditional communities in Brazil, the *caiçaras* believe in forest guardians, the *caipora* or *curupira*, *mãe da mata* (forest mother), and *boitatá*; in spirits that protect the animals (*anhangá*); in spirits that protect reproductive animals *(Tapiora)*: and in spirits of the water, who punish those who fish too hard (*Mãe d'Água*) (Diegues, 1994).

Technology

Fishing is performed with canoes (paddled or motor) and with the use of different types of nets and hooks. For example, from the southern coast of Rio de Janeiro State to the northern coast of São Paulo State, there are: motor and paddled canoes used with encircling nets for fish and shrimp at Gamboa, Itacuruçá Island; boats used with set gill nets at Jaguanum Island and Picinguaba; paddled canoes used with encircling nets at Puruba and Quiririm rivers, in addition to beach seines used at Puruba Beach; paddled or motor canoes used with hook and line at Búzios and Vitória islands.

Following the study by Forman (1970) on a coastal population of north-eastern Brazil, special attention was given to the local knowledge of local people concerning technology changes. Forman claimed that an innovation might be 'rationally' discarded due to inappropriate local features, such as the non-adoption of boats by raft fishermen. Many innovations suited to environmental conditions and beneficial to people tend to be absorbed by communities, as occurred with the *lambreta* at Búzios Island (Begossi and Richerson, 1991). The adoption of the *lambreta*, a jig to catch bluefish, at Búzios Island shows that fishermen had a sharp perception of their needs and of the benefits of this jig. Technological changes imposed by outsiders, without knowledge about the ecological and social context and the needs of the community, have a high probability of failure. Innovations in ecolog-ical and cultural vacuums (Alexander, 1975) may decrease the resilience of a system.

Property rights and institutions among *caiçaras*

The use of resources by the *caiçaras* follows specific local rules. Many rules and practices are based on kinship: for example, manioc flour processing includes an extended family; a crew consisting of brothers is very common in fishing trips. Among the *caiçaras*, informal family rights were observed concerning places to set gill nets at Búzios Island and concerning fishing areas used by artisanal fishers at Sepetiba Bay (Begossi, 1995). Property rights of individuals, families or communities are an important condition for local management, because they depend on the exclusion of outsiders. Groups are able to manage resources if they can exclude other potential users and regulate joint use among group members (Berkes *et al.*, 1989).

Resource tenure systems in fisheries are common in traditional and neo-traditional communities throughout the world (Berkes, 1985). The first to recognize property rights systems among coastal Brazilian fishermen was Forman (1967), who studied the use of secrecy in the management of fishing spots in Coqueiral (Alagoas State). Cordell (1978) observed territoriality among coastal fishermen from Bahia State. Lima (1989, in Diegues, 1994) observed rules and division of spots with regard to net fishing in the beach of Itaipú, Rio de Janeiro State.

The degree of organization of the *caiçaras*, and in consequence the strategies employed by them to make decisions concerning demands (such as fishing or farming) or disputes, are extremely variable among communities. In some, such as at Búzios Island, the lineage system, based on kinship, is dominant. Decisions are usually taken as part of family talks, and leadership is an attribute of the old, who are consulted about the internal and external problems of the community. At one of the most populated harbours on Búzios Island, an octogenarian, the grandfather or father of most inhabitants of this harbour, was very respected and often consulted. Anyone arriving at the harbour was supposed to visit him.

On the other hand, at Sepetiba Bay, fishermen get together to discuss communal problems and leadership is a consequence of local activities performed (fishing, political participation). At Sepetiba Bay, fishers actively defend their bay against intruders, involving the local politicians and the local press (Begossi, 1995). It was suggested that fishermen's tactics and strategies of fishing were influenced by territorial conflicts (Begossi, 1992a). This study at Sepetiba, using optimal foraging analysis, showed that shrimp fishermen from Gamboa (Itacuruçá Island, Sepetiba Bay) were leaving the patches (fishing grounds) later than predicted by the optimal foraging model, trying to get more shrimp out of each patch. Two hypotheses were considered: (a) fishers fished longer because it was difficult to evaluate the density of non-visible prey; (b) competition from industrial fishing was pushing local fishers towards a less conservative behaviour. It is difficult to evaluate whether this practice of fishing longer decreased or increased ecological resilience at Gamboa. If fishers were just getting more shrimp, this was a form of posturing to industrial fishers, i.e. 'marking their spots'. On the other hand, if this behaviour contributed to overfishing, fishers were decreasing their ability to cope with perturbations coming from industrial fishing. Due to the small scale of the artisanal *caiçara* fisheries, and that of its technology, overfishing seemed unlikely.

Government regulation of land and resource use also interferes with neo-traditional communities. Many *caiçara* populations are located in parks (such as the State Park of Serra do Mar – *Parque Estadual da Serra do Mar*), in Biological Reserves (*Reserva Biológica Estadual da Praia do Sul*, Grande Island, Aventureiro) or Biosphere Reserves, where there are sets of state rules for fishing and for agriculture. One of these includes the prohibition of slash-and-burn techniques at the *roça* (Sales 1994). If the *caiçaras* cannot grow their manioc, they would have to buy it in the markets along with other products. Therefore, they will need more cash to be available through the exploitation of natural resources or tourism.

The *caiçaras* are in constant contact with outsiders and with the economic and cultural influences of the Brazilian society. Neo-traditional societies are probably representatives of the Gleasonian concept of communities, because of their different degrees of interaction with society (see Figure 6.2). *Caiçara* economy, despite being based on petty commodity production (Diegues, 1983; 1994; Sider, 1986), has capitalist features. For example, at Búzios Island the lineage system interacts with the capitalist system through the role of different buyers (Begossi, 1996). *Caiçaras* have always been engaged in the economic cycles of the region (Begossi *et al.*, 1993), and have continually changed adaptive strategies and economic behaviour to be able to make a livelihood, even though little change is observed in their day-to-day life.

Caiçaras are also engaged in Pentecostal religions and associations. Even in very small *caiçara* communities, two or three different kinds of churches may be found – God Assembly, Christian Congregation, Adventists (Begossi, 1996). These religious sects change the behaviour of individuals, helping to isolate them from the outside society (some break their radios and TVs after conversion) and are the focus of conflicts within communities.

The Amazon Forest

Amazon supports about 20% of the world's vascular plant species and 2500–3000 fish species (Bennett, 1992). The Amazon rainforest includes vegetation zones ranging from dense forest to areas with sandy soils and sparse vegetation. According to Prance (1978), the vegetation types include the non-flooded forest (*terra firme*), inundated forests, non-flooded and flooded savannas (on *terra firme* and on *várzea*, respectively), *campina*, montane vegetation, coastal vegetation and river beaches. The *terra firme* represents 85% of the Amazon, which includes dense forest with tall trees and large biomass, liana forests, low forests, *campina* (a low, dense type of vegetation on white sand), and bamboo forests, among others. The *várzea* comprises periodically flooded areas by white water rivers, and the *igapó* includes periodically flooded areas by black water rivers or the permanently flooded swamp forest.

Some types of Amazon vegetation are indicators of past human activities. Balée (1989) reviewed such vegetation types, which are the *babaçu* (*cocais*) forests, made up of *Orbignya martiana*, covering 197 000 km²; forests including *Elaeis oleifera* (*caiaué*); forests with a high frequency of palms; the *campina* vegetation, with white sandy soil; bamboo (*Guadua* spp.) forests, occurring in 85 000 km²; the *apête*, or forest islands planted by the Kayapó; forests of nuts (*castanhas – Bertholletia excelsa*), occupying 8000 km², and the liana forests (*mata de cipó*), covering about 100 000 km², which are considered to occur together with fertile soils.

Amazon diversity, especially in the *várzea*, is high, including about 250 species of plants per hectare (Prance, 1978). Setz (1993), sampling forest fragments close to Manaus, in Central Amazon, found 552 species of plants, primarily of the families Leguminosae, Sapotaceae, Chrysobalanaceae, Lauraceae, Annonaceae, Moraceae, Lecythidaceae and Melastomataceae. Salomão *et al.* (1995) sampled 21 sites of 1 ha each, in the States of Pará, Maranhão and Rondônia, and found 790 species of plants of 68 families (10 cm diameter at breast height); among these, 108 species included 20 or more individuals (three palms), showing an enormous diversity.

Soils in the Amazon range from the fertile alluvial soils of the *várzea*, to oxisols to the poor sandy soils. As pointed out by Wambeke (1978), the most frequent soils in the *terra firme* are oxisols. Mammal distribution in the Amazon follows the soil and vegetation variability. Large areas covered by poor soils have a lower mammal productivity (Emmons, 1983). In a classic study, Fittkau and Klinge (1973) showed that the animal biomass in the Amazon is relatively small: in central Amazônia it is 0.02% of the total biomass.

In 1968, the German limnologist H. Sioli classified the Amazon rivers into three groups: white waters (*águas brancas*), such as the Amazon and many tributaries of the west and southwest; clear waters (*águas claras*), such as Tapajós and Tocantins; and black water (*águas pretas*), such as the Negro. Most white water rivers come from the Andes, having a high quantity of sediments in the water and forming the *várzea*. The Amazon *várzea* is renowned for supporting high-density populations, due to the fertile soils and to high concentrations of aquatic animals (McGrath *et al.,* 1993). Clear water rivers come from Guianas and central Brazil, being associated with oxisols (latosols). Black water rivers have a very low quantity of sediments, being associated with podzol soils (Sioli, 1985).

Amazon fisheries include diffuse, small-scale and large-scale fisheries. The first type occurs around small towns and in areas inhabited by scattered riverine fishers, including subsistence fishers. The second is practised in large cities as Manaus and Belém (Bayley and Petrere, 1989) (see Figure 6.1). Among 32 fish species landed in the Manaus market in 1978, 72% were represented by four species: *tambaquí* (*Colossoma macropomum*), *jaraquí* (two species of *Semaprochilodus*) and *curimatá* (*Prochilodus nigricans*) (Petrere, 1989).

In Brazil, the Amazon technically includes nine States, or 5 000 000 km². Estimates of deforestation rates in the Brazilian Amazon have varied because the methods and areas used for estimations also varied. Data from 1991 showed about 10.5% of forest was cleared (426 000 km²), which is almost the size of the State of California (Fearnside, 1993). According to Morán (1993), the conversion of land to pasture, mining and timber activities is the main source of deforestation in the Amazon. Cattle ranches, in particular, have been encouraged by Brazilian government policies, in spite of studies showing the environmental degradation related to them (Fearnside, 1979; 1980). Recent studies (Reiners *et al.,* 1994) have shown the consequences for the soil and vegetation of rainforest lands used for pasture. Shifting cultivation has a low impact because small areas are cleared and there is evidence of management of fallow periods by natives. Ranchers account for about 70% of deforestation in the Amazon (Fearnside, 1993).

The caboclos

Caboclo subsistence is based on small agriculture and fishing. Like the *caiçaras,* the *caboclos* descend from both Indians and Portuguese and, to a lesser extent, they may include African influences. According to Morán (1974), the *caboclo* culture started with the arrival of the Portuguese (1500–1850), following a phase of acculturation and an extractive economy based on rubber (1850–1970). Moran's study shows that the *caboclo* can be a rubber or nut collector, a horticulturist, a canoe paddler, a fisher, usually earning a living from many or some of these activities.

 Caboclo livelihood is based on small-scale agriculture with the cultivation of manioc, maize, rice, beans, watermelon, and papaya, and fishing in the rivers – *igarapés* (small rivers) or *igapós* (flooded forest). River water level is usually an important aspect in the life of the *caboclos*, because their subsistence is managed and adapted to such conditions. When the water is low ('summer'), fishing is an important activity; when the water is high, in the wet season ('winter'), wildlife hunting in the forest tends to be important for subsistence.

 The Amazon includes different kinds of environments, and the *caboclo* living is closely tied to such environments. This study is focused on riverine *caboclos* and small agriculturists from the Tocantins River, the Extractive Reserve of the Upper Juruá and the Transamazon highway.

Local knowledge

Like the *caiçaras,* the caboclos have detailed knowledge of their surroundings. Perhaps unlike the *caiçaras,* the *caboclos* are more heterogeneous, not only because of the high diversity of Amazon environments, but also because of their greater interaction with other communities. For example, fishers who live on the banks of the Tocantins River have a different subsistence and fishing technology than the urban fishers. Fishers from cities such as Imperatriz (Maranhão State) use boats and nets for fishing, besides depending more on industrialized resources, whereas more isolated riverine fishermen use hook and line and paddle canoes and practise small agriculture. Therefore, a stronger or weaker dependence on natural resources may influence the contemporary knowledge of the *caboclo.*

Land cultivation

Like the *caiçaras,* manioc cultivation and the production of manioc flour are typical of the *caboclo* subsistence. Slash-and-burn techniques are used for cultivation and a variety of fruits from trees and from the high diversity of palms is collected in the forest. One of the best examples of how local knowledge forms a basis for successful livelihood strategies in the Amazon Forest is given by studies by Morán (1977, 1979, 1990) among settlers of the Transamazon highway located close to Altamira (Pará State). In the 1970s, southern (23%), northeastern (30%) and west central (13%) migrants, along with *caboclos*, comprised the settlers along the highway in a Brazilian government programme. In spite of the credit given to southerners by the Brazilian government, the local environmental knowledge of the *caboclos* made them more successful as farmers compared to the new settlers.

The Brazilian government's high expectation of success by the southerners was based on their modern farming experience, residential stability, previous credits and initial capital (Morán, 1983). However, the *caboclos* were able to choose the more fertile sites, using as signs or indicators certain plants associated with soil fertility (Morán, 1990). Their main strategy was the planting of manioc and tobacco in small plots using shifting cultivation. Southerners, by contrast, cleared large areas to plant rice, maize and beans. They raised cattle, whereas the *caboclos* raised pigs and chicken. The *caboclo* sites also showed a slower decrease in fertility than did the lands of the new settlers. The *caboclo* produced twice the yield per hectare, and included a diversified cropping pattern which allowed higher incomes than rice farming, less expenditure on consumption, greater use of local technology and of the involvement of family labour (Morán, 1983).

Plants and animals

A variety of plants are used by the *caboclos*, many of them for medicinal practices (Amorozo and Gély, 1988). While there are communities with a detailed knowledge of medicinal plants, other communities seem to have lost part of this knowledge. Despite 300 species being mentioned as used by caboclos (rubber-tappers, etc.) of the Extractive Reserve of the Upper Juruá (Empéraire and Delavaux, 1992), the use of medicinal plants is being replaced by industrial medicine. This is, at least in part, due to the activity of local politicians who in some Amazon areas, distribute medicines as part of their political campaigns, encouraging local people to depend on industrialized drugs and to stop using their traditional medicines. One of the projects in this Extractive Reserve was the restoration of knowledge about plant uses in local health care centres. In these centres, people are trained to teach the cultivation and uses of local medicinal plants to the population.

Animals are an important part of the *caboclo* diet. In the dry season, fish is usually a significant protein source, and in the wet season, game meat is essential. Game among Transamazon settlers includes agoutis (*Agouti* sp.), paca (*Agouti paca*), peccaries (*Tayassu tajacu* and *T. pecari*), deer (*Mazama* sp.) and tapir (*Tapirus terrestris*), as well as many monkey species (Smith, 1976). Game is a very important protein source in the wet season at the Upper Juruá, when deer, peccaries, monkeys, and small rodents were hunted (Begossi, Amaral and Silvano, 1995).

The dry season (June–October) is the time for fishing. Fish consumption includes many catfishes of the Pimelodidae family (one of the most important Amazon fish families) such as *mandi* (species of *Pimelodella*, *Pimelodina* and *Pimelodus*) and surubim (*Pseudoplatystoma fasciatum*), Curimatidae (*Prochilodus nigricans*), as well as species of the family Anostomidae (*piau*). At the Upper Juruá, the Loricariidae (a type of small catfish, with a hard body texture, locally called *bode*) are very important for consumption (Begossi *et al.,* 1995).

Food taboos, related to game and to fish, are part of *caboclo* culture. Tabooed fish may be called *carregado* or *reimoso*. The meats of tapir, collared peccary, nine-banded armadillo and tortoise are strong taboos (*muito reimosos*). Also tabooed are white-lipped peccary, paca, agouti, rabbit and monkeys. Meat not tabooed ('clean', 'mild') includes deer, fowl, chicken and cattle (Smith, 1976).

The fish that are food taboos among *caboclos* of Itacoatiara (Smith, 1981) include, among others, pirapitinga (*Colossoma bidens*), matrinchão (*Brycon* spp.) and curimatá. At the Tocantins River, Begossi and Braga (1992) found that many medicinal fish are avoided as food, such as species of ray, jaú (*Paulicea lutkeni*), and poraquê (*Electrophorus electricus*). At the Upper Juruá, *caboclos* use ray and traíra (*Hoplias malabaricus*) for medicinal purposes, and these are also avoided as food (Begossi *et al.,* 1995).

As with the *caiçaras*, the *caboclos* also avoid eating certain species of fish during illness. These are usually carnivorous fish, such as surubim, barbado (*Pinirampus pirinampu*), jaú and pirarara (*Phractocephalus hemiliopterus*). Some of the taboos mentioned by Smith (1981) refer to this category, when the meat is classified as *carregada* or *reimosa*. The origin of this concept is, according to Smith (1981), still unclear, but it may be related to the Latin *rheum* (thick fluid) and to Hippocratic medicine. A similar relation between tabooed and medicinal or carnivorous fish is found among both *caiçaras* and *caboclos* (Begossi, 1992b; Begossi and Braga, 1992).

Caboclos believe in animal spirits, such as *mães de bicho*, which steal the shadows of hunters that kill too many of each species. *Curupira* is well known in Brazilian folklore. It is a forest guardian represented by a small man with feet turned backward who punishes hunters who are too ambitious, by attracting them into the forest till they lose their way (Morán, 1974). These beliefs may serve as cultural rules that in practice function as conservation measures (Reichel-Dolmatoff 1976) – an *etic* interpretation in Harris' (1976) terms.

Technology

Fishing technology is highly diverse in the Amazon and is related to the different riverine communities. Nets and cast nets are very common in the Amazon rivers, but a variety of jigs, harpoons (*arpão, zagaia, bicheiro*) and traps are found – explosives and fish poisons are also used (Goulding 1979; Smith 1981). Since the beginning of this century, migratory species (such as species of *Semaprochilodus, Colossoma* and *Brycon*) have been caught with seines (a Portuguese introduction) and dynamite (Ribeiro and Petrere, 1990). Paddled canoes are an important technology for both fishing and transportation in the Amazon.

Property rights and institutions among *caboclos*

Contrary to the informality of the *caiçaras* in dealing with internal and external questions, the *caboclos*, especially the *seringueiros* or rubber-tappers, are organized in associations and actively participate in local and environmental politics. Fishers from the Lower Amazon River have developed new management strategies for lake fisheries, involving the exclusion of outsiders and regulating fishing activities (McGrath *et al.,* 1993). According to these authors, territorial fishing in the Amazon is especially related to lakes, for several reasons: they are important for subsistence; they are enclosed bodies of water; and the *caboclos* see a direct link between local fishing pressure and productivity.

Caboclos, influenced by the Liberation Theology and leftist parties, built strong political organizations and movements that culminated in the common management of resources through extractive reserves. In the case of the Alto Juruá, State of Acre, the scattered families along the Juruá River were able to communicate via the radio station *Verdes Florestas*, which transmitted all kinds of messages. Certainly, this was of prime importance in guaranteeing the local organization of rubber-tappers, which are integrated in their National Council.

Extractive reserves, defined as 'forest areas inhabited by extractive populations granted long-term usufruct rights to forest resources which they collectively manage'(Schwartzman, 1989), and legally defined as 'territorial areas used by extractive populations for sustainable exploitation and natural resource conservation'(Decree 98.897, January 30, 1990), are probably the best-known Brazilian example of common property resource management. The first extractive reserve (Alto Juruá) was legally established in 1990. Families were organized into local and national rubber-tapper councils. The organization of the reserve is an activity which involves the local people in meetings, with researchers and representatives of the councils. There is a local council in each rubber-tapper reserve. At the Upper Juruá, the *ASAREAJ* (*Associação dos seringueiros e agricultores da reserva extrativista do Alto Juruá* – Association of the rubber-tappers and agricultors from the Upper Juruá Extractive Reserve) is responsible for local management. Local meetings are attended by residents, after listening to radio *Verdes Florestas*. In our field work (Begossi *et al.*, 1995), we interviewed families located on the banks of the rivers Juruá, Tejo, Bagé, S. João and Breu (see Figure 6.1). Some families already knew about our arrival, even in very distant and isolated places, after listening to their radios. In 1994, as the result of local meetings, a first management plan was proposed by the Rubber-Tapper Council to the Environmental Federal Agency (IBAMA), and it was approved by them.

Caiçara and *caboclo* contributions to ecological resilience

It can be argued that groups that borrow from two or more cultural traditions, as the *caiçara* and the *caboclo* do, have a richer cultural capital and a wider range of adaptive options they can call upon. Thus, such groups may have greater cultural flexibility than either of the parent societies. By contrast, Indian (*Indios*) societies tend to be relatively isolated; their different language also contributes to such isolation.

We assume that neo-traditional communities can be depicted in terms of Gleasonian boundaries, being more open than are Indian communities. Such openness allows neo-traditional communities a higher cultural and economic flexibility. In Figure 6.2, neo-traditional communities do not show sharp edges, in contrast to Indian communities, depicted following the original view of community structure by Clements (Whitaker, 1975). Culture is adaptive because, among other things, it is variable and flexible. Variability or, in other words, *diversity* (genes, species, culture) is the basis for survival and allows communities to cope with environmental changes.

By contrast, Indian societies are less open and hence generally less flexible. The disappearance of many Indian societies in Brazil may be partially explained through their inability to deal with European diseases and institutions. Some Indian societies that 'survived', such as the Guarani, Terena, Guajajara, Tikuna and Macuxi, and that show demographic increases in spite of more than 200 years' contact, did so because of their flexibility. These societies, still maintaining an ethnic identity, show biological adaptations to diseases and cultural abilities to deal with the outside society, including political organizations (Gomes 1988). Thus, the ability of communities to interact with the rest of society and the flexibility of these interactions are factors that may increase their capacity to cope with changes.

For both *caiçaras* and *caboclos* manioc cultivation and the process of producing flour are practices adapted to local soil and rainforest conditions, which come from native Indians. They moved to commercial fishing (after selling manioc), a fundamental channel for cash in both the Atlantic Forest coast and the Amazon. The change from agricultural to fishery production among the *caiçaras,* responding to market pressures in the 1950s, increased their chance of survival (a similar economic change was also observed for the riverine *caboclos* – McGrath *et al.,* 1993). Individuals who entered fishing activities had a better economic performance, and were followed by others in the community. Greater economic flexibility may represent a higher probability of 'community survival' and, apparently, neo-traditional communities benefit from such economic flexibility. Therefore, market participation of *caboclos* is an adaptive strategy – one way to increase their *cultural flexibility*.

The important point is that such cultural flexibility is often related to cultural behaviours which might increase ecological resilience, such as techniques (traditional and neo-traditional). These behaviours include practices of diversified cropping, of increasing manioc diversities, of taboos and beliefs that help conserve medicinally important species.

A relevant aspect of neo-traditional populations is their ability to cope with a mixture of cultures, in which cultural traditions survive alongside innovations. Naturally, the loss of some innovations and knowledge may work against ecological resilience. Examples include the loss of knowledge about medicinal plants observed among *caiçaras* and *caboclos* of the extractive reserve of the Upper Juruá.

The ability of *caboclos* to deal with different groups from society, such as politicians, scientists and government planners, permitted them to 'innovate'. When the rubber-tappers of the extractive reserve of the Upper Juruá started to work together with ecologists from Brazilian universities, they were trying to prevent the intrusion of unsustainable uses into their areas. It was the rubber-tappers themselves who created the extractive reserves, enabling ecologically sustainable practices consistent with increases in ecological resilience (Figure 6.3).

Extractive reserves are a form of common resource management fol-

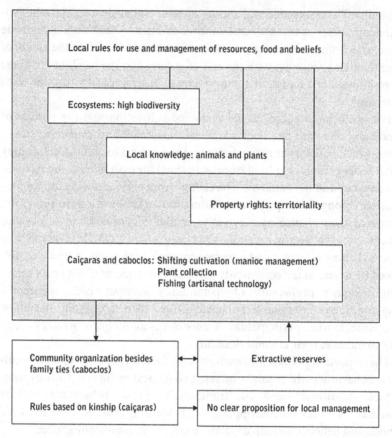

Figure 6.3. Linkages of the ecological and social systems of *caiçaras* and *caboclos*, showing their different outcomes in local management.

lowing the definition by Gibbs and Bromley (1989) in which resources are managed by rules from the user-group and uses are conditional upon the interdependent behaviour of group members. As institutions, extractive reserves include mechanisms for efficiency, stability, resilience and equitability. For example, there are rules to minimize disputes, to enable rubber-tappers to cope with changes, and to ensure that resources are used by all members of the community. The political organization and movement of rubber-tappers in the Amazon, with the creation of extractive reserves, is a case in which strengthening the local social system has paid off in terms of maintaining the sustainability of the local ecological system.

Outcomes

Despite their geographical separation, *caiçaras* of the Atlantic Forest coast and *caboclos* of the Amazon have many interesting points in common regarding the knowledge of the tropical forest environment and cultural rules. In spite of the higher heterogeneity of the Amazon environment compared to the Atlantic Forest, both are highly diverse environments and comprise prime targets for conservation.

Property-rights rules were observed for both *caiçaras* and *caboclos*, but their *outcomes* were completely different. The institutional arrangements of *caiçaras* and *caboclos* involved different interactions to the outside society, probably leading the *caboclos* (rubber-tappers) to a common management through their political organization. Also, the economic activity of rubber-tappers, initially based on extracting products from the forest, helped in an ecologically sound perception of their surroundings.

Differences in the main economic activity (tapping trees to produce natural rubber) of some *caboclos*, in their history of economic exploitation by landowners (*seringalistas*) and in their institutional arrangements contributed to different outcomes, from those of the *caiçaras* (see Figure 6.3). While the *caiçaras* do not seem to be organized politically, or have recently emerging organizations, the *caboclos* are extremely well organized and willing to defend their livelihoods. The results are that *caboclos* seem to be working towards more successful outcomes in terms of a sustainable community.

The successful outcome for the *caboclos* is the creation of the extractive reserve system and their ability to form alliances with native Indians, including the Forest People (*Povos da Floresta*). Extractive reserves were conceptualized and formed by the rubber-tapper movement, which achieved international status through its organizations, reinforced after the killing of one of its leaders, *Chico Mendes*. Extractive reserves are an innovation and a mechanism for local resource management. The concept of an extractive reserve includes concerns about biological and cultural diversity, as well as access and equitability in the use of resources by the commoners. According to Allegretti (1992), the extractive reserve concept is a model for sustainable development because: (a) it enables the forest to be used as a productive resource base; (b) it combines social and economic considerations; and (c) it democratizes the structure of land use.

The Upper Juruá Extractive Reserve was created in 1990 and, following the efforts of the rubber-tappers and researchers, culminated in a preliminary management plan in 1994. These efforts included projects related to biodiversity, social economy, local knowledge (Cunha, Brown and Almeida, 1993; 1994), education and health. The creation of extractive reserves involved major changes in the local economy because the control of land reverted to rubber-tappers, breaking down the old system of economic exploitation of the rubber-tapper (*seringueiro*) by the landowner (*seringalista*). A description of this system is found elsewhere (Rancy, 1986). The creation of such reserves involved conflicts with local politicians and with landowners, as well as the creation of a new economic system for the inhabitants of the reserve area in order to put into practice a local management system for common resources.

Caiçaras, with their lack of genuine political organizations, and *caboclos*, with their celebrated rubber-tapper organizations, show two very different patterns of responses in their dealings with threats to their natural resources. The extractive tradition in the Amazon, pushed by the exploitative system of rubber-tapping, contributed to new ideas and organizations which led to the extractive reserves. By contrast, the *caiçaras* remained strongly tied to tourism and, under the influence of Pentecostal doctrines, ended more isolated and culturally less flexible than the *caboclos*. The creation of mechanisms similar to the extractive reserves in the Atlantic Forest will depend on preliminary steps being taken to build institutions among the *caiçaras* for natural resource management. Extractive reserves are in reality a mechanism to legalize a local management system that already exists in an area.

Acknowledgements

This study is part of the subproject *Linking social and ecological systems for resilience and sustainability,* included in the research programme *Property rights and the performance of natural systems,* directed by Susan Hanna, from the Beijer International Institute of Ecological Economics. I thank the Brazilian Agencies CNPq and FAPESP for grants and research scholarships during this study, and the MacArthur Foundation for financial support for the Upper Juruá projects. I also thank F. Berkes for a detailed revision of the manuscript and very helpful comments.

References

Aizen, M.A. and Feinsinger, P. 1994. Habitat fragmentation, native insect pollinators, and feral honey bees in Argentine 'Chaco Serrano'. *Ecological Applications* 4: 378–92.

Alexander, P. 1975. Innovation in a cultural vacuum: the mechanization of Sri Lanka fisheries. *Human Organization* 34: 333–44.

Allegretti, M.H. 1992. Reconciling people and land: the prospects for sustainable extraction in the Amazon. In *Development or Destruction: the Conversion of Tropical Forest to Pasture in Latin America*, pp.250–4 ed. T.E. Downing, S.B. Hecht, M.A. Pearson and C. Garcia-Downing. Boulder: Westview.

Amorozo, M.C. de M. and Gély, A. 1988. Uso de plantas medicinais por caboclos do baixo Amazonas. Barcarena, PA, Brasil. *Boletim do Museu Paraense Emílio Goeldi, Série Botanica* 4: 47–130.

Balée, W. 1989. Cultura na vegetação da Amazônia brasileira. In *Biologia e ecologia humana na Amazônia, avaliação e perspectivas*, pp. 00–00, ed. W.A. Neves. Belém: CNPQ/MPEG.

Bayley, P. B. and Petrere M. Jr. 1989. Amazon fisheries: assessment methods, current status and management options. *Proceedings of the Large River Symposium, Canadian Special Publication of Fisheries and Aquatic Sciences* 106: 385–98.

Begossi, A. 1989. *Food diversity and choice, and technology in a Brazilian fishing community (Búzios Island, Brazil)*. PhD Dissertation, University of California, Davis, USA.

Begossi, A. 1992a. The use of optimal foraging theory in the understanding of fishing strategies: a case from Sepetiba Bay (Rio de Janeiro State, Brazil). *Human Ecology* 20: 463–75.

Begossi, A. 1992b. Food taboos at Búzios Island (Brazil): their significance and relation to folk medicine. *Journal of Ethnobiology* 12: 117–39.

Begossi, A. 1995. Fishing spots and sea tenure: incipient forms of local management in Atlantic Forest coastal communities. *Human Ecology* 23: 387–406.

Begossi, A. 1996. The fishers and buyers from Búzios Island (Brazil): kin ties and modes of production. *Ciência e Cultura*, 48(3): 142–7.

Begossi, A., Amaral, B.D. and Silvano, R.A.M. 1995. Reserva Extrativista do Alto Juruá. In *A questão ambiental: cenários de pesquisa*, pp.95–106, ed. S. Barbosa. Textos NEPAM, UNICAMP.

Begossi, A. and Braga, F. M. de S. 1992. Food taboos and folk medicine among fishermen from the Tocantins River (Brazil). *Amazoniana* 12: 101–18.

Begossi, A. and Figueiredo, J. L. 1995. Ethnoichthyology of southern coastal fishermen: cases from Búzios Island and Sepetiba Bay (Brazil). *Bulletin of Marine Science* 56: 710–17.

Begossi, A., Leitão-Filho, H.F. and Richerson, P.J. 1993. Plant uses in a Brazilian coastal fishing community (Búzios Island). *Journal of Ethnobiology* 13: 233–56.

Begossi, A. and Richerson, P.J. 1991. The diffusion of 'lambreta', an artificial lure, at Búzios Island (Brazil), *MAST*, 4: 87–103.

Begossi, A. and Richerson, P.J. 1993. Biodiversity, family income, and ecological niche: a study on the comsumption of animal foods on Búzios Island (Brazil). *Ecology of Food and Nutrition* 30: 51–61.

Bennett, B.C. 1992. Plants and people of the Amazonian rainforests. *Bioscience* 42: 599–607.

Benz, B.F., Santana, F.M., Pineda, R. L., Cevallos, J. E., Robles, L. H. and Niz, D. L. 1995. Characterization of mestizo plant use in the Sierra de Manantlan, Jalisco-Colima, Mexico. *Journal of Ethnobiology* 14: 23–41.

Berkes, F. 1985. Fishermen and the 'tragedy of the commons'. *Environmental Conservation* 12: 199–206.

Berkes, F., Feeny, D., McCay, B.J. and Acheson, J.M. 1989. The benefits of the commons. *Nature* 340: 91–3.

Berkes, F. and Folke, C. 1992. A systems perspective on the interrelations between natural, human-made and cultural capital. *Ecological Economics* 5: 1–8.

Boyd, R. and Richerson, P.J. 1981. Culture, biology and the evolution of variation between human groups. In *Science and the Question of Human Equality*. pp.99–152, ed. M.C. Collins, I.W. Wainer and T.A. Bremner. AAAS Selected Symposium, Boulder, Colorado.

Boyd, R. and Richerson, P.J. 1985. *Culture and the Evolutionary Process*, Chicago: Chicago University Press.

Brown, C.H. 1985. Mode of subsistence and folk biological taxonomy. *Current Anthropology* 26: 43–53.

Capobianco, J.P. 1994. Objetivos e atuação da SOS Mata Atlântica. *Mata Atlântica e Imprensa*. Relato do Laboratório Ambiental realizado para a imprensa no Vale do Ribeira/SP. São Paulo: Fundação SOS Mata Atlântica/Fundação Konrad Adenauer.

Coimbra, A.F. 1991. Fauna. In *Atlantic Rain Forest*, pp.129–57, São Paulo: Index/SOS Mata Atlântica.

Cordell, J. 1978. Carrying capacity analysis of fixed-territorial fishing. *Ethnology* 17: 1–24.

Cunha, M.C., Brown, K. and Almeida, M.W.B. 1993. *Can Traditional Forest-dwellers Self-manage Conservation Areas? A Probing Experiment in the Juruá Extractive Reserve, Acre, Brazil*. MacArthur Foundation.

Cunha, M.C., Brown, K. and Almeida, M.W.B. 1994. *Can Traditional Forest-dwellers Self-manage Conservation Areas? A Probing Experiment in the Juruá Extractive Reserve, Acre, Brazil*. MacArthur Foundation.

Cury, R. 1993. *Dinâmica evolutiva e caracterização de germoplasma de mandioca (Manihot esculenta Crantz) na agricultura autóctone do sul do Estado de São Paulo*. Master thesis, ESALQ, Piracicaba.

Diegues, A.C. 1983. *Pescadores, camponeses e trabalhadores do mar*. Rio de Janeiro: Editôra Ática.

Diegues, A.C. 1994. *O mito moderno da natureza* intocada. São Paulo: NUPAUP/USP.

Emmons, L.H. 1983. Geographic variation in densities and diversities of non-flying mammals in Amazonia. *Biotropica* 16: 210–22.

Empéraire, L. and Delavaux, J. 1992. *Relatório de campo, etnobotânica*. Unpublished.

Fearnside, P.M. 1979. Previsão da produção bovina na rodovia Transamazônica do Brasil. *Acta Amazônica* 9: 689–700.

Fearnside, P.M. 1980. Os efeitos das pastagens sobre a fertilidade do solo na Amazônia brasileira: consequencias para a sustentabilidade de produção bovina. *Acta Amazônica* 10: 119–32.

Fearnside, P.M. 1993. Deforestation in Brazilian Amazonia: the effect of population and land tenure. *Ambio* 22: 537–45.

Figueiredo, G.M., Leitão-Filho, H. F. and Begossi, A. 1993. Ethnobotany of Atlantic Forest coastal communities: diversity of plant uses in Gamboa (Itacuruça Island, Brazil). *Human Ecology* 21: 419–30.

Fittkau, E. J. and Klinge, H. 1973. On biomass and trophic structure of the Central Amazonian rain forest ecosystem. *Biotropica* 5: 2–14.

Forman, S. 1967. Cognition of the catch: the location of fishing spots in a Brazilian coastal village. *Ethnology* 6: 417–26.

Forman, S. 1970. *The Raft Fishermen*. Bloomington: University of Indiana Press.

França, A. 1954. *A Ilha de São Sebastião*. Estudo de geografia humana, Boletim 178, Geografia no. 10, Universidade de São Paulo.

Gibbs, C.J.N. and Bromley, D.W. 1989. Institutional arrangements for management of rural resources: common-property regimes. In *Common Property Resources*, pp. 22–32, ed. F. Berkes. London: Belhaven Press.

Gomes, M.P. 1988. *Os índios do Brasil*. Petrópolis: Editôra Vozes.

Goulding, M. 1979. *Ecologia da pesca no rio* Madeira. Manaus: INPA/CNPq.

Habermehl, G. G. 1981. *Venomous Animals and their Toxins*, Berlin: Springer-Verlag.

Hanazaki, N., Leitão-Filho, H.F. and Begossi, A. 19?? Uso de recursos na Mata Atlântica:o caso da Ponta do Almada (Ubatuba, Brazil). *Interciência*, 21(6): 268–76.

Harris, M. 1976. History and significance of the emic/etic distiction. *Annual Review of Anthropology* 5: 329–50.

Holling, C.S. 1992. Cross-scale morphology, geometry, and dynamics of ecosystems. *Ecological Monographs* 62: 447–89.

Joly, C.A., Leitão-Filho, H. F. and Silva, S. M. 1991. The floristic heritage. In Index/SOS Mata Atlântica, *Atlantic Rain Forest*, pp.97–125, ed. Index/SOS Mata Atlântica. São Paulo.

Leitão-Filho, H.F. 1982. Aspectos taxonômicos das florestas do Estado de São Paulo. *Silvicultura em São Paulo* 16: 197–206.

Leitão-Filho, H.F. 1987. Considerações sobre a florística de florestas tropicais e subtropicais do Brasil. *Boletim do Instituto de Pesquisas em Engenharia Florestal (Piracicaba)* 35: 41–6.

Levin, S.A. 1992. The problem of pattern and scale in ecologym. The Robert H. MacArthur Award Lecture. *Ecology* 73: 1943–67.

Lewis, N.D. 1984. Ciguatera – parameters of a tropical health problem. *Human Ecology*, 12: 253–74.

Lino, C.F., ed. 1992. *Reserva da Biosfera da Mata Atlântica*, vol. 1: Referências Básicas. Brazil: Consórcio Mata Atlântica/Universidade Estadual de Campinas.

Malcom, J.R. 1994. Edge effects in central Amazonian forest fragments, *Ecology* 75: 2438–45.

Marcílio, M.L. 1986. *Caiçara: terra e população*. São Paulo: Paulinas.

McGrath, D.G., Castro, F., Futemma, C., Amaral, B. D. and Calabria, J. 1993. Fisheries and the evolution of resource management on the Lower Amazon floodplain. *Human Ecology* 21: 167–95.

McKey, D. and Beckerman, S. 1993. Chemical ecology, plant evolution and traditional manioc cultivation systems. In *Tropical Forests, People and Food. Biocultural Interactions and Applications to Development*, pp.00–00, ed. C.M. Hladik, A. Hladik, O.F. Linares, H. Pagezy, A. Semple, and M. Hadley. Man and the Biosphere Series, Vol. 3. Paris: UNESCO; and Carnforth, UK: Parthenon.

Morán, E.F. 1974. The adaptive system of the Amazonian *caboclo*. In *Man in the Amazon*, pp.136–59, ed. C. Wagley. Gainesville: University of Florida.

Morán , E.F. 1977. Estratégias de Sobrevivência: O uso de recurcos ao longo da Rodovia Transamazônica. *Acta Amazônica* 7: 363–79.

Morán, E.F. 1979. Criteria for choosing sucessful homesteaders in Brazil. *Research in Economic Anthropology* 2: 339–59.

Morán, E.F. 1983. Government – directed settlement in the 1970s: an assessment of Transamazon highway colonization. In *The Dilemma of Amazonian Development*, pp.297–317, ed. E. Moran. Westview Special Studies on Latin America and the Caribbean. Boulder: Westwiew Press.

Morán, E.F. 1990. *A ecologia humana das populações da Amazônia*. Petrópolis, Rio de Janeiro: Editor Vozes.

Morán, E.F. 1993. Deforestation and land use in the Brazilian Amazon. *Human Ecology* 21: 1–21.

Mussolini, G. 1980. *Ensaios de antropologia indígena e caiçara*. Rio de Janeiro: Editôra Paz e Terra.

Myers, N. 1988. Tropical forests and their species, going going . . ., in *Biodiversity*. pp.3–18, ed. E.O. Wilson. Washington DC: National Academy Press.

Oliveira, R.R., Lima, D.F., Sampaio, P.D., Silva, R.F. and Toffoli, D.Di G. 1994. Roça caiçara, um sistema 'primitivo'auto-sustentável, *Ciência Hoje* 18 (104): 44–51.

Paz, V.A. and Begossi, A. 1996. Ethnoichthyology of Gamboa fishermen (Sepetiba Bay, Brazil). *Journal of Ethnobiology* 00–00.

Petrere M. Jr 1989. River fisheries in Brazil: a review. *Regulated Rivers and Management*, 4: 1–16.

Prance, G.T. 1978. The origin and evolution of the Amazon flora. *Interciência* 3: 207–22.

Putman, R.J. and Wratten, S.D. 1984. *Principles of Ecology*. Berkeley: University of California Press.

Rancy, C.M.D. 1986. *Raízes do Acre*. Rio Branco: Editôra Falangola.

Reichel-Dolmatoff, G. 1976. Cosmology as ecological analysis: a view from the rain forest. *Man* 11: 307–18.

Reiners, W.A., Bouwman, A.F., Parsons, W.F.J. and Keller, M. 1994. Tropical rain forest conversion to pasture: changes in vegetation and soil properties. *Ecological Applications* 4: 363–77.

Ribeiro, M.C.L. de B. and Petrere M. Jr 1990. Fisheries ecology and management of the jaraqui (*Semaprochilodus taeniurus, S. insignis*) in Central Amazônia. *Regulated Rivers: Research and Management* 5: 195–215.

Rossato, S.C. 1996. *Uitilizçaõ de Plantas por Poulaçoes do Litoral norte do estado de Saõ Paulo*. Masters thesis, Unicamp, Campinas, Saõ Paulo.

Sales, R.R. 1994. Desenvolvimento e preservação. Relato do Laboratório Ambiental realizado para a imprensa no Vale do Ribeira/SP, pp.41–6. Fundação SOS Mata Atlântica/Fundação Konrad Adenauer, São Paulo.

Salomão, R. de P., Rosa, N. A., Nepstad, D. C. and Bakk, A. 1995. Estrutura diamétrica e breve caracterização ecológica-econômica de 108 espécies arbóreas da floresta Amazônica brasileira. *Interciência* 20: 20–29.

Schwartzman, S. 1989. Extractive Reserves: the rubber tappers' strategy for sustainable use of the Amazon rainforest. In *Fragile Lands of Latin America, Strategies for Sustainable Development*. pp.150–65, ed. J. Biowdev. Boulder: Westview Press.

Setz, E.Z.F. 1993. *Ecologia alimentar de um grupo de parauacus (Pithecia pithecia crysocephala em um fragmento florestal na Amazônia central*. Doctoral thesis, Unicamp, Campinas, Brazil.

Sider, G.M. 1986. *Culture and Class in Anthropology and History: a Newfoundland Illustration*, Cambridge: Cambridge University Press.

Sioli, H. 1985. *Amazônia*. Petrópolis, Rio de Janeiro: Editora Vozes.

Smith, N.J. 1976. Utilization of game along Brazil's transamazon highway. *Acta Amazônica* 6: 455–66.

Smith, N.J. 1981. *Man, Fishes, and the Amazon*. New York: Columbia University Press.

Toft, C. 1986. Communities of species with parasitic life-styles. In *Community Ecology*. pp.445–63, ed. J. Diamond and T. Case. New York: Harper and Row.

Wambeke, A. Van. 1978. Properties and potentials of soils in the Amazon Basin. *Interciência* 3: 233–49.
Whittaker, R.H. 1975. *Communities and Ecosystems.* New York: Macmillan.
Wilson, E.O. 1988. The current state of biological diversity. In *Biodiversity*, pp.3–18, ed. E.O. Wilson. Washington DC: National Academy Press.
Wilson, E.O. 1992. *The Diversity of Life.* Cambridge, Mass.: The Belknap Press of Harvard University Press.

7

Indigenous African resource management of a tropical rainforest ecosystem: a case study of the Yoruba of Ara, Nigeria

D. MICHAEL WARREN & JENNIFER PINKSTON

Introduction

Indigenous knowledge refers to local-level knowledge systems unique to a particular community or ethnic group. Numerous colonial officers expressed great interest in community-based knowledge systems, particularly knowledge about the biological realm (e.g. Dalziel, 1937; Hutchinson and Dalziel, 1954–72; Keay, 1989). Many African governments are now recognizing these systems as a national resource that provides the basis for sustainable approaches to agriculture and natural resource management. By recording the indigenous knowledge and decision-making systems, development workers can understand how community-based organizations facilitate the identification of problems and attempt to deal with them through experimentation and innovation (Warren, Slikkerveer and Brokensha, 1995; Blunt and Warren, 1996; Warren, Adedokun and Omolaoye, 1996a; 1997; Warren and Warren, 1997). Figure 7.1 depicts the dynamic nature of knowledge systems within a problem-solving context.

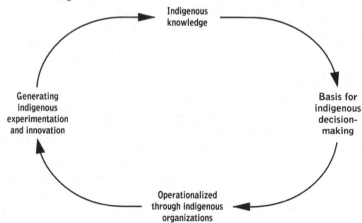

Figure 7.1.   Indigenous knowledge cycle.

The past decade has seen a marked improvement in understanding the various factors that must be addressed when dealing with conservation of natural resources (Fisher, 1995; GRAIN, 1995; Cicin-Sain and Knecht, 1995). These factors are the components of the guiding framework for this study, which was developed by Berkes and Folke to analyse linkages between social and ecological systems in order to understand the resilience and sustainability of a socio-ecological system (Berkes and Folke, 1994; Folke and Berkes, 1995). The importance of and global commitment towards the conservation of biodiversity are no longer questioned. The complementarity of cultural and biological diversity as reflected through local-level indigenous knowledge of natural resources and their management is now documented in numerous case studies (Berkes 1987; 1992; 1995; Oldfield and Alcorn 1991; Rajasekaran, Warren and Babu, 1991; Warren, 1992a; 1993; Berkes, Folke and Gadgil 1993; Gadgil, Berkes and Folke 1993; Berkes and Folke, 1994; Haverkort and Millar 1994; Hyndman 1994; Warren and Rajasekaran, 1994; Rural Advancement Fund International 1994; Phillips and Titilola 1995; Verger 1995). Moreover, the institutional role of indigenous organizations in local-level decision making involving identification and prioritization of community problems and their solution is better understood through recent studies on the management of common property (Warren, 1992c; Blunt and Warren, 1996; Castro, 1995; Lansing and Kremer, 1995; Messerschmidt, 1995). A wealth of data is now available on indigenous knowledge of biological resources (Isawumi, 1990; Berlin, 1992; Martin, 1995).

A growing set of case studies from every region of the world available at the CIKARD Documentation Unit and Library indicates that every community has the capacity to deal with its problems generated by local, regional, national and global influences in both reactive and proactive approaches. Even though this capacity may not always be sufficient to deal effectively with every problem, it is clear that each community has members who play important roles as active experimenters and innovators. Case studies that analyse the nature of local-level experiments in the realm of agriculture and natural resource management indicate that they are often stimulated by the communal identification and discussion of an ever-changing array of conditions and the problems they impose on a community (Warren 1991, 1992b; Prain and Bagalanon, 1994; den Biggelaar, 1995; Prain, Fujisaka and Warren, 1997).

This case study is based on a traditional Yoruba state located in the moist semi-deciduous rainforest in southwestern Nigeria. It represents the initial research stage to describe the interacting forces that have resulted in changes in the ecosystem from pre-colonial times through the colonial era to the present post-colonial period. The study delineates the indigenous knowledge of the natural resources and how this knowledge is reflected in individual and community decision making regarding the utilization of these resources. The natural resources include the flora and fauna as well as physical resources such as soil and water. There is a deliberate maintenance of diversity in domesticated and non-domesticated plants and animals by the Ara community that is characteristic of other agroforestry systems documented in other parts of Africa. This case study represents a social–ecological system that is still resilient but is in an important transition period which could result in a negative set of ecological changes that may in turn result in a non-sustainable system.

The case-study setting

The traditional city-state of Ara includes the community of Ara and 40 smaller towns and villages that are linked historically through their allegiance to the *oba,* or king, whose title is the Alara. Ara is located eight miles west of the Ibadan–Osogbo Expressway on the Ede–Ara–Ejigbo road, which crosses the expressway seven miles south of Osogbo, the capital of Osun State (see Figure 7.2). It is also linked to the expressway by a second route from Ara via Ikotun and Iragberi to Awo, the site of the Egbedore Local Government Secretariat, Awo itself being located within one mile of the expressway. The town of Ara is comprised of 1000 houses, representing 1541 family households and approximately 10 000 inhabitants. According to elderly citizens of Ara, the current numbers of houses and citizens represent an approximate five-fold increase since the turn of this century. The traditional state is located between longitudes 4 and 5 degrees of the Greenwich meridian and between 7 degrees 30 minutes and 8 degrees north of the equator in the drier part of the high forest (Ara Development Planning Workshop, 1991; Omolaoye 1994).

Ara is situated on a highland plateau area ranging from 500 to 1200 feet (152.5 to 366 metres) above sea level. It is bounded on the west by the Gbaga River, on the east by the Ewure River, on the south by the Odo-Igbo River, and on the north by the Oworu River and the Abeyinjona River. Larger rivers include the Aaro River, which flows between Ara and Ejigbo, and the Osun River to the east, which flows through the cities of Osogbo and Ede. There are several large granitic hills in the area, some of which are inhabited.

The traditional city-state of Ara was located in thick tropical rainforest.

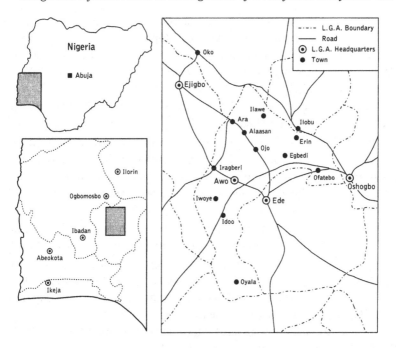

Figure 7.2.   Map of study area.

Within this area there are now only a few isolated stands of primary tropical forest that have never been cleared. These scattered spots are at Obabon
on the edge of the town of Ara, near Ilawe and Ooye, and in areas adjacent
to rivers. Valuable trees typical of the tropical forest of Ara include *àràbà*
(*Ceiba pentandra*), *irókò* (*Milicia excelsa*), *ògànó* (*Khaya senegalensis,*
mahogany), various species of palm trees introduced by early settlers, *arère*
(*Triplochiton scleroxylon*), naturally occurring native species of rubber
(*Funtumia africana* and *F. elastica*), and *afàrà* (*Terminalia superba*).[1] Some
indigenous trees are planted traditionally as boundary markers or fences,
such as *Newbouldia laevis, Moringa oleifera* and *Dracaena spp.* Secondary
tropical forest which emerged after the primary forest had been cleared for
farming purposes is currently the common type of forest over most of the
Alara's lands. The secondary forest, although less abundant in terms of
numbers of botanical species than the primary forest, is also characterized
by the addition of exotic trees such as teak (*Tectona grandis*), neem
(*Azadirachta indica*) – introduced to Nigeria from Ghana in 1928 – Persian
lilac (*Melia azaderach*), *Gmelina arborea*, and *Cassia siamea* (Richard
Lowe, personal communication).

There are two seasons, the rainy (*igba ojo*) and the dry (*igba eerun*)

162   D. Michael Warren & Jennifer Pinkston

(Osunade 1994). The rainy season, characterized by the south-westerly monsoon rains, usually starts in late March/early April and ends in September. The heaviest rainfall is recorded in the months of July and August, when most rivers are flooded. The rainy season has an annual rainfall of between 1143 mm and 1524 mm.

The dry season, influenced by the north-easterly dry trade winds from the Sahara, starts around mid-September and continues until March. Near the end of December, Ara normally experiences the harmattan (*igba oye*), which lasts for about a month. This is the period with the greatest daily temperature range, when the mean maximum daytime temperature is between 32 and 35°, and the mean minimum night-time temperature is between 20 and 22°.

Ara is one of the three oldest towns (along with Awo and Ojo) located within the Egbedore Local Government of Osun State. An ancient Yoruba town, the current Alara is the twentieth king according to oral histories. The founder of Ara was Orira, who migrated from Ile-Ife, the original town of the Yoruba ethnic group of Nigeria. Oral history sets Orira's migration from Ile-Ife possibly as early as the tenth century.

The *àáfin Alara* (Alara's palace) is located at the centre of Ara township. Surrounding the palace are houses grouped into compounds in the traditional Yoruba pattern. A compound is made up of houses or buildings in which patrilineal blood relatives or tenants live. Each compound has a head who is the administrative leader and family head of the compound on behalf of the Alara.

The oldest indigenous organization of Ara is the traditional council comprised of the Alara, his chiefs, and the *molebi*, the representatives of the five ruling families from whom a king is chosen in a rotational system. The Ara Traditional Council performs executive, legislative and judicial functions. It is the overall governing council of the town, settles disputes in the town, receives visitors, enacts laws guiding security and well-being of the citizens, ensures that the laws are obeyed, discusses and endorses development projects, manages the market, and represents the interests of the citizens at the local, state and federal levels of government.

About 90% of the adult inhabitants on the Alara's land are engaged in full-time and part-time farming. Many inhabitants are also engaged in full-time and part-time occupations such as commerce and trade, tailoring, hunting, craft production of baskets, pottery and handwoven cloth, and various trades such as masonry, carpentry and blacksmithing.

The soil of Ara is rich in humus where forest cover remains, with soil types including clay, laterite, loam and sandy soil (Warren, 1992b). The oldest farming system, which still predominates, is an agroforestry system combining arable food crops with domesticated tree crops. Until the twentieth century, the tree crops were limited to oil palm (*Elaeis guineensis*) and some varieties of citrus (*Citrus spp.*). Over the past 75 years, this perennial mixed plantation system of trees bearing cash crops has expanded to include kola (*Cola spp.*), cocoa (*Theobroma cacao*), oil palm (*Elaeis guineensis*), coffee (*Coffea canephora*), coconut (*Cocos nucifera*), cashew (*Anacardium occidentale*), new varieties of citrus such as lemon (*Citrus limon*) and tangerine (*Citrus reticulata*), guava (*Psidium guajava*), Pará rubber (*Hevea brasiliensis*) and mango (*Mangifera indica*), with kola, cocoa and oil palm being of primary commercial value. The main crop season for kola and cocoa is October – February, with smaller crops obtained in the off-season, March – September. Ara is one of the important production areas in Osun State for cocoa, palm oil, kola and rubber. The introduction of cocoa near the turn of this century had a major impact on Ara land tenure. As a perennial tree crop, the cocoa tree lives for at least 40 years. An individual could now sell or pledge his cocoa crop. Since the crop could not be separated from the land, there was a drift toward freehold land tenure (Lloyd, 1962).

The second most prevalent farming system is that based on the biennial and annual mixed cropping of arable crops such as yam (*Dioscorea spp.*), plantain (*Musa paradisiaca*), banana (*Musa sapientum*), cassava (*Manihot esculenta*), cocoyam (*Colocasia esculenta*), rice (*Oryza spp.*), maize (*Zea mays*), papaya (*Carica papaya*), sweet potato (*Ipomea batatas*), beans (e.g. *Phaseolus lunatus*), groundnuts (*Arachis hypogaea*), and vegetables such as okro (*Abelmoschus esculentus*), tomatoes (*Solanum lycopersicum*), onions (*Allium cepa*), melon (*ègúsí*) (*Citrullus lanatus*), pepper (*Capsicum spp.*), and a wide variety of leafy green vegetables (e.g. *Amaranthus hybridus*). Considerable local effort goes into learning about and acquiring genetic material for new varieties of all of these cultivars. Ara citizens living in places such as Cote d'Ivoire and Ghana will send seeds, seedlings and cuttings of varieties that are known not to exist in Ara. Each variety is carefully evaluated according to locally defined positive and negative characteristics related to production, harvesting, storage, processing and consumption (a study on this topic is expected to be completed in 1997 by Dr A.A. Amusan, Obafemi Awolowo University, Ile-Ife; also see Prain *et al.*, 1997). The most common tendency of farmers is carefully to maintain a wide range of the varieties of any given cultivar. One finds, for example, several varieties of yam (*Dioscorea spp.*), the traditional staple crop in Ara, (*isu dagidagi, isu èsúrú, isu ewùrà, isu gidi, isu gbàngí, isu igangan, isu monro*), cassava (*gbokogbala, oko iyawo, ẹ̀gẹ́ odongbo, ẹ̀gẹ́ lalupon, ẹ̀gẹ́ fausa dudu, ẹ̀gẹ́ laani*), and pepper (*átá rodo, átá sonbo, átá were, átá owonso, átá tantase, átá funfun, átá koruko*). In the past these were grown on a subsistence level, but today most farmers in Ara also grow them as cash crops. Cassava (*Manihot esculenta*), introduced to Nigeria from the Americas during the slave trade, became a cash crop in Ara only after about 1970.

Historically, the primary role of farmer fell to Ara men. Women assisted in planting, weeding and harvesting, but their major role was in the processing of agricultural products. This has changed dramatically in the last several decades due to economic imperatives that have pulled many women into production agriculture on their own plots of land, where they make all of the decisions regarding agriculture. Women farmers may hire labourers to assist in the clearing of new fields and in the soil preparation prior to planting, although some female farmers perform these tasks themselves. There is no longer any division of labour of arable food crops related to gender. Both male and female farmers in Ara can and do produce the full range of arable food crops. Master's thesis research on women's role in agriculture was conducted in Ara during the summer of 1995 by Penny Andresen (Andresen, 1996). Ara farmers still rely on human labour with farm tools such as the short-handled hoe, cutlass, axe, knife and the cocoa hook. In addition, some farmers now use limited amounts of fertilizers and pesticides.

Yams, planted January–March and again June–July, are harvested after one year. Maize is planted in both March and August. Many inhabitants also actively rear a variety of domesticated animals including free-ranging sheep (*àgutàn*), goats (*ewúrẹ́*), pigs (*ẹlẹ́dẹ̀*), chickens (*adìẹ*) and ducks (*pẹ́pẹ́yẹ*). Two nearby communities of Fulani cattle herders maintain substantial herds of cattle (*màálúù*). A wide range of wildlife such as various species of antelope (e.g. *ẹtu*), squirrels (e.g. *ikún*), grasscutters (cane rats) (*ọ̀yà*), bush rats (e.g. *ẹmọ́*), bush fowl (*àparò*), forest snails (*ìgbín*) and fish (*ẹja*) are actively trapped and hunted.

Of the approximately 10 000 inhabitants in Ara, about 2050 residents – virtually all women – are involved in the production and processing of palm products such as palm oil, palm kernel oil and palm wine. About 500 farmers produce cocoa, 700 are involved in maize production, 400 produce kola, 50 do vegetable gardening, 80 farmers concentrate on the production of oranges and pineapples, and most farmers produce yam, cocoyam, cassava and maize in intercropped systems.

Historically, the Yoruba have been known for organizing themselves through a myriad of indigenous organizations and associations. In Ara township alone there are more than 100 formally established associations ranging from community-wide organizations such as the Ara Traditional Council, the Ara Development Council, and the Ara Descendants Union (which has 19 branches in major urban areas in Cote d'Ivoire and Nigeria, where large numbers of Ara citizens reside and work), to numerous occupational associations such as the palm oil producers, the Ara Young Farmers Association, the palm oil buyers, and the small-scale traders association, to social clubs and religious associations (Warren *et al.*, 1996a).

Changing land-use patterns in Ara

Numerous discussions with a wide variety of recognized Ara leaders in the areas of agriculture, agroforestry and hunting/trapping indicate changes in land-use patterns and management from the pre-colonial to the colonial and post-colonial periods. In 1914, the United Kingdom united its colonies and protectorates in southern and northern Nigeria into the unified colony of Nigeria – which became independent in 1960.

In pre-colonial Ara, natural resource-use patterns focused on subsistence-level production of food crops and a limited range of naturally occurring tree crops such as oil palm. Hunting and trapping of wildlife were very important. Fishing took place in the rivers and lagoons. The commercial harvesting of timber for lumber had not begun. Aside from oil palm and some species of citrus used in subsistence patterns, the other major tree crops that have assumed great importance in the economy had not yet been introduced. As in other Yoruba communities, some land was set aside for religious and other purposes: 'Forest land was regularly set aside for various purposes: as hunting forests, religious groves, isolation or quarantine forests, and to serve as the abode of fairies and spirits' (NEST, 1991: 251). Osunade has delineated Yoruba categories of forests according to their use (1988a), several of which are found in Ara. Land located at some distance from Ara mainly devoted to hunting is referred to as *igbó ode*. High forest, *igbó egàn*, or abandoned secondary forest when used exclusively for hunting is also referred to as *igbó ode*. Religious groves *(igbó orò)* are usually small, uncultivated forests near the community. They may have specific names depending on their religious function. In Ara, one still finds the *igbó egúngún* where the ancestral masqueraders prepare for annual community rituals. In some Yoruba communities, one finds other religious groves *(igbó ègbèé)* that are reserved for the burial of persons whose death was considered abnormal or mysterious, such as by being struck by lightning. Another term is *igbó iwin*, an area considered to be inhabited by fearful spirits.

Freehold land tenure did not, and still does not, exist in Nigeria. Since the founding of the Ara state, the Alara was seen as the trustee for the allocation of land and assumed the task of land distribution. Given the rich soil, vegetation and wildlife, immigrants were drawn to the area. The Alara allocated specific lands for the migrants whose original farming and hunting villages have grown over the centuries. Of the 40 towns and villages on the Alara's lands, several have been founded as recently as 50 years ago. The migrants to Ara had user rights only, but could not sell the land. Lands allocated to migrants fell under their own control as long as they continued to utilize them, and could be inherited through the patrilineage. Land allowed to go into bush-fallow would be recultivated within 10 years in order to reassert ownership. Otherwise it could be assumed that the owner had abandoned it, and the land might be reallocated to or appropriated by someone else. Even today, the Alara and his chiefs have an indepth knowledge of all lands that fall under the ultimate suzerainty of the Alara, along with the history of use patterns of each parcel of land. Farmers from settled towns and villages could apply to the Alara for allocation of additional unoccupied and unutilized lands for farming. Inhabitants could plant trees freely on their own land as well as on any unoccupied land and control them and their fruits. Hunting and trapping could be carried out both on lands allocated to individuals by the Alara and on unallocated lands in a loosely regulated communal-property system. Population pressure was so low that there was no problem with over-exploiting the wildlife resource. Agriculture was essentially subsistence in nature. Fishing could be carried out by anyone on parts of rivers that did not flow through lands allocated to a given individual. Fishing on rivers flowing through allocated lands was allowed with permission of the person living on the lands.

Religion played a role in assuring the maintenance of biodiversity in precolonial Ara. In the 'Yoruba worldview, the land is a divine phenomenon with celestial status. She is a spiritual woman – a goddess who knows everything about anything in existence. Agriculture is a divine occupation originated by a woman known as *Orisa-oko*' (Opefeyitimi, in preparation). It was forbidden to cut certain trees such as *iroko* (African teak), *ose* (baobab), or *ope* (oil palm); if there were an unavoidable need to cut any of these trees in the past, 'appropriate and elaborate rituals of propitiation were offered for permission of their dwelling spirits before and after cutting' (Opefeyitimi, in preparation). During that period of time there were sacred forests and sacred trees of various types that could not be utilized by the inhabitants in any way. The Alara had a restricted-use forest behind the palace that was also used for his own agricultural purposes. This secondary forest still exists today. There were also areas tabooed through the Yoruba religion. *Ilẹ̀ orò* were areas of forest occupied by spirits that could not be utilized by any inhabitants for any purposes. Sacred areas used for religious purposes or where various Alara were believed to have descended into the earth upon their death were also off-limits to human endeavours. Dr. David Ladipo (personal communication) recently conducted a botanical survey in a Nigerian sacred grove in the rainforest ecozone which yielded 330 botanical species, compared with only 23 on the perimeter of the sacred grove where the ecosystem had been disturbed by human activity. *Igi oro* were trees presumed to be inhabited by ghosts that were taboo for any use.

The custodians for these sacred areas, as well as for all loosely regulated communal property areas, both in the past as well as at the present time, are the Ara hunters. In a study conducted by Gerald Smith in summer 1995, it was determined that the traditional guild of Ara hunters is headed by the chief of the hunters, the *Oluode*, and guided by an Ogun priest, the *abore*, who performs ritual duties (Smith and Osunwole, 1996). As in the past, the hunters continue to provide voluntary security services for the community, play a major role in the supply of bush meat for the community, and provide a font of traditional knowledge about the forest and the management of its resources. Although virtually all contemporary members of the hunters' guild are Muslims and Christians, the rituals carried out to determine if a potential member of the guild can be admitted and to ensure success in hunting still follow Ogun, the traditional Yoruba deity that protects hunters and others whose livelihood is dependent upon metal-based tools. Hunting is normally conducted at night with hand-made Dane guns and acetylene lamps with reflectors, preferably in the dry rather than the rainy season. Since the rainy season makes tracking more difficult, given the lush vegetation, and moisture is believed to bring out more poisonous snakes, this is the annual period when wildlife regenerates its numbers. The hunters' guild, at least 100 years old, provides an organized mechanism to maintain ancient traditions and indigenous natural resource knowledge. This knowledge encapsulates natural resource management with traditional Yoruba religious knowledge, which results in hunters being both feared and respected by the Ara community (Smith and Osunwole, 1996).

The Yoruba religion continues to provide many checks and balances on resource utilization, even though its influence has been diminished since Islam and Christianity were introduced to Ara. Supernatural sanctions can occur at different levels. At the individual level, a person can seek supernatural means to punish someone else who impinges upon one's property and refuses to comply with orders from the owner of the land, from the chiefs and from the Alara. These supernatural sanctions can result in various types of misfortunes, ranging from illness, mishaps and bad luck for the guilty person and his/her relatives. The ancestral spirits can also be called upon to sanction negative behaviour on behalf of the owner. Various deities important within the Yoruba religion such as Ogun and Osun can also be invoked to bring misfortune upon the person unwilling to comply with orders from those in authority. In small-scale communities where everyone is aware of conflicts, people unwilling to comply with those in authority may face public humiliation and shame that can carry over onto their immediate and extended family. These supernatural forces provided the basis upon which the Alara and his chiefs sanctioned undesired behaviour of inhabitants. Although the introduction of Islam to Ara in about 1860, followed by Christianity in 1896, has diluted the impact of these traditional beliefs and sanctions, they remain an important force that can still strongly influence behaviour.

Although Nigeria was formally established as a colony in 1914, it was several decades before the presence of a colonial power was felt in Ara. What was of considerable importance in ecological and economic terms was the colonial emphasis on tree cash crops. In the primary rainforest of Ara, this meant the introduction of a wider array of cash crops based on products from trees. This occurred even as forests were retained as buffers between adjacent traditional states, and also served for hunting and gathering, and for agricultural expansion. Forest reserves were established during the colonial era by the promulgation of forest law (based on that of Burma). Local people accepted the establishment of these reserves because they valued the forest and were allowed to continue their hunting and gathering practices. During the 1920s and 1930s, the future rapid growth in population and the expansion of cash cropping were not foreseen (Richard Lowe, personal communication).

The first kola tree seedlings (*Cola nitida*) made it to the Ara area about 1929 (Akinbode, 1982: 26). The seeds were brought from the Gold Coast by a Yoruba petty trader to a nearby community, from where they diffused rapidly to places like Ara. *Cola nitida* has assumed great economic importance as it features heavily in international trade, while *Cola acuminata*, which is native to the area, is of significant importance locally.

Cocoa (*Theobroma cacao*) was introduced into the Delta area of Nigeria in 1874 from Fernando Po, having been brought there from the Americas by the Spanish. Although one variety of coffee (*Coffea canephora*) is indigenous to Nigeria, it is not produced on a commercial level in Nigeria. Pará rubber (*Hevea brasiliensis*) was introduced to Nigeria by the British around 1912. Since this species requires rainfall of about 2000 mm annually for the commercial production of latex, it is not planted on any scale in Ara. Mango (*Mangifera indica*) originated in the Indo-Burma region and may have spread to West Africa from East Africa. Guava (*Psidium guajava*) and cashew (*Anacardium occidentale*) arrived in Nigeria in the early part of the nineteenth century from the Americas and gradually moved from the coastal areas northward into the Ara area. Sweet oranges and grapefruit were distributed by the colonial Agriculture Department and by Christian missionaries, but probably were not planted extensively until after the First World War (Opeke, 1982). Limes may have been introduced to Nigeria by the Arabs (Dalziel, 1937).

Oral histories from elderly Ara farmers indicate that the first cocoa, rubber and coffee seedlings were brought to Ara as early as 1914 from what is now Imo State in southeastern Nigeria. All three tree crops have had an important impact in the Ara area, although the importance has varied from year to year, depending on the market prices for the three crops. In ecological terms, the introduction of kola, cocoa, coffee and rubber meant that some of the original forest trees were removed for the new cash crop tree plantations. Cocoa thrives best with scattered high forest cover, so cocoa trees were introduced into forest areas without as much impact on the original forest trees as with other exotic tree crops. A cocoa plantation forms a dense canopy which tends to shade out undergrowth. Cocoa is often intercropped with kola trees.

Commercial exploitation of timber began in the coastal areas of Nigeria in the 1880s (Egboh, 1985). During the colonial era, the tropical hardwoods in Ara began to assume greater economic importance. Handsawn harvesting of a limited number of trees began with the production of pitsawn planks for local consumption. During the 1930s and 1940s, the Alara joined forces with adjacent traditional states for the construction of the first roads and bridges by communal labour. Following this action, the Alara's lands were opened to easier access by outside interests, including those of timber contractors, although serious commercial exploitation of timber in Ara did not begin until the 1960s. Tropical hardwood trees became a minor source of income as individuals had the right to sell trees located on their own land. This pattern has continued to the present period when large trees may be sold for as little as the equivalent of US$1–5. Since these trees cannot be processed locally, and there is no knowledge of the international value of some of the hardwood trees, they are not yet viewed as significant economic assets within the community. Communal action took place in 1991 to prohibit these sales, but it has not been effective and the harvesting continues unabated. Today timber contractors locate trees by using motorcycles which can ply the bushpaths. Chain saws are used to fell the trees.

From the colonial era throughout the post-colonial period, the area of mature high forest has been diminishing. Currently, the vegetation is about 95% secondary forest and about 5% derived savanna in the northern portion of the Alara's lands. In addition to the harvesting of tropical hardwoods by outside timber contractors, the second major impact has been the rapidly increasing population. In the past quarter of a century, a system of bush-fallow rotation cultivation has emerged as a new farming system along with sedentary/permanent cultivation with an agroforestry base, and sedentary mixed farming characterized by crops and domesticated animal production. In the bush-fallow system, arable food crops are grown for two or three years, then the land is left to recover under a regrowth of natural vegetation lasting around seven years. Secondary forest developed from coppice growth from stumps and root suckers and from relict trees useful for their fruits and other products retained from the original forest (Richard Lowe, personal communication).

Several factors have influenced the emergence of the bush-fallow cultivation system. Many trees for cash crops such as kola and cocoa have begun to age and decrease in productivity. This has occurred at the same time that prices for many tree crops such as cocoa have fluctuated dramatically, and population has increased in Ara as well as in nearby urban centres, leading to an ever-increasing demand for food crops – with a concomitant continual increase in their prices. One common response to this new set of circumstances has been to cut down the ageing trees, replacing them not with new seedlings but rather with arable food crops. In the agroforestry system, the farmer could rely on continual enrichment of the top soil from the organic material such as leaves from the trees. In areas cleared of trees, not only has the organic material disappeared, but the soil erosion from both water and wind has increased dramatically without the ameliorating influence of trees. Currently, about 85% of the farming areas are still in the bush-fallow system of cultivation based on tree crops mixed with arable crops. Farmland under shifting cultivation has been rotated on a 5–10-year fallow basis. Farmers have already had sufficient experience with the shifting cultivation to understand the long-term negative impact on productivity due to any shortening of the bush-fallow period. They have also learned very quickly that there is a significant advantage in the agroforestry-based farming system which enhances soil fertility by

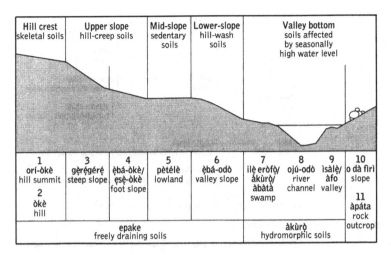

Figure 7.3.   Land forms and major soil types in Ara, Nigeria.

restricting rainwater run-off and by the continual addition of organic material from leaves used as green manure. These new experiences are now being reflected in the indigenous knowledge system within Ara.

Citizens of Ara are well aware of the wide variety of land forms in this part of Nigeria. Each type of land form is named in the Yoruba language. Many land forms are also characterized by particular soil types (Osunade, 1988b; 1992). Figure 7.3 (based on Osunade, 1985) demonstrates a land transect for Ara, while Figure 7.4 shows the three ways in which the Yoruba identify, categorize and classify soils by (a) texture, (b) color and (c) fertility.

Four categories of soil are considered particularly useful for agriculture. These are *ilè dú*, dark humus rich in organic content; *bole*, clay-based soil, found in Ara in two forms – *bole dú*, dark humus mixed with clay, and *bole pupa*, reddish coloured clay; *yanrìn*, sandy soil; and *ilè olókùta*, soil mixed with small stones. Soils are often differentiated by colour terms such as dark-coloured (*dúdú* or *dú*), reddish (*pupa*) and white (*funfun*). The third method of classifying soil is according to fertility. Fertile soil (*ilè olórà*) is known to decline in fertility after years of continual cropping. The term for infertile soils is *asálè*, meaning 'land fading in fertility'. Knowledge about the number of years that a type of soil must be left in fallow (*ìgbòrò*) before regaining full or partial fertility is now extensive.

The farmers in Ara are very conscious of the need for soil management. Efforts are made to reduce the impact of both water and wind erosion. On sloping land, farmers will plant lines of pineapples or trees to reduce the

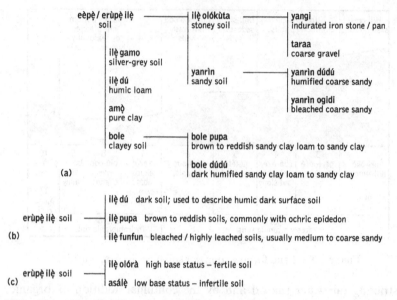

Figure 7.4.   Yoruba soil terms classified by: (a) texture, (b) colour, and (c) fertility.

impact of surface run-off. Farmers carefully observe both annual food crops and tree cash crops for apparent reductions in yield, the most obvious indicator of declining soil fertility. Regeneration of soil is carried out through incorporation of green manure (*iràwe*), household garbage (*idọti*), and ash (*eeru*). Although inorganic fertilizer (*ajilẹ̀:* literally to wake up, *ji*, the soil, *ilẹ̀*) use has increased in recent years, both the high price and unreliable supply have kept traditional approaches to soil regeneration important. Application of fertilizers has also been found to interfere with the keeping properties of yams (Richard Lowe, personal communication).

Farmers also have an extensive understanding of which type of soil is best suited for the wide variety of tree crops and arable food crops grown in Ara. These soil–crop relationships feature heavily in decision making related to acquisition and management of land for agriculture. This indigenous knowledge is outlined in detail in Warren (1992b: ll).

A major shift in the power of the Alara and his chiefs took place in 1978 with the Federal Nigerian Land Use Decree, which greatly diminished the control over the allocation of lands in Nigeria by the chiefs. This decree, incorporated into the National Constitution, transferred the control of land from traditional authorities to the political authorities. A system of state leasehold was created, with the occupier becoming a tenant of the state. Although the law is widely ignored, it has had a major impact in reducing the power base and influence of chiefs at the local level (Lowe 1986). In the past decade, a growing number of individuals have had their lands surveyed and registered with the Federal Lands Office. Many landowners are also going through the lengthy, expensive, and complex bureaucratic procedure of acquiring Certificates of Occupancy which provide the owner and his/her descendants with leasehold renewable rights over the land for 99 years. In the past, the chief was the ultimate authority in the allocation of lands and the reallocation of lands which had fallen into disuse. Today, once a Certificate of Occupancy has been acquired by an individual who has been allocated land by a chief, that individual can leave the land unused or reallocate it to anyone else without even informing the chief. Although this major shift in land-tenure policies is too recent to allow one to make reliable observations, it appears that those individuals who have secured the certificates value these land assets highly and care for them with at least as much diligence as those under the former system.

Historically, trees and tree crops have been controlled exclusively by males. Trees are regarded as an affirmation of land ownership, but the tendency has been to plant exotic species, as these more clearly demonstrate land title. Since only men inherit land, women were not usually permitted to plant trees, but may purchase the tree crops such as palm nuts and process and sell the products. In 1994, the Ara Women's Cooperative Food Products industry set up a small-scale factory with diesel and petrol-driven machines for the production of palm oil, palm kernel oil, and *gari* flour from cassava. The women's co-operative has expressed its desire to plant an oil palm plantation that they would control. If this is approved by the Ara Traditional Council, it will result in a major cultural shift in control of tree-crop products. When the new experience with intermediate technology by the women leads to a more efficient production of palm oil, there will be increased planting of the oil palm, particularly hybrid varieties. An increased volume of available palm nuts coupled with the improved technological basis for processing them will have a major impact on the increased demand for fuelwood requirements for the cooking of the palm nuts. A study on this topic was conducted in Ara during the summer of 1995 by Darci Thelaner (1997).

Land allocated to individuals from the common pool of forested land has increased so fast that there will be no more land for allocation within another decade or so. Hunting, trapping and fishing are rapidly being restricted to a person's own lands. Timber can only be harvested on one's own land. Tree crops are still controlled by the individual who planted them or his descendants. Whenever a person acquires new land, separate negotiations must take place with the owners of the trees on that land should the new owner want access to the trees and their resources. The forces of Islam and Christianity have virtually eliminated the role of sacred groves and trees from the landscape of Ara.

In 1992, an attempt to increase the biodiversity in the area was initiated through the West Africa Multipurpose Tree Project funded by USAID through ICRAF and housed at the International Institute of Tropical Agriculture in Ibadan. Seedlings of numerous tree species were brought to Ara, but interest in protecting the seedlings from the free-ranging goats and sheep was limited to those species already known within the community as having a certain economic value. The unfamiliar species, particularly those with only aesthetic value, were not tended and have not survived. On the other hand, there have been dramatic increases in biodiversity due to the farmers' own efforts to acquire a growing number of varieties of cash crops derived from both trees (such as varieties of hybrid oil palm and guava) and arable crops (such as yams, cocoyams, maize, pepper and cassava). Given the fact that there are still limited numbers of all of the original forest trees and wildlife, the addition of varieties and species of tree crops and arable food crops has actually resulted in an increased incidence of biodiversity on the Alara's lands.

Tables 7.1 and 7.2, based on the analytical framework, summarize the endogenous and exogenous forces influencing the social – ecological system in Ara during the pre-colonial, colonial, and post-colonial periods. In terms of resilience of the system, the inexorable process of social and ecological change has been most influenced by changing patterns of economic opportunities such as the introduction of tree cash crops, the ever-increasing need for cash, and the growing economic role of arable food crops as cash crops required by the ever-increasing population. First-hand experience with changing farming systems is reflected in the changing indigenous knowledge system.

Indigenous knowledge of biodiversity in Ara

The summer and Christmas of 1994 and the summer of 1995 were spent with farmers and hunters in Ara, conducting a biodiversity inventory. Ara residents recognize particular individuals as having specialized knowledge of the flora and fauna of Ara. Most of these individuals are engaged in farming, hunting and the preparation of traditional plant-based medicine. They include both men and women. Since virtually every citizen is involved in farming, the general knowledge of the natural environment is considerable. The Yoruba language is rich in terminology used to classify

Table 7.1. *Changing social and ecological systems in Ara*

Framework components	Time periods		
	Pre-colonial	Colonial	Post-colonial
Ecosystem	Natural resources (land, forest wildlife) exceed demands. Tree crops: oil palm, citrus	Introduction of kola, coffee, cocoa, rubber, mango. Initial demands for tropical hardwoods. Intensive agriculture based on agroforestry. Soil fertility maintained. Increase in bio-diversity through introduction of exotic tree species	Increase in derived savanna. Potential future decrease in biodiversity. First generation of cash-crop trees decrease in productivity; decreasing prices of tree cash crops result in new emphasis on food crops, resulting in decline of agroforestry system; increase in extension of bush fallow system, resulting in declining soil fertility. Increased number of food crop varieties; decrease in the population of some species of forest trees and wildlife. Decreased available land base and increased need for fuelwood in cooking and oil palm production
Population	Low level; in-migration of hunters and farmers	Population increase; continuing in-migration	Population increase results in out-migration to Ondo State (cocoa) and urban areas in Nigeria, Ghana, Cote d'Ivoire. Increased urbanization results in an increased demand for food
Technology	Hand-held iron tools; hand-made Dane guns for hunting	Communal-built infrastructure such as roads and bridges	State Highways (1994–5); chainsaws, provision of pipe-borne water (1988) and electricity (1990). Motorcycles used to locate timber. Intermediate technology production decreases labour and increases palm oil, increasing fuelwood use

Table 7.1. (*cont.*)

Framework components	Time periods		
	Pre-colonial	Colonial	post-Colonial
Local knowledge	Based on subsistence levels of agriculture, hunting, fishing	Based on cash economy – tree cash crops. Males: agriculture production; Females: food processing (e.g. palm oil, *gari*)	Cash economy based on both tree crops and food crops. Women enter production agriculture and continue food processing. Increase IK on crop varieties, and extensive versus intensive farming systems
Property rights	*Oba* controls land distribution; males with freehold rights to planted trees and tree products	*Oba* controls land allocation; males with freehold controls of tree cash crops	1978 Land Use Decree; Certificates of Occupancy (COO); *Oba* controls initial land allocation; COO holders free to re-allocate; women express interest in planting and owning oil palm trees
Social change	Political: Ara Traditional Council; Religion: Yoruba religion; Islam 1860; Baptists 1896. Education: non-formal. Economy: subsistence	Political: ATC, ADU (1947). Religion: Islam, Christianity expand influence; Yoruba religion diminishes. Education: introduction of Western education (1933); Economy: tree cash crops	Political: decrease influence of ATC. Religion: Yoruba religion influence decreases; Education: universal primary education (school fees). Economy: tree cash crops and food crops; increase need for cash – school fees, water and electric rates, Western medicine ADU: links to Ara with cash remittances

Note:
ATC = Ara Traditional Council; ADU = Ara Descendants Union; COO = Certificate of Occupancy; IK = Indigenous Knowledge.

Table 7.2. *Exogenous influences on Ara's social–ecological system*

	Pre-colonial (pre-1914)	Colonial (1914–59)	Post-colonial (post 1960)
Regional	Intra-Yoruba warfare influences migration and settlement patterns in 19th century	Communal-constructed infrastructure (roads, bridges) improves access to Ara	Population and urbanization increase in Osogbo, Ede, Ilobu, Ejigbo, and Ibadan result in increased demand for food cash crops. State highways. Water and electric provision require cash for utility bills
National	Not applicable	Tree cash crops introduced. Western education (1933). Western medicine (1954 maternity, 1956 dispensary)	Universal primary education. Increase in population. 1978 Land Use Decree. Intermediate technology available for women's food processing resulting in decreased female labour loads and increased female income, and possible future increase in need for fuelwood
Global	Global religions reach Ara: 1860 Islam, 1896 Baptist	Uncertain global supply, demand and prices for tree cash crops	Uncertain prices of tree cash crops. High demand for tropical hardwoods

the flora and fauna as well as landforms and soil types (Warren, 1992b). Most Ara citizens are familiar with hundreds of species of animals and plants. The numerous hunters in the area still utilize the hand-made Dane guns for both daytime and night-time hunting. They are very knowledgeable about the behaviour of the animals that are hunted. The farmers are likewise cognizant about the production, processing and consumption characteristics that are considered negative and positive for each variety of crop cultivar (Amusan, 1996; Prain, et al., 1997; Warren and Warren, 1997). These characteristics influence local-level experimentation and the active search for new varieties that enhance the positive traits. For example, the varieties of yams that are actively cultivated include *isu gidi*, *isu igangan*, *isu ewùrà*, *isu gbàngí*, *isu dagidagi* and *isu esúrú*, each of which has different characteristics related to time for maturity, storage qualities, taste, suitability for use in preparing pounded yam (*iyán*), and marketability.

Although many plant and animal species identified in Ara and named in the Yoruba language correspond to the same biological species, one does face several taxonomic difficulties when comparing the two systems. A single Yoruba name, for example, may be applied to more than one botanical species (Abraham, 1958; Olorode, 1985). Among fauna, certain animals are categorized by behaviour, such as animals living in aquatic settings, or as having perceived morphological similarities. Hence, crabs, tortoises and crocodiles may be categorized as types of *jomijòkè* (translated loosely as amphibians), spiders are categorized as types of *kòkòrò* (translated loosely as insects), and earthworms and intestinal worms are categorized as *aràn* (translated as worms). One finds similar problems in categorizing trees (*igi*) and shrubs (*ekè*), as well as grasses (*koríko*), which sometimes include non-grass species regarded as weeds (*èpò*) (Joyce Lowe, personal communication).

The Yoruba differentiate living things (*ohùn eléèmí*) from non-living things (*ohùn àláìlémí*) (Figure 7.5).

Flora (*èdá-ewéko*) is differentiated into domesticated plants (*ògbìn*) and non-domesticated plants (e.g. *igi igboligbè*). Domesticated plants include major categories such as domesticated trees (*igi ògbìn*), fruit trees (*igi eléso*), flowers (*itanna*), vegetables/herbs (*ewébè*, *ohùn ògbìn*), green vegetables (*èfó*), and okra class vegetables (*ilasadò*). Non-domesticated plant categories include trees (*igi*), shrubs (*ekè*), climbers (*ìtàkùn*), grasses (*koríko*), weeds and wild grasses (*èpò*), mushrooms (*olú*), medicinal plants (*igi ogun*), and a wide variety of other types of plants ranging from mildew and moss to various creepers and parasitic plants such as mistletoe (Olorode 1985).

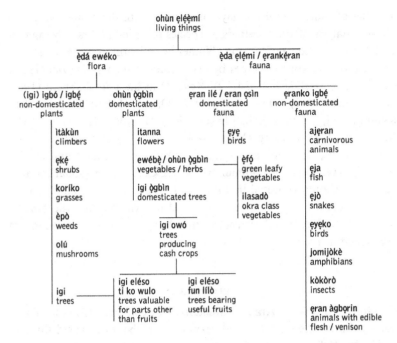

Figure 7.5.   Yoruba terms for living things found in Ara, Nigeria.

Other important categories refer to trees producing cash crops (*igi owó*), which are differentiated into trees bearing useful fruits (*igi eléso fun lílò*) and trees valuable for parts other than fruits (*igi eléso ti ko wulo*). The former category is further divided into trees whose fruits are sold, such as cocoa (*igi eléso fun tita nìkan*), and trees with fruits that are both consumed locally as well as sold, such as oranges (*igi eléso fun tita ati jije*). The latter are divided into those trees whose trunk/stem is valuable, such as the *irókò* forest tree (*igi ti a nla opa rẹ nìkan*), and those that are valuable due to the leaves, such as the neem tree (*igi ti nlo ewé rẹ nìkan*).

Fauna (*èdá-ẹlẹ́mi, ẹrankẹran*) is differentiated into domesticated fauna (*eran-ilé, ẹran-ọsìn*) and non-domesticated fauna (*ẹranko-igbẹ́*). The latter category has as subdivisions of fish (*ẹja*), snakes (*ẹjò*), wild birds (*ẹyẹko*), amphibians (*jomijòkè*), insects (*kòkòrò*), and a wide variety of smaller categories such as bats (*àdán*) and mice (*èkúté*) and single species such as Maxwell's duiker (*ẹtu*) and the bushback (*ìgalà*). Other categories of fauna are animals with feathers (*abìyé*), carnivorous animals (*ajẹran*), birds of prey (*ẹyẹ-ọdẹ*), animals with claws (*ẹran-abekánnáa*), and animals producing venison (*ẹran-àgbọrin*).

Placement of specific varieties and species into the more general Yoruba

categories of fauna and flora may not coincide with the biological taxonomy. An analysis of the hundreds of species recorded in Ara is being prepared for a separate publication. Numerous scientific names have been
identified for the Yoruba names by using sources such as Dalziel (1937);
Hutchinson and Dalziel (1954–72); Keay, Onochie and Stanfield (1960;
1964); Irvine (1969); Onwueme (1982); Opeke (1982); Gbile (1984);
Isawumi (1984; 1993a; 1993b; 1994a; 1994b); Burkill (1985; 1994; 1995);
Oliver-Bever (1986); Keay (1989); Abbiw (1990); Lowe and Soladoye
(1990); Akerele, Heywood and Synge (1991); Sayer, Harcourt and Collins
(1992); J. Lowe, (personal communication); R.G. Lowe, (personal communication). It should be noted that this inventory is still far from complete,
and that there are several species recorded for which either the Yoruba
name could not be remembered at the time the species was identified in Ara
or the scientific name is yet to be identified (Table 7.3).

Conclusions

The traditional state of Ara is characterized by changing patterns of biodiversity during the past hundred years that can be investigated through
oral history. Before the colonial period, population pressure was very low
and land was easily available. There was a loosely regulated communal-
property system for hunting, trapping and fishing. Tree crops could be
planted and controlled by the person planting and tending the seedlings in
any convenient location. The Alara and his chiefs had total control over
allocation of lands for farming. Men controlled trees and tree crops.
Because the population in Ara was perhaps five times less at the turn of this
century than now, agriculture and hunting were based on subsistence patterns. The relative isolation of Ara also limited market transactions with
other traditional states.

The colonial era brought the construction of roads and bridges that
opened the Alara's lands to easier access. Tropical hardwood trees became
a minor cash crop that was harvested by outsider timber contractors, a
pattern that continues to the present time. During this period, several
important tree cash crops were introduced such as kola, cocoa, coffee and
rubber, resulting in the most important farming system that continues
today, that of agroforestry. The cash economy became a growing dominant
force, with cash required to pay new education fees. As the population
began to increase, patterns for hunting, trapping and fishing changed, with
limitations of access to land that was controlled by the hunter or trapper.
This period is also characterized by the growing influence of Islam and

Table 7.3. *Plant and animal species/varieties
identified in Ara with Yoruba names**

A.	Flora	
1.	Domesticated	
	a.	Trees: 41
	b.	Flowers: 6
	c.	Vegetables: 49
	d.	Other: 38
2.	Non-domesticated	
	a.	Trees and shrubs: 114
	b.	Grasses: 7
	c.	Weeds: 7
	d.	Mushrooms: 5
	e.	Medicinal plants: 9
	f.	Other: 19
B.	Fauna	
1.	Domesticated	
	a.	Birds: 6
	b.	Other: 10
2.	Non-domesticated	
	a.	Fish: 9
	b.	Snakes: 8
	c.	Birds: 30
	d.	Amphibians: 6
	e.	Insects: 62
	f.	Other: 44

* Although there are some species double-listed in
different categories, this list represents the number of
species recorded as of July 1995. It is expected that
future field trips will result in refinements and additions
to this inventory.

Christianity, which tended to portray any traditional system related to
Yoruba religion as pagan and involving idol worship. This included the
sacred groves and trees that were regarded as being inhabited by spirits. The
traditional restraints on behaviour in these localities began to diminish.

During the post-colonial period, population pressure increased dramatically at the same time as many of the older first-generation cocoa and kola
trees began to decline in productivity. The demand for arable food crops in
Ara and nearby urban areas resulted in the introduction of a fallow-based
farming system that has begun to influence the growth of derived savanna
into the northern edge of the Alara's lands. At the same time, however,
demand for tree products, particularly palm oil, increased dramatically,

resulting in increased planting of several hybrid tree species. Based on the growing importance of many arable food crops, an actively pursued increase in the varieties of crops such as cassava, maize, yams and coco-yams has occurred. Given the fact that none of the floral or faunal species remembered by the elderly citizens of Ara dating to the nineteenth century has disappeared, the result has been an increase in biodiversity in the area – despite the greatly decreased area of high forest and the diminished wildlife due to hunting. Although the 1978 Nigerian Federal Land Use Decree has greatly diminished the traditional role of the Alara and his chiefs in land use and land allocation, the array of indigenous organiza-tions in Ara, such as the Ara Descendants Union, continue to influence behavioural patterns related to the introduction of new species and vari-eties of plants that have an economic value. At the same time, the role of the traditional sacred groves and trees in terms of biodiversity has virtually disappeared. Gender patterns related to the control of trees and tree prod-ucts are also being challenged by the women of Ara.

The primary driving forces of change appear to be the following:

(a) Social: increasing population and patterns of urbanization, changing gender roles, and changing knowledge that reflects the social–ecological interactions.

(b) Economic: increasing requirement for cash needed for school fees, medical services and utility bills, cash first derived from exotic tree cash crops and more recently from both timber and arable food cash crops; 1978 Nigerian Land Use Decree that changed the basis of land use to long-term leasehold status outside the suzerainty of the traditional polit-ical leaders.

(c) Ecological: shift from primary rainforest to secondary rainforest with the introduction of exotic tree cash crops, and then to derived savanna as more timber has been harvested and agroforestry farming systems have been replaced by bush-fallow.

Although there is likely to be a future loss of indigenous forest tree and wildlife species, and changes in the species and varieties of arable food cash crops, indigenous knowledge will probably continue to be reflected in new experiences and understanding of the important role of forest cover and agroforestry farming systems in maintaining soil fertility.

Reflecting on the indigenous knowledge cycle (see Figure 7.1) within the Ara context, it becomes apparent that knowledge changes and expands in relation to local reactions to an ever-changing set of circumstances per-ceived locally as 'a problem.' This emic or insider's perception of 'a problem' can be at great variance with an etic or outsider's perception of 'a problem'. Very few Ara citizens view the cutting and selling at low prices of tropical hardwood trees as 'a problem' since there are no means locally to process the timber, and many of these trees have no important economic uses. On the other hand, the relationship between tree cover and soil fertility is now much better understood as farmers gain experience with a fallow-based farming system in areas where aged cash crop trees have been felled and replaced only by arable cash crops. The conscious effort to increase the number of species and varieties of arable food crops reflects the indigenous knowledge of the value of biodiversity in reducing agricultural risk related to factors such as crop pests and vagaries in the weather. This is a clear example of resilience – the principles and practices of local resource management systems that enable long-term use of a resource – emerging within the indigenous knowledge of soil fertility and soil regeneration. Clearly, the cultural capital of Ara reflected in its local knowledge system is related to the natural resource capital as it is defined and redefined across time in changing sets of circumstances. Biodiversity is not necessarily valued for its own sake, but is certainly valued and systematically pursued for species that have a high value for the community at a given period of time.

It is difficult to predict accurately future changes in a complex agro-ecosystem such as that in Ara. Although the system has been remarkably resilient over a long period of time, its sustainability in the near future may be in jeopardy. The ever-increasing need for cash to pay school fees, medical expenses and utility bills forces many citizens into short-term economic actions. A worst-case scenario during the next two decades could involve a continuing increase in population and urbanization, increased out-migration, a sharp decline in biodiversity as the number of many species of endogenous trees and wildlife declines to zero, a continuing decline in the number of first-generation productive trees with cash crops, a decline in tree cover and increase in the extent of the derived savanna, a decline in soil fertility and soil tilth with an increased reliance on external inputs, the diminishing influence of the *oba* (the Alara) and his chiefs as the available land base shrinks to zero and the pattern of leasehold land increases due to the 1978 Federal Land Use Decree, and changing gender roles in production agriculture and control of trees and tree crops. If the use of intermediate technology for palm-oil production and the number of oil palm trees continue to increase, there will be an increased need for fuelwood to cook the palm nuts prior to their being processed (Thelaner, 1997).

On the other hand, there are also indications that the current system may continue to be resilient and absorb future perturbations. Natural resource management in Ara has been adaptive, modified and evolutionary from the original settlement period to the present. The indigenous knowledge of Ara men and women has evolved as new opportunites have presented themselves, such as tree cash crops and numerous new varieties of food crops. A major lesson about the relationship between tree cover and soil fertility has been learned by farmers who have moved from an agroforestry bush-fallow system with tree crops intercropped with arable food crops to a bush-fallow farming system where the tree cover is entirely removed. The negative impacts of loss of soil humus and decreased soil fertility due to the bush-fallow system, where the tree cover has been entirely removed and woody regrowth has been replaced by grass during the fallow period, are now part of the indigenous knowledge system. Rule making and enforcement and conflict management continue to be handled in effective ways from the extended family to the Ara Traditional Council.

The role of indigenous knowledge in determining the resilience of the Ara agro-ecosystem should not be underestimated. Problems and problem situations are defined locally (emically) and must be clearly differentiated from exogenous (etic) perspectives. Currently in Ara, biodiversity and its likely decline in the foreseeable future are not regarded as a problem. On the other hand, first-hand experience in the shift from the intensive agroforestry farming system to the extensive bush-fallow system results in a new understanding of the advantages of agroforestry systems in the maintenance of soil fertility. Farmers, both men and women, do experiment with new crop varieties and actively maintain a growing array of these varieties.

Indigenous knowledge is not currently included in the educational curricula in Ara. A visit to the school farm at the Baptist Grammar Secondary School is reminiscent of the typical Iowa monocrop maize field – but it is the only farm of its type in the Ara traditional state. The role of indigenous knowledge in Nigerian educational policy was explored in a conference on the topic held in December 1995, at the University of Ibadan (Titilola *et al.*, 1994; Warren, Egunjobi and Wahab, 1996b). Why not introduce the indigenous agricultural and natural resource management knowledge systems into the curricula so Ara students can appreciate the vitality and resilience of their own dynamic systems as they reflect changing situations and circumstances?

Acknowledgements

The authors are grateful for suggestions provided by the editors and anonymous reviewers that enabled them to improve this chapter. They also greatly appreciate the numerous insights, corrections and additional information provided by Dr Richard Lowe and Dr Joyce Lowe, based on their enormous knowledge about Nigeria gained through lengthy service there as a forester and botanist, respectively.

Notes

1   The orthography for the Yoruba language often indicates a high tone (á) and low tone (à) as well as the use of a subscript dot to represent the 'e' sound as in 'entry' (ẹ), the British pronunciation of 'o' as in 'not' (ọ), and the 'sh' sound as in 'sugar' (ṣ).

References

Abbiw, D.K. 1990. *Useful Plants of Ghana: West African Uses of Wild and Cultivated Plants*. London: Intermediate Technology Publications and The Royal Botanic Gardens, Kew.

Abraham, R.C. 1958. *Dictionary of Modern Yoruba*. London: University of London Press.

Akerele, O., Heywood, V. and Synge, H., eds. 1991. *The Conservation of Medicinal Plants*. Cambridge: Cambridge University Press.

Akinbode, A. 1982. *Kolanut Production and Trade in Nigeria*. Ibadan: Nigerian Institute of Social and Economic Research.

Amusan, A.A. 1996. Indigenous Yoruba agricultural knowledge systems. (Study in progress.) Ile-Ife: Department of Soil Science, Obafemi Awolowo University.

Andresen, P.R. 1996. *Women Farmers of Ara, Nigeria*. MA Thesis. Department of Anthropology, Iowa State University, Ames.

Ara Development Planning Workshop. 1991. *Ara Development Handbook*. Ibadan: Sim-Sim Press.

Berkes, F. 1987.Common-property resource management and Cree Indian fisheries in Subarctic, Canada. In *The Question of the Commons: the Culture and Ecology of Communal Resources*, pp.66–91, ed. B. McCay, and J. Acheson. Tucson: University of Arizona.

Berkes, F., ed. 1992. *Common Property Resources: Ecology and Community-based Sustainable Development*. London: Belhaven Press.

Berkes, F. 1995. Indigenous knowledge and resource management systems: a native Canadian case study from James Bay. In *Property Rights in a Social and Ecological Context: Case Studies and Design Applications*, pp.99–109, ed. S. Hanna and M. Munasinghe. Washington DC: The Beijer International Institute of Ecological Economics and The World Bank.

Berkes, F. and Folke, C. 1994. Linking social and ecological systems for resilience and sustainability. *Beijer Discussion Paper Series No. 52*. Stockholm: Beijer International Institute of Ecological Economics, The Royal Swedish Academy of Sciences.

Berkes, F., Folke, C. and Gadgil, M. 1995. Traditional ecological knowledge, biodiversity, resilience and sustainability. *In Biodiversity Conservation: Problems and Policies*, pp.269–87, ed. Perrings, C.A., Mäler, K-G., Folke, C., Holling, C. S. and Jansson, B-O. Dordrecht: Kluwer Academic Publishers.

Berlin, B. 1992. *Ethnobiological Classification: Principles of Categorization of*

186   *D. Michael Warren & Jennifer Pinkston*

Plants and Animals in Traditional Societies. Princeton: Princeton University Press.

Blunt, P. and Warren, D.M., eds. 1996. *Indigenous Organizations and Development.* London: Intermediate Technology Publications.

Burkill, H. M. 1985. *The Useful Plants of West Tropical Africa.* 2nd edn, Vol. 1: – Families A–D. London: Royal Botanic Gardens, Kew.

Burkill, H. M. 1994. *The Useful Plants of West Tropical Africa.* 2nd edn, Vol. 2: – Families E–I. London: Royal Botanic Gardens, Kew.

Burkill, H. M. 1995. *The Useful Plants of West Tropical Africa.* 2nd edn Vol. 3: – Families J–L. London: Royal Botanic Gardens, Kew.

Castro, A.P. 1995. *Facing Kirinyaga: a Social History of Forest Commons in Southern Mount Kenya.* London: Intermediate Technology Publications.

Cicin-Sain, B. and Knecht, R.B. 1995. Analysis of Earth Summit prescriptions on incorporating traditional knowledge in natural resource management. In *Property Rights and the Environment: Social and Ecological Issues,* pp.105–17, ed. S. Hanna, and M. Munasinghe. Washington DC: The Beijer International Institute of Ecological Economics and The World Bank.

Dalziel, J. McEwen. 1937. *The Useful Plants of West Tropical Africa.* London: The Crown Agents for the Colonies.

den Biggelaar, C. 1995. The use and value of emic and etic perspectives in endogenous agroforestry knowledge research. Unpublished manuscript.

Egboh, E. O. 1985. *Forest Policy in Nigeria 1897–1960.* Nsukka: University of Nigeria.

Fisher, R. J. 1995. *Collaborative Management of Forests for Conservation and Development.* Gland, Switzerland: The International Union for the Conservation of Nature (IUCN) and the World Wide Fund for Nature.

Folke, C. and Berkes, F. 1995. Mechanisms that link property rights to ecological systems. In *Property Rights and the Environment: Social and Ecological Issues,* pp.121–37, ed. S. Hanna and M. Munasinghe. Washington, DC: The Beijer International Institute of Ecological Economics and The World Bank.

Gadgil, M., Berkes, F. and Folke, C. 1993. 'Indigenous knowledge for biodiversity conservation.' *Ambio* 22 (2+3): 151–6.

Gbile, Z. O. 1984. *Vernacular Names of Nigerian plants (Yoruba).* Ibadan: Forestry Research Institute of Nigeria.

GRAIN 1995. *Towards a Biodiversity Community Rights Regime.* Barcelona: Genetic Resources Action International.

Haverkort, B. and Millar, D. 1994. Constructing diversity: the active role of rural people in maintaining and enhancing biodiversity. *Etnoecologica* 2 (3): 51–64.

Hutchinson, J. and Dalziel, J.M. 1954–72. *Flora of West Tropical Africa,* 2nd edn, revised by Keay, R.W.J. and Hepper, F. N. Vols. I–III. London: Crown Agents.

Hyndman, D. 1994. Conservation through self-determination: promoting the interdependence of cultural and biological diversity. *Human Organization* 53 (3): 296–302.

Irvine, F.R. 1969. *West African Crops.* Oxford: Oxford University Press.

Isawumi, M.A. 1984. The peppery fruits of Nigeria. *The Nigerian Field* 49 (1–4): 37–44.

Isawumi, M.A. 1990. Yoruba system of plant nomenclature and its implications in traditional medicine. *The Nigerian Field* 55 (3+4): 165–71.

Isawumi, M.A. 1993a The common edible fruits of Nigeria – Part I. *The Nigerian Field* 58 (1+2): 27–44.

Isawumi, M.A. 1993b. The common edible fruits of Nigeria – Part II. *The Nigerian Field* 58 (3+4): 157–71.

Isawumi, M.A. 1994a. The common edible fruits of Nigeria – Part III. *The Nigerian Field* 59: 57–70.
Isawumi, M.A. 1994b. The common edible fruits of Nigeria – Part IV. *The Nigerian Field* 59: 111–22.
Keay, R.W.J. 1989 *Trees of Nigeria*. Oxford: Clarendon Press.
Keay, R.W.J., Onochie, C.F.A. and Stanfield, D.P. 1960 1964. *Nigerian Trees*, 2 vols. Ibadan: Department of Forest Research.
Lansing, J.S. and Kremer, J.N. 1995. A socioecological analysis of Balinese water temples. In *The Cultural Dimension of Development: Indigenous Knowledge Systems*, pp.258–68, ed. D.M. Warren, L.J. Slikkerveer, and D. Brokensha. London: Intermediate Technology Publications.
Lloyd, P.C. 1962. *Yoruba Land Law*. Oxford: Oxford University Press.
Lowe, J. and Soladoye, M.O. 1990. Some changes and corrections to names of Nigerian plants since publication of *Flora of West Tropical Africa*, Edition 2. *Nigerian Journal of Botany* 3: 1–24.
Lowe, R.G. 1986 *Agricultural Revolution in Africa?* London: Macmillan.
Martin, Gary J. 1995 *Ethnobotany: a Methods Manual*. London: Chapman and Hall.
Messerschmidt, D.A. 1995. Local traditions and community forestry management: a view from Nepal. In *The Cultural Dimension of Development: Indigenous Knowledge Systems*, pp.231–44, ed. D.M. Warren, L.J. Slikkerveer and D. Brokensha London: Intermediate Technology Publications.
NEST 1991. *Nigeria's Threatened Environment: A National Profile*, Ibadan: Nigerian Environmental Study/Action Team.
Oldfield, M.L. and Alcorn, J.B., eds. 1991. *Biodiversity: Culture, Conservation, and Ecodevelopment*, Boulder, Col.: Westview Press.
Oliver-Bever, B. 1986. *Medicinal plants in Tropical West Africa*. Cambridge: Cambridge University Press.
Olorode, O. 1985. Aspects of plant naming and classification among the Yoruba. *Odu* 27: 82–95.
Omolaoye, O.A. 1994. Rural-urban interaction and community development: a case study of Ara and its surrounding villages in Egbedore Local Government of Osun State. Unpublished MSc thesis, Department of Geography, University of Ibadan, Nigeria.
Onwueme, I.C. 1982. *The Tropical Tuber Crops*, New York: ELBS and John Wiley.
Opefeyitimi, A. in preparation. Sustaining agriculture in an era of ecosystem abuse: insights from Yoruba culture and worldview. In *Strategies and Tactics of Sustainable Agriculture in the Tropics*, ed. M.A. Badejo, and G. Tian.
Opeke, L.K. 1982. *Tropical Tree Crops*, New York: John Wiley.
Osunade, M.A.A. 1985. Criteria used in traditional land evaluation in southwestern Nigeria. *Journal of West African Studies* 28: 57–70.
Osunade, M.A.A. 1988a. Nomenclature and classification of traditional land use types in south western Nigeria. *Savanna: A Journal of the Environmental and Social Sciences* 9: 50–63.
Osunade, M.A.A. 1988b. Soil suitability classification by small farmers. *The Professional Geographer 40*: 194–201.
Osunade, M.A.A. 1992. Identification of crop soils by small farmers of Southwestern Nigeria. *Journal of Environmental Management* 35: 193–203.
Osunade, M.A.A. 1994. Indigenous climate knowledge and agricultural practice in Southwestern Nigeria. *Malaysian Journal of Tropical Geography* 25: 21–8.

Phillips, A. and Titilola, T., eds. 1995. *Indigenous Knowledge Systems and Practices: Case Studies from Nigeria.* Ibadan: Nigerian Institute of Social and Economic Research.

Prain, G. and Bagalanon, C.P., eds. 1994. *Local Knowledge, Global Science and Plant Genetic Resources: Towards a Partnership.* Los Banos: UPWARD.

Prain, G., Fujisaka, S. and Warren, D.M., eds. 1997. *Biological and Cultural Diversity: The Role of Indigenous Agricultural Experimentation in Development.* London: Intermediate Technology Publications.

Rajasekaran, B., Warren D.M. and Babu S.C. 1991. Indigenous natural-resource management systems for sustainable agricultural development – a global perspective. *Journal of International Development* 3 (4): 387–401.

Rural Advancement Fund International 1994. *Conserving Indigenous Knowledge: Integrating Two Systems of Innovation.* New York: United Nations Development Programme.

Sayer, J.A., Harcourt, C.S. and Collins, N. M., eds. 1992. *The Conservation Atlas of Tropical Forests: Africa.* New York: Simon and Schuster.

Smith, G. and Osunwole, S.A., 1996. Urban and rural case studies of the transmission of the Yorùbá therapeutic system. In Fairfax, F., Wahab, B., Egunjobi, L. and Warren, D.M., eds. *Alàáfìà: Studies of Yorùbá Concepts of Health and Well-being in Nigeria,* pp.45–52. Studies in Technology and Social Change No. 25. Center for Indigenous Knowledge for Agriculture and Rural Development, Iowa State University, Ames.

Thelaner, D. 1997. The effects of mechanized palm oil production on fuelwood availability in Ara, Nigeria. MA Thesis. Department of Anthropology, Iowa State University, Ames.

Titilola, T., Egunjobi, L., Amusan, A. and Wahab B. 1994. Introduction of indigenous knowledge into the education curriculum of primary, secondary and tertiary institutions in Nigeria: a policy guideline. Unpublished report. CIKARD, Ames.

Verger, P. F. 1995. *Ewé: The Use of Plants in Yoruba Society.* Sao Paulo: Odebrecht–Companhia das Letras.

Warren, D.M. 1991. Using indigenous knowledge in agricultural development, *World Bank Discussion Papers No. 127.* Washington, DC: The World Bank.

Warren, D.M. 1992a. Indigenous knowledge, biodiversity, conservation and development. Keynote address, International Conference on Conservation of Biodiversity in Africa: Local Initiatives and Institutional Roles. August 30–September 3, 1992. National Museums of Kenya, Nairobi. In 1996. *Sustainable Development in Third World Countries,* pp.81–9, ed. J. Valentine. Westport, Conn.: Greenwood.

Warren, D.M. 1992b. A preliminary analysis of indigenous soil classification and management systems in four ecozones of Nigeria. *Discussion paper RCMD 92/1.* Ibadan: International Institute of Tropical Agriculture and the African Resource Centre for Indigenous Knowledge.

Warren, D.M. 1992c. Strengthening indigenous Nigerian organizations and associations for rural development: the case of Ara community. *NISER Occasional Paper No. 1.* Ibadan: Nigerian Institute of Social and Economic Research.

Warren, D.M. 1993. Indigenous knowledge and sustainable agricultural and rural development in Africa: policy issues and strategies for the twenty-first century. Paper presented at the annual meeting of the African Studies Association, Boston, December 4–7, 1993.

Warren, D. M., Adedokun, R. and Omolaoye, A. 1996a. Indigenous organizations and development: the case of Ara, Nigeria. In *Indigenous Organizations and Development*, pp.43–9, ed. P. Blunt and D.M. Warren London: Intermediate Technology Publications.

Warren, D., Egunjobi, L. and Wahab, R., eds. 1996b. *Indigenous Knowledge in Education: Proceedings of a Regional Workshop on Integration of Indigenous Knowledge into Nigerian Education Curriculum*. Indigenous Knowledge Study Group, University of Ibadan, Ibadan.

Warren, D.M. and Rajasekaran, B. 1994. Using indigenous knowledge for sustainable dry-land management: a global perspective. In *Listening to the People: Social Aspects of Dry-land Management*, pp.89–100, ed. D. Stiles. Nairobi: Desertification Control Programme Activity Centre, United Nations Environment Programme. Republished in Stiles, D., ed. 1995. *Social Aspects of Sustainable Dryland Management*, pp.193–209. New York: John Wiley and Sons.

Warren, D.M., Slikkerveer, L.J. and Brokensha, D., eds. 1995. *The Cultural Dimension of Development: Indigenous Knowledge Systems*. London: Intermediate Technology Publications.

Warren, D.M. and Warren, M. 1997. Local-level experimentation with social organization and management of self-reliant agricultural development: the case of Ara, Nigeria. In *Biological and Cultural Diversity: the Role of Indigenous Agricultural Experimentation in Development*, ed. G. Prain, S. Fujisaka and D.M. Warren. London: Intermediate Technology Publications.

8

Managing for human and ecological context in the Maine soft shell clam fishery

SUSAN S. HANNA

Introduction

Soft shell clams (*Mya arenaria*) are distributed throughout the intertidal zone of Maine's coast and offshore islands. A community-based fishery for this species has existed for over 200 years, managed with varying degrees of sustainability. Over time, the fishery has maintained many of its traditional characteristics against the backdrop of an industrializing economy, supporting subsistence, recreational and commercial uses. Uses of the clam resource have reflected the larger ecological and economic context, whose health and productivity have varied over time and area. Management of the clam resource also varies by area, reflecting objectives of social efficiency and equity, as defined by coastal communities. The co-evolution of management within the larger economy has been made possible by the nesting of property rights in community and state levels of governance, and by the co-operation of communities and the state in research and management. Management authority at each level involves responsibilities as well as rights. This chapter addresses the key aspects of property rights and responsibilities in this fishery as they exist within the context of the ecosystem, the mix of technology and people, local knowledge, and the interaction of markets and management. Maine's clam management is a unique example of local-level management adapted to the ecological and cultural context. It is also an example of a system struggling to maintain itself in the face of economic and political change. Signs of strain such as falling clam harvests and rising management costs have signalled a need for management to continue to adapt to its changing environment.

The ecosystem

Maine's coastline extends for over 7200 km in linear length, with intertidal ecosystems that vary in sediment type, current strength, salinity and water temperature. Sediment types include rocky rubble, clay and fine sand. Tides periodically cover and uncover the bottom with salt water in currents that change with bottom topography. Salinity levels vary according to the proximity of freshwater sources. Water temperatures rise and fall with the time of year and topography: in summer, steeply sloped intertidal areas are covered with deep cold water at high tide; mildly sloped intertidal areas have shallow, warm water cover. The reverse is true in winter. These different physical attributes result in variation in the species compositions of benthic communities along the coast.

Throughout its life history, the soft shell clam uses a variety of intertidal habitats. Spawning takes place in warm water over a four-month period. The larval stage is free drifting, subject to water currents and winds. At the end of the larval stage, juveniles settle to the ocean bottom and attach to the sediment by byssal threads. The bottom-dwelling stage lasts about a year, during which the animals are semi-mobile. At onset of the adult stage, clams burrow into the sediment, becoming sedentary. From that period on, clams are limited to the small range of vertical movement necessary to siphon-feed from surface water. Their preferred location is at the lower end of the intertidal zone where nutrient-bearing salt water coverage is at its maximum. Once settled in sediment, clams grow at different rates, taking between three and six years to mature to recruitment size. The warmer the summer water, the faster the growth (Dow, 1972).

Soft shell clams adapt to a range of environments, including tidal estuaries, coves, open pebble beaches, and island intertidal areas. Their preferred environment is fine sediment ('mud flats') with a high salinity level. Co-resident in the intertidal ecosystem are other clam species, crabs, mussels, snails, barnacles, seaweeds and marine worms (Crowder, 1931; Teal, Teal and Fish, 1969). Other intertidal species that are harvested for commercial use in Maine include marine worms, used as bait; blue mussels, sold as food; seaweed, sold as a source of carageenan; and rock crabs, eaten as subsistence or sold as food.

The survival and productivity of soft shell clams depend on a number of physical and biological factors: water temperature, salinity, currents, competition, predation, and the presence of toxins. Growth rates increase with water temperature up to a threshold temperature level, beyond which warm water inhibits growth and causes mortality (Dow, 1972). Clams can adapt to different sediment types but can be smothered by abrupt changes in sediment type or level. If water circulation is sluggish, growth is slow; if too rapid, setting of juveniles on the bottom may be inhibited. Soft shell clams compete for sediment space with blue mussels (*Mytilus edulis*), which attach themselves to the bottom surface. Clams are most vulnerable to predation by green crabs (*Carcinides maenas*), which grow best in warm water. Rock crabs (*Cancer irroratus*), moon snails (*Polynices*), horseshoe crab (*Limulus polyphemus*), winter flounder (*Pseudopleuronectes americanus*), herring gulls and ducks also prey on soft shell clams (Bigelow and Schroeder, 1953; Dow and Wallace, 1957). Clam mortality is also caused by algal blooms, pesticide runoff, oil spills and other pollutants (Dow and Wallace, 1961).

The structure of the intertidal ecosystem is relatively well known, but the system's function presents several areas of scientific uncertainty. Factors which determine the productive variability of system components are only partially known, as are the drivers of predator–prey interactions and community interactions. The system as a whole appears to be resilient to perturbations that occur in the form of changes in water temperature, pollution and predation. Individual system components vary in their relative magnitude over time, but system structure has remained fairly stable to a range of disturbances. Subsurface species in particular are protected from many sources of environmental variability by the layer of sediment above them (Teal *et al.*, 1969). The ecosystem is well adapted to the gradual seasonal changes in water temperature and salinity levels.

Although the intertidal ecosystem appears to be robust to disturbances at the system level, individual species are sensitive to the external effects of various human activities. Pesticide and manure runoff from farms, salt, sand and oil runoff from roads, and other industrial and municipal wastes all affect the productivity and health of individual species. Because human population concentrations and industrial activities vary by location along the Maine coast, ecosystem health also varies by area. As a sedentary species in the nearshore zone, soft shell clams are particularly subject to contamination from anthropogenic pollution such as bacteria (*Escherichia coli*), heavy metals and radionuclides. The intertidal ecosystem has suffered chronic accumulations of various pollutants, which have caused a gradual degradation of ecosystem health that, while not leading to large-scale shifts in system structure, have increased requirements for biological monitoring and remediation.

People and technology

Soft shell clams are harvested for food and commercial sale by residents of Maine coastal communities. Coastal communities are small, ranging in population from 500 to 10000 people, and many are isolated (Anon 1989). The isolation fosters independence and an ethic of self-sufficiency. At the same time, isolation creates social and economic interdependencies within the community. As population density gradually decreases moving from west to east along the coast, economic activity and employment opportunities also decrease, and isolation increases. Community economies in the eastern end of the coast tend to be more dependent on the natural resource base for employment and subsistence.

With minor exceptions, methods for harvesting soft shell clams have stayed the same for several hundred years. The scene at low tide of clam diggers bent over mud flats could as easily be 1696 as 1996. Clam harvesting is limited by state law to hand implements, and is difficult physical labour (State of Maine, 1994). The preferred harvest tool is a long-tined clam fork, a 'rake', and wooden slat basket, a 'hod'. Mechanical dredges are allowed in exceptional cases to harvest clams on mildly contaminated flats for depuration (holding clams in clean water until bacteria are filtered out) and sale.

Flats are exposed twice a day by the tides, for about four hours each tide in the lower near-water end of the intertidal zone. Harvest time is limited to low tides during daylight hours, which means a single low tide in winter and sometimes two in summer. The digging process can cause incidental mortality to undersized clams by breaking shells and distributing them on the surface (Glude, 1954), but these sources of damage vary widely with the skill of individual clam diggers. Many clam diggers combine clam harvesting with other resource-based occupations such as worm digging, mussel or seaweed harvest, and wood cutting. Employment is seasonal, stopping in winter when severe temperatures cover the surface of clam flats with ice. The number of people working as clam diggers and their average harvest per tide vary widely over time, depending on resource supply, market conditions, and employment alternatives. Community residents of all ages harvest clams for subsistence, but the more physically difficult commercial harvest tends to be concentrated among men from 20 to 35, although some continue until age 65 (Maine Department of Marine Resources, 1994).

Soft shell clams spent years in relative economic obscurity before emerging as a major seafood industry of Maine, although they were historically important as food and bait. Midden remains indicate that clams were a major food source for coastal tribes of Native Americans (Maine Department of Sea and Shore Fisheries, 1918). In the 1600s, European settlers used clams for subsistence:

The clams in the flats, and the fish in the sea, were the two great factors that enabled our forefathers to obtain a foothold on these shores, living on them until they could clear the land and push back the interior

(Maine Department of Sea and Shore Fisheries 1906).

As other sources of food production were developed in the Colonies, clams became a food of last resort, a 'poor farm' fishery (Maine Department of Sea and Shore Fisheries, 1954). By the mid-1800s, clams had become a common food source for Maine residents. Salted clams were a major source of bait for offshore Grand Banks fishermen until 1875, when they were replaced by fresh bait. By the late 1800s, clams were being shipped to Boston and New York in barrels for hotel and retail markets (Maine Department of Sea and Shore Fisheries, 1906), followed in the early 1900s by the development of an export market to Portugal. Clams were canned from the early to mid-1900s, after which they were sold as a fresh product, in-shell and shucked (Dow and Wallace, 1957), the product form that continues to the present.

The size of the soft shell clam harvest is influenced by a number of factors, most importantly market conditions, weather, climate cycles, ecosystem health and anthropogenic pollutants. Ex-vessel prices vary with the size of landings, seasons, and the supply of substitutes. Weather affects the survival of both spat and mature animals, for example when storm-driven sediment shifts smother clams. Weather also limits flat availability to days without ice cover. Climate cycles affect water temperature, which influences spawning time, larval survival, animal growth rates and survival. Large algal blooms compete with clams for oxygen. Periodic toxic algal blooms such as the 'red tide' (*Gonyaulax tamarensis*), a dinoflagellate causing paralytic shellfish poisoning (PSP) in humans, result in the closure of areas to harvest. Pollution in the forms of bacteria (*E. coli*), heavy metals, and radionuclides also remove areas of clams from use.

The production of soft shell clams is well adapted to the social and economic context of Maine. In most Maine coastal communities, labour and resources are abundant relative to capital. Because hand harvesting is slow and less technically efficient than mechanical devices, it employs larger numbers of people. Capital investment in hand harvesting is low, adjusting the use of productive inputs towards labour and away from capital. As a labour-intensive technology, the clam rake is appropriate to the relative labour abundance, and as a low-cost harvest technology it allows access to a larger number of people than would be the case with high-cost technology. Since the technology's availability does not depend on the clam digger's access to capital markets, equity of access is created. Because it is a hand tool, the extent to which the technology can be used to enhance the marginal productivity of some clam diggers over others is extremely limited. Productivity differences in clam harvesting result only from individual skill and effort levels. The extent to which clam rakes are also the most appropriate technology for minimizing clam breakage is unclear, but probably depends on the skill of the individual digger. As with technology constraints in other fisheries, the clam rake appears to have been kept as legal gear in recognition of the dual community objectives of employment and equity.

Expectations of full community access to the resource are historical, built on the early role played by soft shell clams in the subsistence and survival of the first European settlers. Restricting harvest technology to hand tools is consistent with the traditional values of Maine's coastal communities which emphasize equal access to resources. These values are similar to those in other traditional resource-based communities which emphasize security of livelihood and equity of access (Berkes and Farvar, 1989). The use of hand technology for Maine clams sets technology at appropriate levels for resource endowments and has promoted social objectives that have remained important to coastal communities over time.

Local knowledge

A unique aspect of Maine's soft shell clam fishery is the integration of informal local knowledge with locally generated formal scientific information. The coast of Maine is geographically diverse, offering wide scope for site-specific ecological knowledge. Information about the intertidal ecosystem is acquired by coastal community residents in both formal and informal ways. Informal knowledge has probably existed for as long as people have harvested clams on a regular basis; harvest areas are part of the home communities of clam diggers, many of whom have lived in the same community throughout their lives. Repeated exposure to the intertidal ecosystem imparts information about ecosystem structure, preferred habitat, causes of mortality, and interannual variability. In earlier times of low human population density and little market demand for soft shell clams, ecosystem knowledge was probably passively acquired but little used, since exploitation pressures were not severe enough to present conservation problems.

Increasing human population densities and strong market demand have subjected the intertidal ecosystem to a number of pressures requiring monitoring and remediation, where informal knowledge must be supplemented with scientific knowledge. Scientific knowledge of the soft shell clam and its ecosystem is generated at both the state and community levels. Local knowledge may include information about habitats, distribution, growth rates, general ecosystem properties, and the effects of harvest on population. Informal local knowledge, acquired through the experience of clam harvesters and other coastal community residents, is fed into the design of scientific experiments generating information on a systematic basis. Local community residents co-operate with and are assisted by representatives of the Maine Department of Marine Resources in the production of site-specific scientific information on the intertidal ecosystem (Hanna Dearborn, 1977).

The type of knowledge required to *use* the soft shell clam resource is limited to rules about state and local regulations, legal gear, and gear operation. A certain level of skill is required to harvest the resource with minimal damage to the ecosystem by avoiding shell breakage, exposure of undersized clams, and destruction of other species. But because soft shell clams are a sedentary species, their location is predictable and in fact signalled by siphon holes in the mud. Once clams are burrowed, they are fairly easily located, making search time minimal.

The knowledge required to *sustain* the resource over time is considerably more complex. The easy access to clams on exposed tidal flats makes them particularly vulnerable to exploitation. Moreover, to sustain soft shell clam populations over the long term requires ecosystem information on population size, growth rates and predation as well as on the physical dangers to the resource from siltation, salinity levels, bacteria and other contamination. In addition, sustainability requires a system of controls adapted to both ecological and harvest variability.

The gap between the level of knowledge needed to use the resource and that needed to sustain the resource is wide. Attempts have been made to bridge this gap by locating the production of scientific information at the site of resource harvest. The community is formally tied to conservation responsibilities through statutory provision for local involvement in conservation and management. The state regulatory regime explicitly recognizes the value of local knowledge in providing for clam conservation plans. Community residents work in conjunction with state biologists to develop plans and conduct resource surveys in a co-operative production of scientific knowledge. Through shellfish conservation committees, community members participate in the definition of needed research, the design of experiments, and the conduct of research. With the technical advice of state biologists, residents conduct resource surveys, test water quality, monitor for contaminants, and conduct experiments in predation control, growth, hatchery and nursery operations and reseeding of flats (Gouldsboro, 1993; Islesboro, 1993; Jonesboro, 1993; Waldoboro, 1993; Scarborough, 1993).

The co-operative generation of scientific information is also reflected in the allocation of monitoring responsibilities between the community and the state. Coastal communities monitor environmental information specific to their ecosystem and to harvest activities in that ecosystem. The information includes growth and productivity of soft shell clam populations, predation, areas needing enhancement, seasonal and area closures, local harvest and access. The state monitors environmental information related to overall productivity, resource status and the health of Maine citizens.

For people in coastal communities, the costs of developing and maintaining a local knowledge base can be high. Costs include the direct costs of co-ordination, information gathering, experimentation and monitoring, and as coastal populations have increased, the need to monitor water quality, non-point pollution and harvest effort has also increased. The expected benefits to the community of maintaining the knowledge base are returned in the economic, recreational, and aesthetic values of the intertidal areas, but whether the benefits from the local production of information exceed the costs depends in large part on the social and economic alternatives to the use of the intertidal zone.

Property rights

Both informal local knowledge and the local production of scientific knowledge are enabled and enhanced by the particular structure of property rights that exist for the soft shell clam resource. Property rights are based on a 1641 ordinance of the Colony of Massachusetts Bay, which included the territory that later became the State of Maine:

Every inhabitant who is a householder shall have free fishing and fowling in any great ponds, bays, coves, and rivers so far as the sea ebbs and flows within the precincts of the town where they dwell, unless the freemen of the same town, or of the general court, have otherwise appropriated them.

(Anon, 1970)

This ordinance reflects the importance of fisheries as a means of subsistence to the early European settlers. The Colonial Ordinance assigned rights of access to all 'householders', who were later identified by Maine's first legislature as all residents of a town:

That very inhabitant of the said towns . . . shall have a right to take such other shellfish from their beds therein for the use of his or her family.

(State of Maine, 1821)

The first Maine law endorsed the apportionment of access rights by town boundaries, allocating to town residents the right to take clams for their

198   *Susan S. Hanna*

Table 8.1. *Property rights and responsibilities related to soft shell clam resources of Maine.*

	Property rights	Property responsibilities
State	Final approval on conservation Gear regulations Supplemental conservation measures	Sustainability Co-ordination Public health Enforcement Development
Community	Access: resident/non-resident leasing Regulation fees tools	Conservation research implementation Regulation monitoring enforcement
Individual	Access Appropriation	Support management Legal behaviour

own use and to town officers the right to regulate clam harvesting (Anon, 1970). Appropriation rights of coastal community residents, practised for 180 years, were thus formally established as law. As economic and demographic conditions changed over time, state law further strengthened the rights of coastal towns to establish allowable harvest times, set licence fees, restrict quantities harvested for commercial use, and favour residents over non-residents in rights of use (Hanna Dearborn, 1977). A similar pattern of development was followed in Massachusetts, where property rights to shellfish that began with the same Colonial Ordinance now parallel those of Maine (Muse, 1976).

Attitudes toward rights of access to the intertidal zone are consistent with traditional views of settlers throughout the north-east. McCay (1989) notes that sentiments of early Euro-Americans towards natural resources were towards open access, a denial of class-based privileges which had existed in the countries from which they emigrated. According to the American Public Trust Doctrine, which specifies public ownership of the intertidal zone, the state holds intertidal resources in trust for its citizens, managing resources for their benefit (McCay, 1993). Judd (1992) points to the influence of the pioneer past in establishing a tradition of rights to hunt and fish. The tradition finds its modern expression in conservation efforts which Judd calls '..a blend of everyday, practical observation and dogged traditionalism.'

In the early 1900s the state, to encourage increased clam production, authorized coastal towns to lease up to one-quarter of their clam flats to private cultivators. The idea of private property rights over clam flats was in such conflict with traditional views of equal access by all residents that despite the continuance of legal authorization for private leasing, none has ever occurred. Instead, by the 1950s, community objectives of equitable local access were strengthened by the establishment of rules limiting harvest to hand gear, a technology which assured access to the soft shell clam resource to the maximum number of people.

By the early 1960s, coastal communities had acquired property responsibilities along with their property rights. Towns with clam ordinances were responsible for their enforcement and for the development and support of conservation programmes, replacing state functions in these areas. Access rights could be restricted by individual towns, but only if the restrictions were based on state-approved conservation plans. Access rights continue to be secondary to conservation objectives. The law divides resource management authority between the state and coastal communities; the clam resource is still held by the state in trust for all citizens, but the privilege of managing that resource is assigned to coastal towns if they accept the corresponding responsibilities. To qualify for authority to manage its soft shell clam resource, a town must establish a shellfish conservation committee, develop a conservation plan, and appropriate funds to pay for resource surveys and enforcement. Communities with management plans approved by the Maine Department of Marine Resources have the power to restrict entry, require town residency, restrict the quantity of harvest, establish open and closed areas, set seasons, and charge user fees. They may also lease up to a quarter of their intertidal surface area, although none has done so (Maine Department of Marine Resources, 1994).

The state retains control over licensing of clam diggers, allowed clamming gear, and public health. Anyone harvesting clams for sale must be a Maine resident and purchase a state licence. The state monitors clam processing, pollution levels and toxin levels. The superseding position of the state over the community in some matters means that the clam resource is owned as restricted community property.

Resource appropriators are resident commercial harvesters, resident subsistence gatherers, or non-resident recreational licence holders. Appropriators have various bundles of rights and responsibilities, depending on their status as commercial or subsistence users, or on the community in which they live. Rights of licensed individuals include rights of access to harvest and rights of sale. Communities may attenuate harvest rights by setting harvest quotas, closed areas and closed seasons. The state attenuates harvest rights by restricting harvest method, size of animals harvested and harvest from contaminated areas.

Table 8.1 summarizes soft shell clam rights and responsibilities at the levels of the individual, the community and the state. Individual residents have rights of access and use of the resource, with corresponding responsibilities to fund resource management through tax appropriations and to comply with regulations. Coastal communities have rights of access over intertidal areas within town boundaries, which they may allocate to residents and non-residents, and may lease to private operators. Communities may set regulations and assess use fees. The corresponding responsibilities are the development and implementation of conservation plans, the conduct of research to support plans, ecosystem monitoring, and the monitoring and enforcement of regulations. The state retains the right of final approval on towns' conservation plans, gear regulations and supplemental conservation measures such as minimum legal size. The state bears the ultimate responsibility to sustain the resource over time, coordinate community conservation plans, monitor for public health, supplement enforcement, and develop the resource. Moving from the citizen to state level, the bundles of rights are nested in ascending levels of authority. In general, state rights supersede community rights, which supersede individual rights.

Other components of the intertidal ecosystem also have appropriation rights assigned. Licences are required for the commercial harvest of marine worms, seaweed and mussels, and hand tools are required for all intertidal harvest.

Markets, management and outcomes

Nothing less than the most minute of regulations of particular waters; the most excessive penalties, adjusted to the locality affected; their enforcement by local officers inspired by local pride and personal interest warmly backed up by local sentiment, can suffice, even among a comparatively scanty population, in preserving their fisheries . . .

(Maine Department of Fisheries and Game, 1892).

The structure within which soft shell clams are managed has evolved over time in response to political, economic and ecological forces, but has consistently reflected the sentiment toward local power expressed in the quote above. Political support to maintain and strengthen local property rights to coastal resources has for the most part remained strong. As soft shell clams have ascended in economic importance, interest has heightened in developing practices designed to sustain the flow of harvest over time. In addition, some rent-seeking moves by excluded parties have taken place.

Markets

One of the important drivers of human interaction with the intertidal ecosystem is the market for its resources. The development of markets for any natural resource introduces strong pressures on resource appropriators to maximize short-run gains at the expense of long-run sustainability. Overuse can easily result. Because they are sedentary, soft shell clams are particularly vulnerable to market pressures, and require management that can prevent overuse. Over time, markets have created both neutral and strong pressures on the sustainability of the resource. For example, up until 1885, markets for Maine soft shell clams were limited, exploitation pressure was low, and management needs were minimal:

> ... By leaving these, and the smaller ones down to the seed, to grow until another year, there was an abundance of them every fall. Under these conditions, the flats to the north of Fort Popham, comprising some two hundred acres, were dug over by sixty or one hundred men every winter for twenty-four or five years, without apparently diminishing the clams.
> *(Maine Department of Sea and Shore Fisheries, 1906).*

But after 1885: '. . . orders began to be received from Boston for clams in the shell to supply hotels and clambakes in the summer, and the public at large in the winter. Then commenced the depletion of these bivalves.' (Maine Department of Sea and Shore Fisheries, 1906). In the 1990s, strong market demand continues, ex-vessel prices are high, and the pressure to exploit for short-term gain still exists. Past experience with depletion, the expectation of continued high prices, and the institutional structure which promotes local conservation efforts are providing some countervailing pressures for sustainability. Strong markets provide incentives for conservation and enhancement to maintain the flow of resource services over time. At the same time, they also create short-term pressures to overexploit the resource.

In addition to markets, the degree of economic diversification in coastal communities also plays a role in determining a community's approach to management. The degree of economic dependence on a resource influences the level of management attention people give to it. Coastal communities with diversified economies providing employment alternatives to soft shell clams were traditionally among the ones least active in soft shell clam management. For years, towns on the western end of the coast with chronic intertidal pollution problems were not motivated to address those problems because other labour markets existed and the costs of investing in clean-up and management exceeded the expected benefits of a productive clam resource. In addition, until the mid-1970s, the local authority to control pollution was absent. The enactment of federal pollution control measures lowered or removed the clean-up costs to towns. Strong markets combined with lower management costs to create the incentives that led to restoration efforts in some communities with depleted or damaged clam-growing areas. In these areas, the expected benefits from investment in management now appear to exceed the costs (Islesboro, 1993; Scarborough, 1993; Waldoboro, 1993). Interestingly, towns in the less-developed sections of the coast have been slower to adapt to this change, and this may be due to the costs associated with management.

Management

The combined state/community management regime over soft shell clams is implemented through a co-management process that divides management authority and power between the state and the community. Within the nested structure, authority varies according to the type of decision to be made. Communities develop site-specific conservation plans with technical assistance from the state. Communities with shellfish conservation plans set appropriation limits, set aside areas for regeneration, allocate the resource and enforce local rules. The state retains final approval authority over the conservation component of local management plans, maintains a state-wide size restriction on clams, and restricts the method of harvest. The state also monitors various types of contamination and retains the rights to close areas to harvest when necessary to protect public health. In this way the local management institutions are nested within the wider and superseding state authority over minimum size, legal gear, conservation and public health.

Decentralization of some management decisions leads to management diversity at the local level. Despite this diversity, however, there are certain patterns of interaction among appropriators and between appropriators and the ecosystem which tend to characterize management in all communities with shellfish conservation plans. The appropriators of the soft shell clam resource, primarily residents of coastal communities, interact with other resource users by co-operation and reciprocity, restrictions on free riding, and explicit controls on competition.

Co-operation and reciprocity

The institutional structure and the co-management process are based on co-operative interactions and mutual give-and-take. The values of co-operation and reciprocity, rooted in tradition, are at the heart of negative attitudes towards private leasing of the intertidal area or limiting access of community residents. Although many of the coastal towns have shellfish ordinances that restrict harvest access to residents only, none has a plan that directly limits access of residents through caps on the number of commercial licences. Towns have preferred to manage exploitation through indirect measures such as open and closed areas, open and closed seasons, and enhancement of production, a preference that appears to be based in the traditional desire to keep resource access open to all. The indirect effort–control measures have had mixed success in balancing harvest effort with ecosystem productivity.

Free riding

To some extent, all residents of a coastal community benefit from the actions of those who develop and implement a shellfish conservation plan. In towns with conservation plans, free riding – obtaining the benefit of others' actions without paying a proportional share of the cost – is minimized by the structure and process of management. Monetary costs of management are appropriated from local taxes and local user fees levied on commercial users. Expenditures of time and effort to develop and implement plans are proportionally shared by the major beneficiaries of the resource through inclusion of users in resource surveys and other experiments, and rotating membership on shellfish conservation committees. Some towns require that each commercial licence holder contribute a specified amount of 'conservation hours' as part of the licence privilege (cf. Scarborough, 1993). But despite the mechanisms that minimize free riding, many towns do not develop shellfish conservation plans.

Competition

When marine resources become owned only after capture, exploitation is competitive. Two common forms of competition used by resource appropriators are *scramble competition* and *interference competition* (Hirschleifer, 1978). In scramble competition, each user behaves autonomously to capture resources in a race against other users; for example in the harvest

of large quantities of clams over as short a time period as possible. Free
access to a resource promotes this type of competition, known in fisheries
as 'the race for fish'. Scramble competition is promoted in Maine clam
management by the lack of direct controls on entry or production. It is at
the same time limited in the case of small towns where geographic isolation
places a premium on interactions between residents, and social sanctions
can be applied against behaviour considered destructive to the collective
good. In contrast, interference competition is based on strategies that interfere with users' abilities to compete; for example, manipulating a management process to create rules that work to one's advantage and others'
disadvantage. The system of municipal clam management excludes all but
citizens of coastal towns from resident use privileges and decision making.
Previously, coastal residency as the basis of ownership limited the scope for
interference competition to within the local management system, which,
because of the small number of appropriators in each town and the
participation of those appropriators in resource management plan
development, minimized opportunities for actions designed to manipulate
rules or access. Recently, the residency requirement has given rise to
between-town interference competition, as inland towns attempt to garner
political support for 'resident' access to the clam resource (Weegar, 1996),
and clam diggers from towns with depleted resources attempt access to flats
in other towns.

Feedbacks

Interactions between appropriators of soft shell clams and the ecosystem
that supports the fishery are importantly influenced by the feedbacks
promoted by the system of local rule making. In towns that have enacted
shellfish conservation plans, adaptation to biological variability is
enhanced through local oversight and on-site monitoring of feedback
signals from the environment. For example, early signals of increased
predation on soft shell clams by green crabs are registered by clam
harvesters finding empty shells on the flats. Quick response is possible
because decisions are made locally. Flexible local rule making allows
revision of management decisions that do not lead to the desired
outcome. Rules can be revised without the costly and time-consuming
co-ordination process that would ensue from a more hierarchical decision process. Rapid response and continual monitoring are possible
because action is local. As long as communities are willing to absorb the
cost of management responsibilities, small-scale management helps
develop controls that contribute to a resilient interface between

appropriators and the ecosystem, a necessary condition for sustainability (Berkes and Farvar, 1989; Folke and Kåberger, 1991).

Outcomes

Soft shell clam management within the overriding state constraints results in a wide range of management approaches, from sound management to little or no management beyond the basic state regulations. Management actions take place on a small scale, and are directly translated into outcomes. Coastal communities that have adopted shellfish conservation plans are addressing problems typical of common pool resources: maintaining exclusion and reconciling individual and social incentives.

The exclusion problem is solved by defining exclusion authority over the same geographic area as municipal authority. Management authority over the resource is established at well-defined municipality borders. Each community can decide whether to limit appropriation to residents or to allow harvest by non-residents. Communities may also restrict access to a subset of town members and are responsible for monitoring and enforcing access to the resource. Since towns are small, monitoring and enforcement proceed without reaching diseconomies of scale. Towns have an incentive to protect the productivity of the intertidal area. As Townsend (1985) notes, towns with exclusion provisions for soft shell clams have more output per unit of effort.

The problem of reconciling individual and group incentives is handled most directly by the state requirement that towns wishing to manage the soft shell clam resources within their borders form a Shellfish Conservation Committee (SCC). The composition of the committee varies by town (State of Maine, 1994).

The extent to which communities can maintain exclusion and reconcile individual and social incentives depends on a third problem characteristic of all natural resource management: holding transactions costs to reasonable levels. Although decentralized management for soft shell clams would appear to incur higher costs to the state than a single state-mandated approach, the state realizes certain cost efficiencies by a co-management process (Hanna, 1994; 1995). Allocating decision-making rights to individual communities lowers overall costs of management, most notably costs incurred for describing and monitoring the ecosystem, designing regulations, co-ordinating users and enforcing regulations. The investment in human capital needed to perform resource assessments requires both state and local resources, but from the state's perspective, however, the net cost of local efforts per quantity of information produced is lower than if Maine Department of Marine Resources scientists were responsible for assessing the entire coast. Once the investment in training is made, average costs of resource assessments decline over time.

Even more important to management outcome than the total costs of management is the distribution of those costs. Maine's system shifts many of the transactions costs of management to coastal communities through the assignment of management responsibilities. From the community's perspective, management costs are absorbed in expectation of better management and greater resource productivity. The extent to which communities are willing to invest in management depends critically on the relative magnitude of costs compared to the expected return.

From the community's perspective, there are several attributes of the management system that favour investment. There is greater confidence in the quality of data produced locally because its collection has been overseen by those with the greatest vested interest in the assessment outcome. The legitimacy of regulations is increased by the active participation of clam diggers in the management process, which in turn increases the probability of smooth implementation and lowers the costs of monitoring and enforcement (Jentoft, 1989; 1993). There are fewer surprises in management programmes that reflect the structure of the ecosystem, resulting in more enforceable regulations. Enforcement takes place at the level of face-to-face interaction characteristic of small towns. Uncertainty about collective actions is reduced because a stake in the outcome is created by including users at the design and implementation stages.

The strength of the co-management system for Maine clams derives from the small size and unique character of Maine coastal communities that precludes the areas in which co-management is traditionally vulnerable: the incomplete representation of interests, failure to craft participation around clearly specified rules, corruption of the process, and conflicts of interest. The user fragmentation which undermines management processes is uncommon in small, relatively homogeneous communities. Participation processes are small scale and have their basis in a strong tradition of participatory democracy. Expectations for outcome are clearly specified by state requirements. Corruption of the management process by strategic individual rent seeking is difficult to pursue in the face of close interactions between community members. Conflicts of interest in enforcement would be quickly revealed because monitoring is small scale and widespread.

Despite its strengths, Maine's system is also vulnerable to weaknesses associated with the distribution of many management costs to small communities. For some communities, for example those with a small population combined with a large number of environmental monitoring needs, or a small proportion of the population benefiting directly from the harvest of clams, the higher average costs of management can lead to an unwillingness to accept management responsibility. Data collection, plan development, monitoring and enforcement represent real costs of time and money that may exceed the expected benefits. The absence of a shellfish conservation plan in some towns may reflect an unwillingness to bear the costs of management, or a fear that free riders will erode the benefits of those who do.

There is evidence that in recent years management costs have been rising to the extent that many towns have considered the distribution of costs to be inequitable. A December 1995 announcement of a conference on improving municipal clam management in Maine describes municipal clam management as suffering overload, in particular with regard to the need to meet water quality control mandates and control harvesting effort (Anon, 1995; S. Hoyt, 1995, personal communication, Georges River Clam Project, University of Maine Cooperative Extension Service, 375 Main Street, Rockland, Maine).

The combination of strong markets and increased coastal populations means that these costs would exist regardless of the form of management; the fact that, under co-management, monitoring and effort control are the responsibilities of coastal communities has shifted those costs to the local level.

Conclusions

Maine soft shell clam management is a system in transition. It demonstrates the importance of the human and ecological context to resource production and to the structure and function of resource management. Clam production and clam management have been well adapted to the social, economic and ecological context of the coast of Maine, although there is evidence that increasing costs of biological monitoring and harvest control have thrown the system out of balance. Management that works through a nested system of rights and responsibilities is intended to provide checks and balances against overuse. It may need to be adjusted to realign the distribution of management costs, either through a change in local responsibilities or through the formation of larger management districts that encompass an entire intertidal ecosystem and more than one town.

A major factor influencing the behaviour of people who work in the marine environment is the uncertainty borne of biological variability and the lack of control over collective behaviour. The allocation of management authority over soft shell clams to the local level has addressed the uncertainty associated with collective behaviour, providing communities with the means to control collective actions. In addition, the regulatory regime creates wide access to an employment opportunity in an area where labour is relatively abundant. Management also meets a social goal of open access to all community members, although the effectiveness of that access is limited by the restrictions on harvest technology.

The approach to resolving biological uncertainty is a unique aspect of Maine's soft shell clam management. Knowledge of the ecosystem is generated through the integration of informal local knowledge and locally generated formal scientific information. Locating the production of scientific information at the site of resource harvest helps bridge the gap between knowledge needed to use the resource in the short term and knowledge needed to sustain the resource over the long term. The co-operative generation of scientific information offers the advantage of using a broad spectrum of experts to represent and explain the resource context. It is also proving to be, in some cases, a greater burden than individual communities are willing to bear.

Soft shell clam management has evolved over time in response to political, economic and ecological forces, and there are signs that such forces are playing a role in the current re-evaluation of municipal clam management. Political support to maintain and strengthen local property rights to coastal resources has for the most part remained strong along the coast, although it is now being challenged by inland residents as well as by some coastal residents.

For local-level management to succeed, the expected benefits to the community must exceed management costs. Under the Maine system, most of the costs of management are borne by those who will benefit from management, and the question is whether in certain areas management costs are exceeding the benefits, at least in the short term. A particularly interesting aspect of this resource and its management has been the external benefit to soft shell clam appropriators provided by federally mandated pollution mitigation efforts. In more developed areas in which contamination of the intertidal zone had previously made management problematic, mitigation efforts are lowering the cost of management to acceptable levels. The combination of improved ecosystem health and strong markets for soft shell clams is providing incentives for the expansion of local management into previously abandoned areas.

The institutional structure operant in this case provides constrained local autonomy for coastal communities to manage the soft shell clam resource. Together with local power is a high level of local responsibility, associated in recent years with increasing costs of monitoring and effort control. Some towns have embraced the responsibility; others face costs associated with the responsibility that exceed their expected benefits. The disparity of outcomes re-emphasizes the importance of context in designing institutions for resource management.

Acknowledgements

This work was partially sponsored by Grant No. NA 36RG0451 (Project No. R/PPA-39) from the National Oceanic and Atmospheric Administration to the Oregon State University Sea Grant College Program and by appropriations made by the Oregon State Legislature. The views expressed herein are those of the author and do not necessarily represent the views of NOAA or any of its subagencies. Parts of this research were sponsored by the Beijer International Institute of Ecological Economics, The Royal Swedish Academy of Sciences, Stockholm, Sweden, with support from the World Environment and Resources Program of the John D. and Catherine T. MacArthur Foundation and the Environment Division of the World Bank. The research was conducted under the research programme Property Rights and the Performance of Natural Resource Systems.

References

Anon 1970. *Maine Law Affecting Marine Resources, Vol. 2*. Portland: University of Maine School of Law.
Anon 1989. *The Maine Atlas and Gazetteer*, 14th edition. DeLorme Mapping Company, P.O. BOX 298, Freeport, Maine 04032.
Anon 1995. *Improving Municipal Clam Management in Maine*. Announcement of a conference held December 8, 1995, Rockport, Maine.
Berkes, F. and Farvar, M.T. 1989. Introduction and overview. In *Common Property Resources: Ecology and Community-Based Sustainable Development*, pp.1–17, ed. F. Berkes. London: Belhaven Press.
Bigelow, H.B. and Schroeder, W.C. 1953. *Fishes of the Gulf of Maine*. Fishery Bulletin of the Fish and Wildlife Service Volume 53. Washington, D.C: US Government Printing Office.
Crowder, W. 1931. *Seashore Life Between the Tides*. New York: Dover Publications, Inc.
Dow, R. 1972. Fluctuations in Gulf of Maine sea temperature and specific molluscan abundance, *Journal du Conseil International Pour l'Exploration de la Mer* 34(3): 532–4.

Dow, R. and Wallace, D. 1957. *The Maine Clam*. Augusta, Maine: Maine
    Department of Sea and Shore Fisheries.
Dow, R. and Wallace, D. 1961. *The Soft Shell Clam Industry of Maine*. Augusta,
    Maine: Maine Department of Sea and Shore Fisheries.
Folke, C. and Kåberger, T. 1991. Recent trends in linking the natural environment
    and the economy. In *Linking the Natural Environment and the Economy:
    Essays from the Eco-Eco Group*, pp.273–301, ed. C. Folke and T. Kåberger.
    Dordrecht: Kluwer Academic Publishers.
Glude, J.B. 1954. *Survival of Soft-Shell Clams, Mya Arenaria, Buried at Various
    Depths*. Research Bulletin No. 22. Augusta, Maine: Department of Sea and
    Shore Fisheries.
Gouldsboro, Town of 1993. *Town of Gouldsboro Annual Report*. Maine:
    Gouldsboro.
Hanna, S.S. 1994. Co-Management. In *Limiting Access to Marine Fisheries:
    Keeping the Focus on Conservation*, pp.233–42, ed. K.L. Gimbel. Washington,
    DC: Center for Marine Conservation and World Wildlife Fund.
Hanna, S.S. 1995. Efficiencies of user participation in natural resource
    management. In *Property Rights and the Environment: Social and Ecological
    Issues*, pp. 59–68, ed. S. Hanna and M. Munasinghe. Washington DC: World
    Bank.
Hanna Dearborn, S.S. 1977. *The Soft Shell Clam Industry of Maine: Its
    Institutions of Regulation and Management*. Unpublished MS Thesis,
    University of Maine, Orono, Maine.
Hirschleifer, J. 1978. Competition, cooperation and conflict in economics and
    biology. *American Economic Review* 68(2): 238–43.
Islesboro, Town of 1993. *1992 Annual Report*. Maine: Islesboro.
Jentoft, S. 1989. Fisheries co-management: delegating responsibility to
    fishermen's organizations, *Marine Policy* 13(2): 137–54.
Jentoft, S. 1993. *Dangling Lines*. St. John's, Newfoundland: Institute of Social and
    Economic Research, Memorial University of Newfoundland.
Jonesboro, Town of 1993. *Annual Report of the Municipal Officers of the Town of
    Jonesboro for the Year 1992–1993*. Maine: Jonesboro.
Judd, R.W. 1992. Searching for the roots of the conservation movement: fish
    protection in New England 1865–1900. In *Transactions of the 57th North
    American Wildlife and Natural Resources Conference*, pp.717–23, ed. R.E.
    McCabe. Washington DC: Wildlife Management Institute.
Maine Department of Fisheries and Game 1892. *Biennial Report of the
    Commissioner of Sea and Shore Fisheries*. Augusta, Maine: Kennebec Journal
    Print.
Maine Department of Marine Resources 1994. *Maine Marine Resources Laws*.
    Augusta, Maine: State of Maine, Office of the Governor.
Maine Department of Sea and Shore Fisheries 1906. *Report of the Commissioner
    of Sea and Shore Fisheries of the State of Maine*. Augusta, Maine: Kennebec
    Journal Print.
Maine Department of Sea and Shore Fisheries 1918. *First Biennial Report*.
    Auburn, Maine: Merrill and Webber Company.
Maine Department of Sea and Shore Fisheries 1954. *State of Maine 18th
    Biennial Report*. Augusta, Maine: Maine Department of Sea and Shore
    Fisheries.
McCay, B.J. 1989. The culture of the commoners: historical observations on old
    and new world fisheries. In *The Question of the Commons*, pp.195–216, eds.
    B.J. McCay and J.M. Acheson. Tuscon: The University of Arizona Press.

McCay, B.J. 1993. The making of an environmental doctrine: public trust and American shellfishermen. In *Environmentalism: The View from Anthropology*, pp.85–96, ed. K. Milton London: Routledge.

Muse, B. III. 1976. *The Shellfisheries of Massachusetts: the Political Economy of a Common-Property Resource*. Unpublished MS Thesis, University of Massachusetts, Amherst, Massachusetts.

Scarborough, Town of 1993. *Annual Report of the Town of Scarborough 1992*. Maine: Scarborough.

State of Maine. 1821. *Maine Public Law*. Chapter 179, Section 3.

State of Maine. 1994. *Maine Marine Resources Laws*. Augusta, Maine: State of Maine Office of the Governor, Maine Department of Marine Resources.

Teal, J., Teal, M. and Fish. R. 1969. *Life and Death of the Salt Marsh*. Boston: Little Brown.

Townsend, R.E. 1985. An economic evaluation of restricted entry in Maine's soft-shell clam industry. *North American Journal of Fisheries Management* 5: 57–64.

Waldoboro, Town of. 1993. *1992 Annual Report*. Maine: Waldoboro.

Weegar, A.K. 1996. Class wants clam flat license equity, *Maine Times*, 29 February.

Part III
Success and failure in regional systems

Introduction

Community-based studies are important for the investigation of local social adaptations in the context of the ecosystems in which they are located. Most of the chapters in Parts I and II fall into that category. By contrast, the chapters in Part III are based on distinct ecological regions and deal with social–ecological linkages peculiar to the region or resource area. They have another characteristic in common: they all deal with embedded systems, emphasizing the point that local social systems are not isolated but are subject to national and regional-level disturbances. The chapters consider Mexico's forest ecosystems, the semi-arid ecosystem of the African Sahel, the mountain ecosystems of Hindukush–Himalayas, and the Northwest Atlantic. Two of the chapters are about agriculture and agroforestry systems, one is about rangelands and pastoral peoples, and one is about fishing.

Chapter 9, by Alcorn and Toledo, assesses the hypothesis that national support for community-based property rights systems enables locally adapted systems to persist, and communities to participate in renewal cycles within larger national and global systems. Cast in the idiom of renewal cycles, the authors show how ecological renewal cycles have been affected by their linkage to social renewal cycles. The chapter explores the property rights and resource management systems of communities in two of Mexico's forest ecosystems – the lowland humid tropical forest and the sub-humid temperate forest. It deals with Mexico's recognition of community-based property rights, and its effect on traditional resource management systems in these two forest ecosystems where renewal cycles have depended on social organizational strength, cultural support, and local knowledge systems. The Mexican revolution and radical changes in forest concession policy are evidence for a renewal cycle operating in Mexican society. Both events can be interpreted as having helped avert a pending 'flip' to altered ecological and social systems. With recent globalization policies (NAFTA reforms), Mexican society once again stands at the edge of a creative-destruction phase that could either lead to renewal, or trigger a 'flip' into a different state.

Chapter 10, by Niamir-Fuller, asks the questions: how do pastoralists manage ecosystem variability? What are the social and institutional mechanisms by which pastoralists use their semi-arid environment without causing a deterioration of ecosystem functions? And can these mechanisms still be used today? Pastoral populations of the Sahel have faced a number of challenges in recent decades, including the privatization of the grazing commons, encroachment of cultivation, degradation of fallow lands, and the breakdown of traditional institutions for land management, in addition to recurrent droughts. Niamir-Fuller argues that present populations of livestock cannot be maintained by settled, intensive systems by the majority of pastoral households. Extensive production systems, relying on large tracts of common land, are the only short- and medium-term strategies. This requires a reversal of both 'benign neglect' policies and those biased against pastoralists. It also requires identifying and understanding how traditional practices of pastoralists, such as the rotation of grazing areas and the use of range reserves to provide a 'savings bank' of forage, help manage the variability of the Sahelian ecosystem, and keep it from 'flipping' into a state of permanent degradation.

Chapter 11, by Jodha, presents a synthesis of over 50 studies from several countries in the Hindukush–Himalaya region. It focuses on natural resource-friendly traditional systems of resource management, their decline, and possible approaches for their restoration. In mountain areas, the circumstances created by relatively high degrees of inaccessibility, fragility and diversity compelled communities to adapt their requirements and resource-use systems to the limitations of the resource base, generating positive social system–ecosystem links. Jodha observes that, with increasing administrative and market integration and rapid population growth over the years, these linkages have weakened or disappeared. Based on a closer look at the above processes, the chapter advocates a search for possible present-day functional substitutes for the various elements of traditional systems, in order to restore positive social system–ecosystem links. These include restoring the stake of the community in the health and productivity of its natural resources base; enhancing the sensitivity of decision-makers and resource users to the characteristics of mountain resources; and restoring community involvement in local resource management through decentralized and participatory measures.

In Chapter 12, Finlayson and McCay analyse the collapse of the cod fishery of Newfoundland in the early 1990s. They show that the collapse did not occur because of lack of regulations or excessive numbers of fishers, but because of the introduction of offshore fishing technologies in the context of weak international policies. The chapter argues that Canada's own policies to encourage offshore fisheries, and the establishment of a powerful centralized scientific authority, actually contributed to the demise of the fishery. The separation of science from management, the centrality of science in the process of determining the total allowable catch, an offshore bias that led to overly optimistic estimates of total allowable catch, and uncritical reliance on mathematical models in a complex and variable environment were, in the view of the authors, some of the major factors that lead to the disaster. The inshore fishers, who possess many generations of accumulated knowledge, were marginalized, and their warnings were ignored: 'the day-to-day operational reality of the inshore fishing community increasingly diverged from that of the science-based construction of reality'. The authors see the collapse of cod and of fishing communities as a system 'flip', and explore the question of why there is so little evidence of a similar 'flip' in resource management institutions and thinking.

9

Resilient resource management in Mexico's forest ecosystems: the contribution of property rights

JANIS B. ALCORN & VICTOR M. TOLEDO

Introduction

Holling and Sanderson (1996) have offered a theoretical model for a renewal cycle that describes the dynamic nature of natural and social systems based on Holling's earlier ecosystem model (Holling, 1992). The sustainability of larger-scale systems depends on renewal cycles in the dynamic, local systems that maintain the fine-scale variability arising from adaptations to local environmental and social conditions. In this chapter, we assess the hypothesis that strong, national policy support for community-based property rights systems plays a critical role in enabling locally adapted systems to persist and participate in local renewal cycles within larger national, regional and global systems.

Property rights systems establish relationships between people, not relationships between people and things as is often assumed (Crocombe, 1971; Macpherson, 1978). Property is 'a political relation between persons' (Macpherson, 1978: 4); it is 'an enforceable claim to some use or benefit' (Macpherson, 1978: 6) held by some persons and not by other persons. Property rights systems do not simply define and grant rights; rather, they establish the rights and responsibilities of system participants vis-a-vis each other. Property rights systems are part of communities' law, and as such are core elements that differentiate communities – law being 'the central art by which community and culture are established, maintained and transformed' (White, 1985: 684). When states sanction community-based tenurial systems, they are devolving political power and recognizing the rights of culturally diverse communities to co-exist within their borders. While community-based property rights systems exist within the borders of many nation states, they are seldom recognized as legitimate. On the contrary, local property rights systems are usually ignored or formally extinguished so that the state can claim the lands or declare them open for claim by entrepreneurs, often to generate revenues for the state-linked elites.

In this chapter, we focus on Mexico, an unusual example of a country which has tested a mixture of private individual and community-based property rights within a modern capitalist environment for over 70 years. The Mexican mix includes: corporate community-based land-holdings (66.3% of the production units and covering 59% of the land area of Mexico); private individual holdings (comprised of 30.8% of the production units and covering 40.9% of the land area); and mixed systems (including 2.9% of the production units and covering 0.1% of the land area) (National Census, 1990). Mexico has also supported the world's largest experiment with community-based forestry for the past 20 years (Bray 1995). The Mexican Revolution that returned lands to traditional community-based management (see below) and the more recent revolution in forest concession policy could be viewed as evidence for a renewal cycle operating in Mexican society – both events were radical changes that averted a pending 'flip' to altered ecological and social cycles. These larger-scale cycles have enabled the local-scale cycles nested within them to continue; at the same time, the resilience of the local-scale cycles probably contributed to the renewal phase of the larger-scale cycles.

Different systems of property rights over land are usually compared by studying annual agricultural production trends in relation to progress in formal titling to individuals. However, this approach ignores ecological impacts (e.g. Porter, Allen and Thompson, 1991). In this chapter, property rights over land are assessed in terms of their effect on the resilience of ecosystems (i.e. land, forest, water, and other associated resources that produce more than annual crop yields).

We argue that although sustainable resource management was not the objective of Mexico's land reform legislation enacted some 80 years ago,[1] Mexican recognition of community-based tenure has enabled locally adapted agro-ecosystems to evolve in the face of changes.[2] The Mexican experience offers lessons for other countries. Group property rights are not legally recognized in most countries, although vestiges of pre-existing, customary property rights systems persist in many bio-diverse areas. Legal support for community-based, corporate tenure is an attractive policy option where indigenous peoples and other rural communities continue to use locally adapted resource management systems.

Property rights 'shells'

The impact of property rights systems is much more far-reaching and mul-
tifaceted than is usually appreciated. Property rights systems provide a
basic structure from which spring opportunities and avenues for resource
extraction and protection. Property rights systems vary within a nation
state and between nation states, and their local implementation is affected
by local patterns of decision making within the context established by sub-
regional and national environments.

Property rights systems create 'shells' in the computer sense, in that they
create a structured interface between the inner environment (where a given
system operates) and the outer operating system (with which the inner
system must interact, but which may either support or conflict with the
inner system). A shell functions as a constraining and enabling structure for
the inner environment by linking it in very specific ways to the larger 'operating system' in which the shell is embedded. Specific shell interfaces
respond to cultural, ecological, economic and social factors – both internal
and those arising from externally generated stresses or opportunities.
Transactions across the boundary are limited by the structure of the shell.

Property rights shells are often nested within a hierarchy of shells, each
outermost shell shaping the operating environment inside the next level of
shells. Broad national policy on property rights interfaces with the various
property rights systems present within a nation's borders. National support
for local tenurial systems strengthens the shells around them, while policy
that denies their existence *de facto* creates a more labile shell.[3] The unrecog-
nized shell continues to exist as long as local systems continue to operate, but
it is weakened when individuals bypass it and make local land-use decisions
that are unacceptable under customary law but supported by national law.

In ecological modelling terminology, different local tenurial systems
create discontinuities at the local scale, and in turn contribute to processes
operating at other scales (such as small-scale household-level decisions and
larger-scale landscape-level processes). Tenurial shells are invisible edges
between systems that are often also recognizable on the basis of other
characteristics, such as different political and cultural systems, and
different values that underlie land-use decisions. Most analyses have focused
on fuzzy cultural differences between systems and have failed to focus on
specific types of culture-based decisions with critical impacts, such as
assessing the functional role of different property rights systems in maintaining the functioning of local systems and, in turn, the functioning of
larger-scale systems. Local shells and tenurial systems are invisible to those

who don't participate in local political activity or directly manage local resources. For this reason, few national natural resource managers, econo-mists or ecologists have recognized or assessed the role of tenurial systems in ecological sustainability.

Tenurial shells are created at the interface between competing social and political systems and their associated institutions. In pre-modern times, shells were formed between local communities. The borders and shells changed with trade and war. Remnants of this type of situation can be found in the South Pacific today, where neighbouring communities with different prop-erty rights systems co-exist. In times of stress, the political units inside different local shells have forged organizational links between themselves to resist the destruction of their shells and their property rights. These resistant organizational links take the form of military or political alliances. During colonial and neo-colonial times, the shells around many local communities were disrupted as states refused to recognize pre-existing property rights, and the more common interface became isolated local shells abutting state-based property regimes (Alcorn, 1995; Alcorn and Molnar, 1996). Tenurial shells are created within nations by the state's recognition of local systems; these local systems function within a particular national framework created by that nation *vis-a-vis* other nations. The historical trend has been increasing loss of shells around local tenurial systems and the locally based resource manage-ment systems they support. Yet, communities inside local shells continue to forge organizational links with national support organizations and networks in order to resist legal and illegal efforts to dismantle shells.

States can create or recognize many types of tenurial shells that affect ecological resilience. In this chapter, we specifically focus on the ecological impact of indigenous, community-based tenurial systems embedded within a nation state. Communities interact with outside economies, markets and organizations through a shell interface that facilitates selected interactions across the shell while blocking other types of interactions. Hence, these tenurial shells function as protective borders around subsystems that could not remain viable if fully exposed to the outer environment in which they are embedded.[4]

Each community-based tenurial shell is constructed of linkages into institutions that pervade the lives of community members. The term 'insti-tution' is used here to mean the invisible bodies of rules, regulations, and processes that guide decision-making (Ostrom, 1990; Ostrom, Walker and Gardner, 1992). Such decision making is often carried out within organiza-tional structures – organizations being groups of people acting in relation-ships governed by and legitimized by institutions.[5] Local institutions include rules about use and acceptable distribution of benefits, means by which tenure is determined, and conflict-resolution mechanisms (Berkes *et al.*, 1989; Bromley and Cernea, 1989).

In healthy local subsystems, local feedback loops can lead to recognition of overexploitation of a resource and failure of ecosystem functions.[6] In response to feedback and tensions between individuals seeking access to resources, local institutions have arisen to ensure community members' continued access to resources while restricting access by outsiders, as well as to manage the differentiated access rights of insiders. These institutions result from a political process of trade-offs between members of a community who must work together because of their interdependence in many other spheres. The institutions may be old, but they are dynamic and respond to change according to interpretations of local law as it is applied in specific situations.[7]

While local systems are often referred to as common-property systems to contrast them with the predominant system of individual private property freely exchanged in the market, the term common property is a misleading label. Under these local systems, the community, through the authority of its own institutions, defines and allocates individual and group rights to particular resources within the lands held by the community. These traditional corporate systems derive strength from a cultural and social integrity which, on the one hand, reinforces a unified approach to management decisions and yet, on the other hand, offers individual households the freedom to benefit from differential, individual access to specific resources held within the community.[8] All individuals in a given community do not share equal rights over all the resources held by that community. Within community-based systems, agricultural lands are often held by individual families who pass their rights to their descendants through locally shared rules. Our survey of the literature from many regions of the world indicates that agricultural lands are often held privately in community-based tenurial systems, while rights to forest or pasture lands are more likely to overlap (although private forests and private pastures often co-exist alongside those which are shared by all). Hence, community lands are better described as private property belonging to a specific group of individuals who allocate rights among themselves (Lynch and Alcorn, 1994).

Ejidos and *comunidades*: Mexican shells

The Mexican state formally created *ejidos* and *comunidades* as tenurial shells for communities after the Mexican Revolution, which was born, fought and won on the demand for the return and redistribution of land to peasant communities (Sanderson, 1984). The 1917 Constitution supported land reform and recognized community ownership of land under Article 27.

Community-based tenurial systems in Mexico are similar to those elsewhere in the world. Tenurial rights and responsibilities are defined by local communities within the basic framework established by the state. We refer to these systems as community based because the primary legitimacy of community-based tenure systems is drawn from the community and not from the nation state which recognizes them (Lynch and Alcorn, 1994). In other words, the local community, not the national government, is the primary allocator and enforcer of rights to resources within the boundaries of the community. Responsibilities are defined by the community, and the role of the national government is to defend a community's rights to its resources against the claims of those who are not community members. At the same time, however, in the Mexican case, the state retains ultimate rights over the resources and places restrictions on rights to sell, lease or rent community properties.

Unlike Mexico, most countries[9] do not recognize the traditional tenurial shells created by communities that have long lived among, named, managed and used the resources of their homelands. Instead, the state claims those lands and resources as state property; and those who dwell on and use those lands are labelled as squatters, poachers or other criminal epithets. In these and many other less subtle ways, traditional systems are often weakened by lack of state support. The community's traditional resource management systems and related institutions are slowly undermined by changes made in response to new rules and the confusion created by new resource users whose access was sanctioned by the state but not by the community.

Unsustainable resource use often increases as the old system is weakened and replaced by a new property rights system. The new system is often an aberrant version of the legally specified system as it is interpreted and locally implemented by the politically powerful, including the military. Deforestation is a frequent outcome. The resilience of the forest ecosystem is insufficient to absorb the new disturbance to which it is subjected, and the system flips to a less diverse and less stable system.

In remote areas, on the other hand, traditional tenurial systems often continue to operate and evolve in response to market and cultural changes without legitimation by the government. Alternatively, communities in remote areas may have rights that have been legitimized by the government but be uninformed of their rights and therefore fail to seek state assistance to fend off illegal extraction of their resources (e.g. Cortez Ruiz, 1992).

In Mexico, two forms of community-based corporate ownership are currently recognized and supported by law: *ejidos* and *comunidades*. The *ejido* is a creation of the Mexican revolution that enables groups of people to petition for access to resources to which they have no prior claim. The *comunidad*, on the other hand, is defined as a pre-existing corporate entity whose rights are recognized if its members can demonstrate prior, long-standing, community-based use of the land and waters. The stated objective of legally establishing the post-Revolution *comunidad* was to return to the corporate tenurial system originally recognized by Spanish colonial administrators and based on similar European traditions of corporate land use (Sanderson, 1984; Sheridan, 1988).

During the Porfirian period (1876–1910) prior to the Revolution, the state withdrew its historical support for the communities' tenurial shells. Federal laws eliminated communal property rights that had been sustained from the 1500s when the Spanish colonized the region and recognized local community rights. The state claimed as state property all lands without official titles (Stresser-Péan, 1967; Sanderson, 1984; Barthas, 1994). It in turn gave rights to those same lands to capitalists and owners of *haciendas*, leaving communities to depend on wages for their survival. The impact of these policies varied in different regions of Mexico, but nationwide, by 1910, nearly half of the rural population had become debt peons on *haciendas* and *ranchos*, 82% of all communities were located on *haciendas* and *ranchos*, and free agricultural villages held very little land (Sanderson, 1984:16–18). As a result of the export-oriented policies, prices for food rose significantly, while profits from growth in the export sector primarily accrued to foreign investors. Wages remained low, 'verging on slavery' in some areas (Sanderson 1984). These conditions gave rise to the Mexican Revolution.

Under the post-revolution land reforms, despite the legally specified difference between *comunidades* and *ejidos*, most pre-existing communities were not recognized as *comunidades* on the basis of documented prior claims but were instead granted rights as *ejidos*. In practice, the distinction between communities called *ejidos* and those called *comunidades* is seldom significant; indigenous and mestizo[10] communities are found in both *ejidos* and *comunidades*. As of 1995, all *ejidos* and *comunidades* are functioning as longstanding communities with prior rights.

Under both *ejido* and *comunidad* shells, each household in the community has the right to exploit the community's natural resources necessary for livelihood. The household is, in effect, a user–manager of a set of resources that belongs to everyone in the community. At the same time, the shell

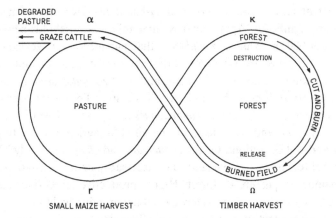

Figure 9.1.   The *milpa* cycle as disrupted prior to the Mexican Revolution, when traditional tenure was lost. The forest renewal cycle was flipped into degraded pasture.

limits transactions with society outside the shell – the household cannot sell or rent community lands to anyone outside the community (but see 1992 revisions below). This prevents the kind of loss of lands to those outside the community that occurred prior to the establishment of *ejido* and *comunidad* shells when families lost their lands to ranchers when they purchased needed goods on credit at ranchers' stores and then the ranchers seized their land for payment. Piece by piece, communities were dismantled, and slowly the forest changed into pasture (Figure 9.1). The *ejido* and *comunidad* shells stopped that trend and prevented communities from being dismantled through the loss of their lands.

Inheritance and membership in *ejidos* and *comunidades* are regulated by communities. Resources are allocated to members of the community who exploit and manage these resources on an individual basis within the limits set by the community. Communities are heterogeneous, dynamic entities containing subunits that form shifting alliances within shared institutions and guided by shared ethics. Their tenurial shells create protective crucibles within which local conflicts and differing strategies can bubble together without being destabilized by external factors. Land disputes within communities are common, but they are generally resolved at the community level and do not become a burden for state agencies (DeWalt and Rees, 1994). Land disputes between communities and ranchers[11] are also common, particularly in forested areas (e.g. Sandoval, 1994), and the state apparatus offers the only recourse for justice in situations where ranchers have enormous political influence and sometimes their own private armies.

The extent and impact of community-based resource management in Mexico are significant. Approximately three million households belong to the nearly 30000 *ejidos* and *comunidades*[12] that manage 59% of Mexico's land area (103 million hectares) and 66% of the total rural production units. Most indigenous communities operate *ejidos*, and mestizo *ejidos* often retain the pre-Hispanic traditions of their indigenous ancestors. For these reasons, it is appropriate to assess *ejidos* and *comunidades* as a group. Most of the land operated by *ejidos* and *comunidades* is marginal for agriculture due to poor climatic and soil conditions. Of the *comunidad* and *ejido* lands, only 22% is agricultural (defined as arable lands, not necessarily under cultivation) and the remainder is pasture or forest. Highly productive lands (particularly those that are irrigated) are privately owned under individual title.

Within the protective and enabling shells created by *ejidos* and *comunidades* in Mexico, communities apply an incredible range of innovative, sustainable, locally adapted natural resource management systems in a wide variety of ecosystems, ranging from desert to rainforest (e.g. Zizumbo Villarreal and Colunga García-Marín, 1982; Toledo and Barrera-Bassols, 1984; Toledo *et al.*, 1985; Wilken, 1987; Gómez-Pompa and Kaus 1990; Mora López and Medellín-Morales, 1992; Nahmad, González and Vásquez, 1994). Indigenous peoples live within the borders of 80% of Mexico's protected areas, an indication of the level of biodiversity maintained by their land-use patterns. In recognition of this situation, in 1992, the Mexican government announced a new special shell that was available to communities – the *campesino* (peasant) ecological reserve (Grupo para la Conservación del Trópico en México, 1992). Under this shell, communities retain their authority to plan and implement activities, but in return for their conservation activities they receive additional protections above those offered to *ejidos* and *comunidades*. To receive this new designation, communities are required to develop their own management plans for conservation and sustainable development. In turn, they are protected from environmentally damaging projects, and they are eligible for special programmes that offer technical advice from universities and non-governmental organizations. An archipelago of communities linked as a network of *campesino* ecological reserves could effectively cover Mexico's biodiversity (Toledo, 1992b; 1994b).

Forest management inside the shells

Every ecological zone in Mexico supports rich reserves of biodiversity, but the forested areas are especially rich (Ramamoorthy *et al.*, 1993). Between

70% and 80% of Mexico's forests is under management by some 7000 to 9000 *ejidos* and *comunidades* (A. Molnar, personal communication, World Bank).[13] From a cultural perspective, it is also noteworthy that 4.8 million indigenous people[14] reside in *ejidos* and *comunidades* in forested areas (National Census, 1990).

We explore community-based property rights and resource management by long-established communities in two of Mexico's forest ecosystems – the lowland humid tropical forest and the sub-humid temperate forest – some 15 million hectares of which remains under the management of *comunidades* and *ejidos* (National Census, 1990). Over the past few decades, ranchers[15] have converted 20 million hectares of lowland and temperate forests into pasture (Toledo, 1992a), and they continue to press on the edges of forested *ejidos* (e.g. Sandoval, 1994), putting external stress on their tenurial shells and their forests. Given the importance of these two forest zones and the high percentage of indigenous communities living in them it is appropriate to select case-study examples from indigenous communities within them.

Patterns from tropical forests around the world suggest that the key elements of sustainable management strategies adapted to the dynamics of tropical forest ecosystems are: (1) patchy disturbance;[16] (2) controls over placement of disturbance; and (3) active development of crops and crop varieties adapted to the local agro-ecosystem patches within the forest matrix. These elements are found in indigenous resource management systems. Patchy disturbance (spatially and/or temporally) creates patches of different types of habitat. For ecosystem integrity to be maintained by patchy disturbance, patches must include undisturbed areas of sufficient size and coverage, and appropriate distribution and composition to: (a) ensure regeneration of the species and the communities; and (b) maintain ecosystem services essential for habitat maintenance of the entire matrix and adjacent ecosystems. The management system's ability to create and maintain patches that meet these criteria is challenged by changes in human population density, political organization, market values, in-migration, intensity of resource extraction, economic opportunities, and other non-biological factors.

Although most case studies contain insufficient information to ascertain the key factors that determine whether a society can create or adapt a resource management system to meet the ecological criteria for forest maintenance, existing evidence suggests that several things need to be shared within the user group, including: (1) cultural values, traditions, and socio-political organizations; (2) controls and incentives; (3) accurate ecological knowledge; and (4) attention to monitoring for negative changes.

Strong cultural traditions, social organizations and institutions have evolved in many forest-dwelling societies which have experienced forest loss and then reacted in an effort to manage or reverse the change. Values and institutions (such as *milpa*, described below) evolved to support agricultural management systems adapted to the tropical forest ecosystem's limits. Controls (including tenurial rights and responsibilities defined by a community) and incentives to encourage community members to respond to evidence that forest is being damaged are also important.

Finally, monitoring to recognize that the forest is being harmed or helped by certain changes may be a group or individual activity, but it must be linked through a feedback mechanism into an institution that can bring the community together to wrestle with a problem if it is detected. If loss of forest is not perceived or not recognized as a problem, then no conscious choice is made to keep or lose the forest, and the process proceeds as an accident. It is also possible that the community may perceive that forest is being lost, but opt for the new land use. If loss is perceived and recognized as a problem, then choices can be made to hold losses to an acceptable level or arrest/reverse the process. However, if loss is perceived, but there is no strong organization to support open discussions, resolve conflicts between different points of view, and enforce restrictions to stop the loss, then forest loss will continue. The successful implementation of community-wide decisions depends on the society's shared values, appropriate organizations, and political power *vis-a-vis* outsiders (or powerful insiders) who may be causing the changes.

The specific cases from Mexico summarized below include: (1) strong tenurial rights held by individual families within a strong tenurial shell supported by the state; (2) some resources under communally shared tenure; (3) evidence that potential negative impacts of land use options are considered in making choices; and (4) shared cultural values, institutions and organizations developed over centuries of changes *in situ*.

The in-situ changes have included shifts from subsistence production to involvement with cash crop production, resistance to outside efforts to eradicate their cultural traditions, and efforts to counter increasing marginalization within the political economy. The specific resource management practices in the two ecoregions differ. In both zones, individual households and community-wide land-use patterns are constructed from core elements (Figure 9.2) that include forest, fallow cycled fields, corridors of wild vegetation within agricultural areas, water bodies, house gardens, and permanent fields including plantations and pastures.

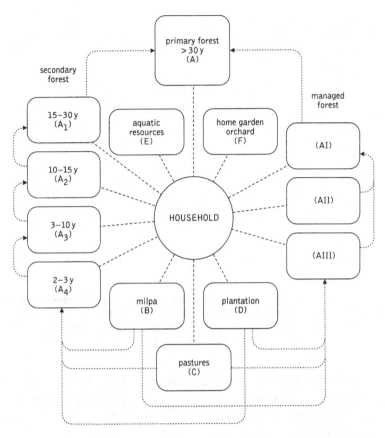

Figure 9.2. The core subunits created and used by tropical humid lowland indigenous communities in Mexico.

The specific type of tenurial rights within a community are probably less important for ecological success than are the legitimacy of the tenurial shell and the strength of the institutions which reinforce tenurial responsibilities and provide the capacity to take action on the basis of feedback from monitoring. In both cases, community institutions (both state-imposed institutions and traditional cultural institutions) influence property rights interpretation and resource management. The local institutions created by the state to regulate activities on *comunidad* and *ejido* lands in accordance with state law include the General Assembly, to which all households are represented by one person, and two important elected three-person committees: the *comisariado* (which represents the community to outside authorities and settles land disputes) and the *consejo de vigilancia* (which monitors the activities of the first committee). Community decisions are

Figure 9.3.   Geographic location of Huastec, Totonac and Purepechan lands.
Huastec and Totonac primarily live in tropical humid forest in the Mexican states
of San Luis Potosi, Hidalgo, Veracruz, and Puebla. Purepechans primarily live in
temperate subhumid forests in Michoacan.

made in General Assembly meetings or special meetings by majority vote;
representatives of all households must attend these meetings or be fined. To
varying degrees, elders and traditional leaders influence the functioning of
these institutions.

    In addition, religious institutions also reinforce community cohesion in
both indigenous and mestizo rural communities across Mexico. Ritual
obligations, rights to community resources, and obligations to manage
those resources are linked. Families affirm their commitment to the community and their relationship to other nearby communities by participat-
ing in a hierarchical *cargo* system involving the care and worship of
particular community saints and the sponsorship of community festivals
that form an annual cycle. These systems and their associated organizations
are Catholic but build on pre-Hispanic traditions (e.g. Vogt, 1976; Hunt,
1977) González Rodrigo, 1993.

Case one: lowland tropical moist forest

Twenty-two indigenous groups (population: 1.56 million) operate *ejidos*
and *comunidades* in the tropical humid zones of Mexico (National Census,

1990). The case-study site is a typical example of indigenous communities occupying tropical humid forest. It is located in northeastern Mexico, on the Gulf Coastal slopes of the Sierra Madre Oriental, in the states of San Luis Potosi and Veracruz (Figure 9.3), where rainforests reach their northernmost range in the Americas (Rzedowski, 1978). Prior to the arrival of the Spanish, the area was occupied for thousands of years and supported complex civilizations. From the time of the earliest written documents, the tropical moist forest region has been characterized as a 'hell' or a 'paradise', depending on the viewer (e.g. Vetancourt, 1689; Tapia Centeno, 1960). If this ecosystem is managed properly, it is a paradise, because it provides a wealth of short-term and long-term benefits. If its special resources, created under hot, humid conditions, are misused for short-term extractive gains or if conversion is attempted, then this ecosystem degenerates into a less valuable ecosystem requiring external inputs to maintain production.

We focus on two indigenous groups located in *ejidos* and *communidades* in contiguous areas (southeastern San Luis Potosi and northern Veracruz) of this ecosystem who use similar resource management systems: the Huastec Maya (population 121 000; 1990 census Table 10, Cuadro 8) and their southern linguistically unrelated neighbours, the Totonac (population 208 000; 1990 census Table 10, Cuadro 8). Totonac and Huastec both retain their respective languages and strong cultural traditions, but at the same time have participated in economies linked to the global economy for several centuries.

Economic differences do exist between families, but only a few families in any given community hold significantly greater resources than the rest. Huastec and Totonac communities occupy *comunidad* and *ejido* lands where population densities average around 100 persons per square kilometre. The communities vary in size from 500 to several thousand hectares, and contain no significantly distinct subgroups[17] of resource users.

Huastec and Totonac communities are spatially distributed as distinct islands in a sea of lands operated by a different group of resource users, the mestizos. Mestizos' political power and domination of the economy influence the technical and organizational options available to indigenous resource users. Mestizos occupy towns, ranches and citrus/sugarcane plantations in the more level lands and areas along roadways, while the islands of indigenous territories tend to be aggregates of communities grouped on steeper, less desirable agricultural lands. There is continued tension over borders between mestizo and indigenous lands. Occasionally, powerful mestizos still assert their rights over these resources, without any legal basis (e.g. Briseño Guerrero, 1994).

The land-use patterns of the indigenous people and the mestizos who own private lands are quite different. Mestizo land use outside *ejidos* generally tends to follow the standard Eurocentric model of monocrops and pastures with intensive herbicide and pesticide use. Mestizos dedicate most of their lands to cattle. Their pastures are largely degraded and unproductive; soil on slopes is eroded and vegetation consists primarily of unpalatable species. This general pattern has been in place for several hundred years (Barthas, 1994; Aguilar-Robledo, 1994).

Property within the borders of the *comunidad* or *ejido* is recognized, used and inherited according to local institutions. Almost all forested land is under family ownership. Small patches of communally shared forest generate income to pay school expenses and maintain other community buildings, and provide subsistence goods for the poor.

A community-elected official adjudicates over land disputes and inheritance decisions in consultation with other community members. Community members understand the state's legal apparatus establishing *ejidos* and *comunidades* as an extension of traditional institutions that control human behaviour in order to protect the community and the land and resources for which the community is collectively responsible. Mesoamerican cultural concepts of ownership extend beyond the usual Western legal considerations. The real owners of the land and forest are divine beings and spirits (including ancestors). Another way of expressing this relationship is that the Earth (with its resources) is a member of the community, and the community has the obligation to treat the Earth and all other community members with respect and concern for their continued well-being (Briseño Guerrero, 1994). In other words, ownership means that the human community has a moral responsibility to maintain the land, its resources, and society in good condition. Hence, despite the apparent clearcut borders between Huastec families' lands, members of one family have the right to ask another family to borrow land or harvest forest products to meet their subsistence needs. This system provides a social safety net for the poorer members of the community.

Disputes over land borders and harvest rights are common and can disrupt congenial relationships between families within a community. Accusations of witchcraft are made against those who attempt to appropriate others' resources for private gain. A belief in witchcraft provides a strong social sanction against actions that go against conservative use of resources and a commitment to the corporate group. Traditional curers reinforce socially appropriate behaviour during their interactions with patients, looking for illness causes in the patient's or others' misuse of resources. Here the importance of the relationship between the divine powers and the land comes into play, as well as the relationship between people. Clearing a private forest along a community watershed, for example, would result in strong pressure (including witchcraft accusations) against the family as well as being interpreted by the curer as causing illness or misfortune because the person went against religious sanctions about protecting water (ecologically unwise). Hence, ecologically sound land use is supported by cultural values and belief in the ethical commitments made between people and spiritual powers when people make land-use decisions. The tenurial shell created by the state supports the traditional belief structure, which in turn supports ecologically sustainable land use.

The effects of these moral commitments and beliefs are visible in the stark contrast between land use on either side of the border where indigenous *ejidos/comunidades* abut mestizo lands. The tenurial shell that reinforces community and cultural values is physically visible at the border. Standing at the border where degraded pasture of large private holders meets the community's patchwork of *milpa* and forest, people tell stories of how their way of life and forests were threatened before the revolution, and how they were unable to reclaim parts of their territory (now outside the border). They say that the revolution was terrible, but they acknowledge that saved their forests and their way of life. Without the *ejido* and *comunidad* arising from the revolution, there would be no borders and no islands, only a sea of pasture.

Both Huastec and Totonac apply a high level of knowledge about species and ecosystems (Alcorn, 1984; 1989a; Barrera-Bassols, Medellín and Ortiz Espejel, 1991; Toledo, Ortiz and Medellin-Morales, 1994). Huastec use 679 plant species and specifically 'manage for' 349 of those species. Totonacs use and manage at least 355 species of plants and animals. Useful species are harvested from lands managed by risk-spreading strategies to make multiple use of available resources while maintaining the natural processes on which agricultural and forest-based systems rely. A survey of indigenous communities in the Mexican lowland humid tropics revealed that 1052 species are used for consumption and sale (Toledo *et al.*, 1995a).

The Huastec and Totonac agro-ecosystem is a fluid mosaic of various resource zones: permanent planted fields, periodically planted fields, fallows, dooryards, orchards, forests and streams. People manage the natural ecosystem for crops, wild plants, wild animals and ecological services. Huastec agriculture relies on a core consisting of three elements: *milpa* cycled fields (cornfields, gardens and fallows), sugarcane fields and managed forests. The Totonac core is similar, except that instead of sugarcane for cash, Totonac have expanded pasture lands for cash generation. Simply put, the two systems create a shifting mosaic of replicates of three standard pieces: forest patches, swidden patches, and cash crop patches. Managed forests, especially along streams, on ridges and steep slopes, have not been cleared in living memory. Approximately 25% of an average Huastec community's land will be under forest; 50% in *milpa*-fallow cycled land; and 25% in sugarcane. In a typical Totonac community, 30% of the land is under forest, 36% under *milpa*, 10% under cash crops (aside from vanilla), and 23% in pasture (Toledo *et al.*, 1994). The *milpa* cycled fields are the most 'mobile' and the managed forests the least mobile part of the shifting mosaic of land use.

The Huastec and Totonac land-use system is generally similar to that of other Mesoamerican *milpa* agriculturalists (Alcorn, 1990). *Milpa* is the Mesoamerican version of integral swidden agriculture (Warner, 1991) applied in most tropical forest areas of the world. As a production system, *milpa* is characterized by its central focus on maize cultivation. From a property rights point of view, *milpa* is an institution which reinforces reciprocity and community-based control of natural resources. Making *milpa* requires reciprocal labour exchange because cutting, burning and planting are done by work parties. *Milpa* also requires decisions made following specific rituals and culturally determined rules of behaviour.

Milpa is the key subsystem of the agro-ecosystem from a social organizational and property rights point of view, as well as from an ecological point of view. If a household stops making *milpa*, it is cut out of reciprocity networks that can be relied on for other types of assistance. At the same time, *milpa* offers strong rights to individual families over the resources produced from their own labour. One analyst of Mesoamerican land tenure has argued that the *milpa* system requires corporate ownership (Rees, 1974). The *milpa* swidden system does not allow continuous use of land for agriculture. *Milpa* requires periodic release of possession, and rights to land without possession can only be vested in a corporate group.

Many familiar with Mesoamerican agriculture think of *milpa* as meaning 'cornfield', but it is not primarily a spatial concept; the spatial meaning is secondary. *Milpa* is an institution and a process. It is a 'script', not a patch in space. A script is an internalized plan used by people carrying out and interpreting routine activities (Schank and Abelson, 1977). Its basic structure is a series of routine steps with alternative subroutines, decision nodes, and room for experimentation (Gladwin and Murtaugh, 1980; Gladwin and Butler, 1984). Ecological knowledge is encoded in the local variation of the *milpa* script derived from the experiences of farmers who have adapted to the local environment over generations (Alcorn, 1989a). The script is passed on to children and supported by cultural beliefs, mythologies and yearly festivals.

Culture plays a very significant role in *milpa* management, and in turn *milpa* has affected culture. Researchers have often commented on the integral role *milpa* plays in Mesoamerican life. For example, Redfield and Villa Rojas (1934) noted for the Yucatec Maya '[t]o abandon one's *milpa* is to forsake the very roots of life'. Rodas *et al.* (1940) note the Quiche do not raise maize to live, they live to raise maize. Nigh (1976), in his study of highland Mayan *milpa*, states that '. . . the making of *milpa* is the central, most sacred act, one which binds together the family, the community, the universe,' and that '*milpa* forms the core institution of Indian society in Mesoamerica and its religious and social importance often appear to exceed its nutritional and economic importance'. Each stage of the *milpa* cycle is named and marked by ritual activities. Tales of a maize culture hero are associated with all stages of the *milpa*. To mistreat maize, or to fail to carry out proper *milpa* steps, is believed to bring misfortune. Periodic sightings of the maize hero have supported cultural revival, because the hero warns people of impending doom if they stop making *milpa* properly.

This strong cultural support buffers the script from total disruption by new economic demands or the introduction of new technologies. Adjustments may be made but a core plan is conserved. For example, Huastec today purchase most of their maize with income from the sale of raw sugar made by cottage industry. Sugar was integrated into their land-use system in the eighteenth century (Mandeville, 1976), but they still do *milpa* agriculture as well – largely because of the cultural importance of 'making *milpa*'. The context in which the *milpa* script has been carried out in the Huasteca and the Totonac area has varied for several thousand years (Alcorn, 1984). The script continues to be modified to fit the context. It enables people to manipulate a forest ecosystem's renewal cycle to produce a crop of maize without disrupting the renewal of the forest (Figure 9.4), but it probably has influenced the structure of the mature forest and the species that inhabit it (c.f. Gomez Pompa, 1971).

As is true of other swidden systems (Warner, 1991), *milpa* is adapted to make ecologically sound use of tropical moist forest. When farmers chop down trees in a plot prior to planting, Westerners think this means the forest is gone. Farmers, however, realize they are actually planting in the forest, because the forest is a process, not a stand of trees. In effect, forest processes are slowed for a few years in a particular space so that a crop can be grown. The farmers recognize that forests are micro-organisms, plants,

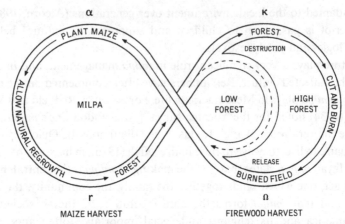

Figure 9.4.   The traditional Mesoamerican *milpa* cycle in the tropical humid low-lands of Mexico as it follows the forest renewal cycle of Holling (1992), supported by traditional tenure, cultural beliefs and institutions. Flip to a less-complex, low-forest ecosystem is prevented by balance between areas allowed to regrow for longer and shorter periods of time.

animals, successional regenerative processes, and nutrient cycling processes – all useful to the farmer, especially the regenerative processes (Alcorn, 1989a). Farmers use their script to regulate the dynamics of their agro-ecosystem. Their script orchestrates many natural processes for the produc-tion of subsistence items and benefits when and where the farmer wants them.

Farmers' *milpa* scripts manipulate natural regrowth to manage regenera-tion processes. Wild plants growing in a field may be managed for future use as items or for the processes in which they participate (Alcorn, 1981) – making agriculture a sort of sequential cropping of crops and non-crops. Farmers know that in the ideal *milpa* cycle, new fields are cleared in high forest, but they also know that the milpa cycle can be done in secondary forest regrowth approximately 3–4 metres in height (three to five years regrowth), provided that only one crop is taken so that the system's regener-ative capacity is not disturbed. If more than one successive crop of maize is taken in a short fallow *milpa*, short-lived weedy species come to dominate the plot and forest regeneration may not occur. The land may come to be dominated by grasses and low shrubs and subject to erosion – as has occurred in mestizos' pasturelands. If the farmer chooses to follow a shorter cycle in some areas, then he or she also sets aside patches of older forest to be areas where *milpa* is not to be made. Farmers who cut all their forest, or who mismanage their *milpa*, are labelled as witches – a strong social sanc-tion to operate within the range of resilience tolerated by the forest.

There is insufficient information available to assess how indigenous management of forests influenced the development of *milpa* over the thousands of years since maize cultivation was introduced to the lowland tropical forest. Older forest management institutions may have formed the foundation for *milpa* and associated religious beliefs that maintain forests. Today, farmers manage their forest patches through selective removal of unwanted individual trees/plants/vines, selective encouragement of desirable species, and attention to spatial distribution of forest patches to ensure that they cover ridges, shelter waterways, and are linked to create large units from individual contiguous patches. These systems have been detailed extensively elsewhere (Alcorn, 1981; 1983; Medellín-Morales, 1986).

Community forest care, patrol and decision making are undertaken as part of *faena* – required work for community benefit which also includes work parties for clearing along major paths, maintaining community buildings, etc. Management decisions are made at Assembly meetings. There are no written rules about the use of community forest, nor any fixed fines for unauthorized use. For example, allegations of collecting firewood for sale (from either individual forests or the community forest) are handled through discussions with the accused party. Peer-pressure derived from a shared value system is generally effective at preventing private gain from community resources that would jeopardize long-term sustainability.

Despite the relatively high population density, approximately 25% of the area is still forested by choice. Instead of clearing more forest for increasing staple production, the *milpa* system was modified to use short fallow periods before all high forest was cleared. People have chosen to increase cash crops and take outside jobs instead of clearing more forest because they value multiple benefits from the forest (Alcorn, 1989b). Increased dependence on cash from outside jobs is associated with increased area under forest.

One other important process essential for ecologically sustainable land use is happening in farmers' homes and *milpa* fields – crop evolution that maintains locally successful varieties (Oldfield and Alcorn, 1987).

The Huastec resource management system makes ecological and economic sense. A typical Huastec community's lands yield a net benefit of cash and subsistence goods valued at $598 per hectare per year (Alcorn, 1989b). This compares favourably with the benefits generated by other systems (Godoy, Lubowski and Markandaya, 1993).

The major stresses that specifically threaten Huastec and Totonac communities' adaptation to the ecosystem include: increasing population, drop in prices for coffee or cacao (the major crops in managed forests), frost damage to forests, price fixing by middlemen, and land grabbing by ranchers and coffee estate (*finca*) owners. The tenacity of community-based management of biodiverse areas under these stresses stands as a monument to communities' commitment to maintain their natural resource base. This commitment is directly related to their resolve to maintain their own cultural and community-based identity. The community secures the economic and social safety net necessary for social reproduction by maintaining direct, long-term access to essential subsistence resources through socially sanctioned management systems based, in turn, on the maintenance of healthy ecosystem function. The communities' health depends on the ecosystem's health. If a community depends on local resources, a more resilient ecosystem means the community can also sustain greater disturbances.

Case two: sub-humid temperate forest

In the sub-humid temperate forest ecosystem (a zone that covers 33 million hectares that is occupied by 1.55 million indigenous people), community-based systems are also adapting to changes. The case study covers two areas managed by Purépechan people (Tarascans) in the state of Michoacan (see Figure 9.3). The oak–pine forests and the intervening grass and shrubland areas support an estimated flora of some 1000 species. Archaeological research documents human settlements in the area beginning approximately 3500 BP and intensive agriculture during prehispanic times (O'Hara, Street-Perot and Burt,1993), and Spanish records indicate that the area supported a large population during the 1500s. Hence, these forests have been disturbed for thousands of years, and local communities have persisted and adapted to a changing series of stresses over time.

The first site is located in the Lake Pátzcuaro basin which includes lake islands, shore, hillsides, mountains and intermontane valleys. Purépechan communities and many of the mixed mestizo settlements around Lake Pátzcuaro retain their language and a strong Purépecha cultural heritage, including tenure systems, resource management systems and social organization. All major basin settlements were in place and occupied by Purépecha people at the time of the Spanish Conquest around 1500. Purépechans make multiple use of the ecosystems available to them. They recognize and name 400 plant species, 53 mushrooms, 31 mammals, 21 reptiles and amphibians, 75 birds and 11 fish species. Two hundred and twenty-four plant and mushroom species have multiple uses for food, medicine and utilitarian values. Purépecha economy is based on a combination of seed, tree and vegetable agriculture, hunting, fishing, gathering, cattle raising, forest management, handicrafts (including weaving based on aquatic plants, wheat and palms), bakeries, and textile weaving. Purépechans recognize 14 different management systems and agricultural landscapes: three rainfed, one dryland and six irrigated agricultural types; two silvicultural systems; and two homegarden types. Based on fishing instruments, there are ten different types of fishing systems. Many of the products from these systems are sold in local markets.

The second site is the community of San Juan Nuevo in pine–oak forest on the high plateau of western Michoacan. In contrast to the Lake Pátzcuaro communities, San Juan Nuevo Parangaricutiro (population 10 000) has used its forest resources and organizational connections to acquire modern machinery for a vertically integrated forest products industry, including factories for mouldings, parquet, furniture, packing crates, charcoal and sawn wood for export markets (Alvarez Icaza, 1993). Although forestry is the main activity at San Juan Nuevo, families also rely on *milpa* fields, homegardens, forestry activities (wood, resin and gathering of medicinal and food plants, mushroom species, lumber, etc.) and cattle raising. While nationally some 65% of forested *ejido* and *comunidades* exploit their forests for commercial sales, San Juan Nuevo is among the few internationally recognized for its successful and profitable forest management. Since 1983, San Juan Nuevo's forestry enterprise has grown in both size and scope. In the last ten years, profits have increased 2,000%, and the personnel from 100 to 1000, with salaries well above the minimum wages for the region. Part of this administrative and economic success lies in the community decision to continue re-investing all profits, rather than distribute them.

San Juan Nuevo illustrates a process of entrepreneurial efficiency and modernization within the traditional tenurial shell. Tenurial rights create a delicate balance between family rights, communal responsibility and enterprise efficiency. Family rights to land and natural resources have been respected as the exploitation of tracts of forests (for wood and resin extraction) by the communal enterprise affects portions of household parcels.

Purépecha communities, like the Huastec and Totonac, have communal ownership of their lands and resources protected under *ejido* and *comunidad* shells, and individual households exercise ownership over their own agricultural lands. Community members may rent or mortgage their lands to other community members. Forest, pasture and lake resources are considered community property, with rules regulating their access and use. Different communities have managed their communal resources in

238   *Janis B. Alcorn & Victor M. Toledo*

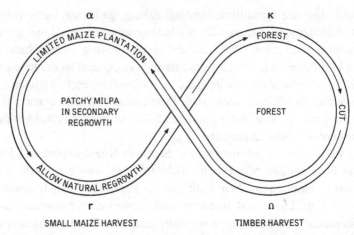

Figure 9.5.   The *milpa* cycle supported by the local *ejido* shell and strong regional organizations in San Juan Nuevo, Michoacan. Tenurial authority exerted by community linked with commercial incentives has resulted in forest increase and enhanced resilience.

different ways. In Pichataro, for example, 4000 ha of pine–oak forests have been divided evenly between eight subdivisions of the community, thereby giving each of the 559 households equal access to forest resources for resin, wood, firewood and food. The lake is used by 700 fisherman from 21 settlements, 19 of which are Purépechan. The lake territory has been divided into sections to be exploited by each community. Each community, in turn, has divided the lake into fishing grounds and shore areas for each fisherman through collectively established rules. Shore areas are physically divided into territories by artificial channels lined by tule reed plants.

In San Juan Nuevo, forests were divided into family patches for exploitation on an individual basis, for resin extraction and small-scale woodworking shops. Until 1970, marketing was controlled by middlemen and much of the forest eventually became degraded from overextraction. But during the 1970s, the *comunidad* joined the Union of Forest Ejidos and Comunidades, and worked for government authorization of community-based forest management and production. By 1981, the community's General Assembly approved the formation of a community enterprise, which successfully competed with middlemen by offering a better price. Sale to the community mill requires sharing rights with the enterprise; the participants enter into co-management arrangements so that the community's forest has slowly come under stronger community control. Forest recovery (Figure 9.5) has occurred because of the tenurial authority exerted by the community. The community as a whole moved to reduce individual rights in order to sustain the forest. It is unlikely that similar state-level action could have prevented clearcutting through zoning or harvest regulations, given the poor record of state-level interventions.

The San Juan Nuevo Purépechans have developed a new local institution associated with the operation of the community's forestry enterprises and the forest co-management rules linked to sustainable extraction for the enterprises. A Communal Council was established, which includes ten representatives from San Juan Nuevo's six subunits, the enterprise directors, property administrators, and a technical committee (Alvarez Icaza, 1993). This group oversees and directs the community's projects, and serves as a forum for developing concensus. The *comunidad* has agreed to re-invest all profits into the enterprise, rather than distributing the profits.

As among the Huastec and Totonac, Purépechan culture supports values placed on reproduction of the community, conservative use of resources, protection of natural processes, economic equity among community members, consensus building, and collective resistance to intrusion by outsiders. Equitable distribution of the communities' land and water resources among individual families prevents over-exploitation by any one family, while communally-shared core values and institutions maintain the community's overall resource use within acceptable bounds.

Discussion

Neither community-based tenure nor state-based tenurial shells are, by themselves, recipes for ecologically sustainable resource management.[18] Knowledge of and commitment to ecologically sustainable management regimes and state implementation of supportive policies are both required elements of the recipe. For example, new *ejidos* of people resettled into tropical forest areas from other ecological zones during the 1960–70s were encouraged by government policy to use capital-intensive inputs and heavy machinery for commercial agriculture (Ewell and Poleman, 1980; DeWalt and Rees, 1994). These ill-advised agricultural schemes failed, and resulted in massive deforestation followed by unproductive cattle ranching. On the other hand, in some cases, spontaneous migrants into forests have attempted (with no government support) to apply the locally adapted, low-input agricultural systems found in long-established communities and successfully established new communities that recreated the land-use patterns of the older communities (Ewell and Poleman 1980). While Mexican policy support has been sufficient to enable ecologically sustainable systems to persist, support has been insufficient to enable those systems to spread and prosper.

New challenges

Challenges to tenurial shell integrity affect ecological sustainability. Over the past several thousand years, community-based management of forest ecosystems has faced a variety of stresses, from colonization, massive depopulation, incorporation of tribute and then cash cropping, forced concentration of populations, introduction of foreign crops and livestock, loss of land rights, and other problems. Yet the system has proven resilient under the protection of the state's authority; indigenous resource management systems continued to adapt to their changing context within the space provided by the protective shells of *ejidos* and *comunidades*. Today, however, there is a new threat – the *ejido* shell is being altered.

In 1992, in preparation for the North American Free Trade Agreement (NAFTA), then-President Salinas revised Article 27 of the Constitution to change the tenurial shells of communities, and Congress passed enabling legislation in the form of the new Agrarian Law, using haunting echoes of the reasoning used during the Porfirian period when the government moved to eliminate the 'unproductive' community-based landholdings (Briseño Guerrero, 1994: 45) in order to make land available to politically powerful élites. Salinas' actions created sweeping changes in the rules regulating *ejidos*, although *comunidades* were technically exempted from these changes. Among the changes are the following: *ejido* members can now rent, sell or mortgage their lands; they no longer have to work the land to retain rights to it; and they can enter into joint ventures with outside entrepreneurs to exploit their resources (DeWalt and Rees, 1994). These recent changes have great potential to undermine the community-based sector and expand the rights of private individual property to mine resources in ecologically fragile areas, instead of seeking ways to support ecologically-sustainable agricultural systems (Toledo, 1995). In effect, the new Agrarian Law tacitly recognizes the existing illegal largeholdings ('latifundios') of politically powerful ranchers, supporting the nationally infamous political bosses of the Huasteca and other indigenous areas (Briseño Guerrero, 1994) and expansion of an inefficient and ecologically damaging land use.

These changes may weaken the recent strength shown by communities that are using traditional communal values to compete in the marketplace. New peasant movements during the past decade have been using collective organization based on traditional values of reciprocity, communal property and voluntary labour to create business corporations that provide quality products at competitive prices in the open market, despite resistance from local elites (Briseño Guerrero, 1994). In Oaxaca and Chiapas, for example, Mam Mayan coffee producers developed an organization to export organically grown coffee (Nigh, 1995). Some 90 organizations, including unions of *ejidos* and other pan-community organizations, have organized to represent the rights of *ejido* and *comunidad* members at the national level, including their rights to defend their territories against encroachment by ranchers and their rights to export products.

It is too early to evaluate the impact of the 1992 changes on forests, but negative ecological and social impacts can be predicted. One can expect an expansion of the situation on the borders of the Lacandon Forest today, where big ranches expand and drive landless poor to clear forest in nature reserves. Outside entrepreneurs are putting pressure on communities to cut their forests for immediate sale, and in some cases replace standing forests with eucalyptus plantations. It will be very difficult for politically weak communities and weak community members to resist pressure from politically powerful people who seek to gain personally from such deals. Local flips to less resilient systems with lower diversity are likely to increase.

Political movements in rural Mexico are seeking a route that involves control of productive processes, including marketing, and ecologically sustainable use of their natural resources as a means to maintain their social and ecological systems (e.g. Bray, 1991; 1992; Declaración del Foro Nacional sobre el Sector Social Forestal, 1992; Merino, 1992). A shell of community-based property rights is critical for the ecological and socio-economic success of this fledgling strategy. Without strong organizations to protect communities' rights and develop supportive policies under NAFTA reforms, the Mexican experiment with community-based tenurial shells will be terminated, and an opportunity for ecologically sustainable development will be lost. Ninety years after the last Mexican Revolution, Mexican society once again stands at the edge of a creative–destruction phase that could either lead to renewal (as before), or take a flip into a different system.

Conclusions

Property rights and tenurial shells, in and of themselves, do not guarantee ecologically sustainable development. Tenurial shells can shelter unsustainable use as well as promote sustainable management.[19] State protection of community-based tenurial systems, however, can be a necessary condition for ecological sustainability in certain situations. The Mexican case demonstrates that state-supported tenurial shells offer a way to protect existing, resilient community-based resource management systems in biologically diverse and ecologically fragile areas.

Priority should be given to supporting community-based tenure in areas rich in biodiversity and forests.[20] Steps for such a programme include: (a) locate existing communities that shelter locally adapted resource management systems; (b) assess how such systems can be best supported within the existing state framework; (c) assess how current policies are hurting these systems; and (d) enact supportive policies. Lynch and Talbott (1995) offer some basic, practical legal steps toward supporting community-based tenurial shells. We have offered other, more specific policy recommendations elsewhere (Toledo, 1994a; Alcorn and Toledo, 1995).

If steps are taken to strengthen state recognition of community-based tenurial systems in countries where the lands of indigenous peoples overlap with the distribution of threatened forests, the resilience of forests and communities may be strengthened so that a flip to a new, less diverse system is averted. This will depend, however, on whether community-based interest groups can scale-up to form politically effective social organizations that shape new interfaces not only with their own national government, but also with powerful blocs of nations and other manifestations of the larger global politico-economic system.

Acknowledgements

We thank Miguel Aguilar, David Bray, Guillermo Castilleja, Mac Chapin, Alejandro de Avila, Mark Freudenberger, Augusta Molnar and Nancy Peluso for their comments. We thank *campesinos* across Mexico for sharing their insights about property rights. Portions of this chapter are reprinted with permission from a World Bank publication (Alcorn and Toledo, 1995). We thank Carl Folke and Fikret Berkes for their inspiration and challenges; and for the opportunity to participate in the research programme in Beijer International Institute of Ecological Economics. We, however, retain responsibility for this chapter and any shortcomings thereof.

Notes

1 The political purposes of Mexico's land reform programme are evident in the slow and sporadic implementation over a period of many decades. The process 'fostered dependency on the apparatus of the state' (Powelson and

Stock, 1987: 29). Programmes to support agriculture and marketing services were designed to build political patronage and power bases for the ruling party rather than to assist communities to develop and market their products. These factors affected productive performance and the outcome of judicial processes when land reform laws were violated.

2 This is not to say that all *ejidos* and *comunidades* have been able to sustainably manage their natural resources, particularly in cases where an insufficient land/resource base was available for the population – either because too little land was granted to the community, or because politically powerful neighbours continued to seize land gradually from the edges and thereby slowly reduced the communities' lands/resources. There is no available survey of *ejidos* and *comunidades* from which one could attempt to determine what percentage are sustainably managing their resources.

3 Even if customary land tenure systems or customary laws are recognized at the national level, conflicting sectoral policies can undermine that recognition. The lack of laws and regulations to support the exercise of specific community rights also undermines the intent of the national recognition of customary rights (Fingleton, 1993; Lynch and Talbott, 1995).

4 State recognition of local systems is not necessarily sufficient to maintain local systems which can fail from the inside if the authority of local institutions is undermined by cultural or political changes.

5 Examples of organizations include families, clans, co-operative societies, community organizations, the church, local government councils, unions, and state agencies.

6 Ventocilla *et al.* (1995a; 1995b) are documenting an on-going process of local ecological damage recognition, community discussion, and local institutional evolution in Panama. Panama has supported tenurial shells around *comarcas* (Herlihy, 1990) – semi-autonomous regions operated by indigenous communities that have functioned to control unsustainable development and conserve biodiversity (e.g. Chapin and Breslin, 1984; Denniston, 1994). The Kuna have formally modified the legislation that recognizes and outlines the internal *comarca* government in order to add a provision for terrestrial and marine resources in co-ordination with national authorities.

7 Hutchins (1980) provides a detailed example of how local institutions use detailed local knowledge and cultural values to interpret evidence to resolve tenurial disputes in changing circumstances in the South Pacific.

8 Land-holdings of individual farm families may be fragmented in order to provide each family with access to available soil types and microhabitats. This fragmentation strengthens conservation of traditional crop varieties by ensuring that families maintain and select varieties adapted to the full range of local conditions (Oldfield and Alcorn, 1987; Brush and Bellon, 1994).

9 The general pattern described in this paragraph is based on an extensive review of published and grey literature from Asia, Latin America, Africa and the Near East, and the authors' fieldwork in Latin America, South Pacific and Asia.

10 People of mixed ancestry who claim Mexican national identity.

11 See note 15.

12 Although the *comunidad* was expressly created for indigenous communities, due to the political reasons described earlier there are only 1231 *comunidades* covering some 9 million hectares (Sheridan, 1988).

13 A recent World Bank sector review has found wide variation in published estimates of forest held by *ejidos* and *comunidades* and in the numbers of *ejidos* and *comunidades* who hold forest.

244   *Janis B. Alcorn & Victor M. Toledo*

14 Mexico's total indigenous population is estimated at 10.5 million (1990 Census) people over the age of five years. There are 54 major indigenous groups speaking up to 240 languages – the number of languages is debated and depends on the criteria used.
15 The term rancher might be somewhat misleading to those unfamiliar with the Mexican situation. Ranchers, or *ganaderos*, are generally powerful elites who do not reside permanently on their ranches. They are analogous to powerful extractive interests in other countries where cattle raising is less common.
16 The ecological importance of patchy disturbance for resilience and patch dynamics in forests is summarized in a seminal book edited by Pickett and White (1985).
17 The term subgroup is used to mean a group which specializes in a profession that requires a particular resource – as in India, where different castes rely on distinct sets of resources.
18 We are not claiming that the *ejido* and *comunidad* system, as administered in the past, is a perfect system – either ecologically or institutionally speaking. These Mexican shells, however, have allowed ecologically-sustainable management and have offered opportunities for economic development in a subset of cases with the shared characteristics described in this chapter.
19 For example, shells offered private corporations in order to shelter foreign investment often result in unsustainable use of resources and environmental pollution.
20 Community-based forest management systems are widespread globally (Messerschmidt, 1993). Other countries offer examples of successful corporate tenure systems and other tenurial options (e.g. Herlihy, 1990; Davis and Wali, 1993; Fox, 1993), and the basic elements of strategies for supporting community-based forest management have been offered from experiences in many countries (Legal Rights and Natural Resources Center, 1994; Lynch and Talbott, 1995; Poole, 1995).

References

Aguilar-Robledo, M. 1994. Reses y ecosistemas: Notas para una evaluación del impacto ambientál de la ganadería bovina en la Huasteca Potosina. *Cuadrante* 11/12: 134–61.
Alcorn, J.B. 1981. Huastec noncrop resource management: Implications for prehistoric rainforest management. *Human Ecology* 9: 395–417.
Alcorn, J.B. 1983. El te'lom huasteco: Presente, pasado, y futuro de un sistema de silvicultura indígena. *Biótica* 8: 315–31.
Alcorn, J.B. 1984. *Huastec Mayan Ethnobotany*. Austin: University of Texas Press.
Alcorn, J.B. 1989a. Process as resource: The traditional agricultural ideology of Bora and Huastec resource management and its implications for research. In *Resource Management in Amazonia: Indigenous and Folk Strategies*, pp.63–77, ed. D.A. Posey and W. Balée. New York: Botanical Garden, Bronx.
Alcorn, J.B. 1989b. An economic analysis of Huastec Mayan forest management. In *Fragile Lands of Latin America: Strategies for Sustainable Development*, pp.182–203, ed. J.O. Browder. Boulder, Col.: Westview Press
Alcorn, J.B. 1990. Indigenous agroforestry systems in the Latin American tropics. In *Agroecology and Small Farm Development*, pp.203–20, ed. M.A. Altieri and S.B. Hecht. Boca Raton: CRC Press.

Alcorn, J.B. 1995. Economic botany, conservation, and development: What's the connection? *Annals of the Missouri Botanical Garden* 82: 34–46.

Alcorn, J.B. and Molnar, A. 1996. 'Deforestation and human – forest relationships: What can we learn from India?' In *Tropical Deforestation: The Human Dimension*, pp.99–121, ed. L. Sponsel, T. Headland and R. Bailey. New York: Columbia Univerity Press.

Alcorn, J.B. and Toledo, V. 1995. The role of tenurial shells in ecological sustainability. In *Property Rights in a Social and Ecological Context*, pp.123–40, ed. S. Hanna and M. Munasinghe. Washington DC: The World Bank.

Alvarez Icaza, P. 1993. Forestry as social enterprise. *Cultural Survival Quarterly* Spring 1993: 45–7.

Barrera-Bassols, N., Medellín S. and Ortiz Espejel. B. 1991. Un reducto de la abundancia: el caso excepcional de la milpa en Plan de Hidalgo, Veracruz. In *Reestructuración Económica y Subsistencia Rural*, pp.163–82, ed. C. Hewitt de Alcántara. Mexico City: United Nations Research Institute for Social Development (UNRISD), Centro Tepoztlán and El Colegio de México.

Barthas, B. 1994. Sistemas de producción y conflictos agrarios en la Huasteca potosina (1870–1910). *Cuadrante* 11/12: 30–42.

Berkes, F., Feeny, D., McCay, B.J. and.Acheson, J.M. 1989. The benefits of the commons. *Nature* 340: 91–3.

Bray, D. 1991. The forests of Mexico: Moving from concessions to communities. *Grassroots Development* 15(3): 16–17.

Bray, D. 1992. La lucha por el bosque: Conservación y desarrollo en la Sierra Juárez. *El Cotidiano* 48, June 1992: 21–7.

Bray, D. 1995'. Peasant organizations and 'the permanent reconstruction of nature'. *Journal for Environment and Development* 00:00–00.

Briseño Guerrero, J. 1994. Tapabocas dos: El control del acceso a la tierra comunal ante el 'nuevo' articulo 27 constitucional. *Cuadrante* 11/12: 43–52.

Bromley, D. and Cernea, M. 1989. *The Management of Common Property Resources: Some Conceptual and Operational Fallacies.* Discussion Paper No. 57. Washington DC: World Bank.

Brush, S.B. and Bellon, M.R. 1994. Keepers of maize in Chiapas, Mexico. *Economic Botany* 48: 196–209.

Chapin, M. and Breslin, P. 1984. Conservation Kuna-style. *Grassroots Development* 8(2): 26–35.

Cortez Ruiz, C. 1992. El sector forestal mexicano ante el TLC [NAFTA]. *El Cotidiano* 48, June 1992: 79–85.

Crocombe, R. 1971. An approach to the analysis of land tenure systems. In *Land Tenure in the Pacific*, pp.1–17, ed. R. Crocombe. Melbourne: Oxford University Press.

Davis, S.H. and Wali, A. 1993. Indigenous territories and tropical forest management in Latin America. *Policy Research Working Paper, WPS 1100.* Washington DC: World Bank.

Declaración del Foro Nacional sobre el Sector Social Forestal 1992. *El Cotidiano* 48, June 1992: 49–52.

Denniston, D. 1994. Defending the land with maps. *World Watch* 7(1): 27–31.

DeWalt, B.R. and Rees, M.W. 1994. *The End of Agrarian Reform in Mexico: Past Lessons, Future Prospects.* San Diego: Center for US–Mexican Studies, University of California.

Ewell, P.T. and Poleman, T.T. 1980. *Uxpanapa: Agricultural Development in the Mexican Tropics*, New York: Pergamon Press.

246   *Janis B. Alcorn & Victor M. Toledo*

Fingleton, J.S. 1993. Conservation, environment protection, and customary land tenure. In *Papua New Guinea Conservation Needs Assessment*, Volume 1, pp.31–56, ed. J.B. Alcorn. Washington DC: Government of Papua New Guinea & Biodiversity Support Program, World Wildlife Fund.

Fox, J., ed. 1993. Legal frameworks in forest management in Asia. *Occasional Papers, Program on Environment, Paper No.16*. Honolulu: East–West Center.

Gladwin, C.H. and Butler, J. 1984. Is gardening an adaptive strategy for Florida family farmers? *Human Organization* 43: 208–16.

Gladwin, C.H. and Murtaugh, M.M. 1980. The attentive/pre-attentive distinction in agricultural decisions. In *Agricultural Decision Making*, pp.115–36, ed. P. Barlett. New York: Academic Press.

Godoy, R.A., Lubowski, R. and Markandaya, A. 1993. A method for the economic valuation of non-timber forest products. *Economic Botany* 47(3): 220–33.

Gómez Pompa, A. 1971. Posible papel de la vegetación secundaria en la evolución de la flora tropical. *Biotropica* 3: 125–35.

Gómez-Pompa, A. and Kaus, A. 1990. Traditional management of tropical forests in Mexico. In *Alternatives to Deforestation*, pp.45–64. ed. A.B. Anderson. New York: Columbia University Press.

González Rodrigo, J. 1993. *Santa Catarina del Monte, Bosques y Hongos*. Colección Tepetlaóstoc 3. Mexico City: Universidad Iberoamericana.

Grupo para la Conservación del Trópico en México 1992. *Compromisos con el Trópico Mexicano*. Mexico City: Grupo para la Conservación del Trópico en México.

Herlihy, P. 1990. Panama's quiet revolution: Comarca homelands and Indian rights. *Cultural Survival Quarterly* 13(3): 17–24.

Holling, C.S. 1992. Cross-scale morphology, geometry, and dynamics of ecosystems. *Ecological Monographs* 62(4): 447–502.

Holling, C.S. and Sanderson, S. 1996. Dynamics of disharmony in ecological and social systems. In *Rights to Nature*, pp.00–00, ed. S. Hanna, C. Folke and K-G. Mäler. Washington DC: Island Press.

Hunt, E. 1977. *The Transformation of the Hummingbird*. Ithaca, New York: Cornell University Press.

Hutchins, E. 1980. *Culture and Inference*. Cambridge, Mass.: Harvard University Press.

Legal Rights and Natural Resources Center/Kasama sa Kalikasan. 1994. '*Baguio Declaration.*' NGO Policy Workshop on Strategies for Effectively Promoting Community-Based Management of Tropical Forest Resources. Manila: Legal Rights and Natural Resources Center (LRC)/Kasama sa Kalikasan (KSK).

Lynch, O.J. and Alcorn, J.B. 1994. Tenurial rights and community-based conservation. In *Natural Connections: Community-Based Conservation*, pp.373–92, ed. D. Western, M. Wright and S. Strum. Washington DC: Island Press.

Lynch, O.J. and Talbott, K. 1995. *Balancing Acts: Community-based Forest Management and National Law in Asia and the Pacific*. Washington DC: World Resources Institute.

Macpherson, C.B. 1978. The meaning of property. In *Property*, pp.1–13, ed. C.B. Macpherson. Toronto: University of Toronto Press.

Mandeville, P.B. 1976. *La Jurisdicción de la Villa de Santiago de los Valles en 1700–1800*. Mexico: Academia de Historia Potosina, San Luis Potosi.

Medellín-Morales, S.G. 1986. *Uso y Manejo de las Especies Vegetales Comestibles, Medicinales, Para Construcción y Combustibles en una Comunidad Totonaca de la Costa (Plan de Hidalgo, Papantla, Veracruz, México)*. Xalapa, Veracruz: Programa Formación de Recursos Humanos, INIREB.

Merino, L. 1992. La experiencia de la Organización Forestal de la Zona Maya. *El Cotidiano* 48, June 1992: 40–43.

Messerschmidt, D.A., ed. 1993. *Common Forest Resource Management: Annotated Bibliography of Asia, Africa, and Latin America*. Rome: Food and Agriculture Organization of the United Nations (FAO).

Mora López, J.L. and Medellín-Morales, S. 1992. Los núcleos campesinos de la Reserva de la Biosfera 'El Cielo': Aliados en la conservación? *BIOTAM* 4(2): 13–40.

Nahmad, S., González, A. and Vásquez, M.A. 1994. *Medio Ambiente y Tecnologías Indígenas en el Sur del Oaxaca*. Mexico: Centro de Ecología y Desarrollo.

National Census 1990. *Estados Unidos Mexicanos, Resumen General XI Censo General de Población y Vivienda*. Mexico City: Instituto Nacional de Estadística, Geografía e Informática.

Nigh, R. 1976. Evolutionary Ecology of Maya Agriculture in Highland Chiapas, Mexico. PhD dissertation, Stanford University. Ann Arbor: University Microfilms.

O'Hara, S., Street-Perot, F.A. and Burt, T.P. 1993. Accelerated soil erosion around a Mexican highland lake caused by prehispanic agriculture. *Nature* 362: 48–51.

Oldfield, M.L. and Alcorn. J.B. 1987. Conservation of traditional agroecosystems. *BioScience* 37: 199–208.

Ostrom, E. 1990. *Governing the Commons: The Evolution of Institutions for Collective Action*. New York: Cambridge University Press.

Ostrom, E., Walker, J. and Gardner, R. 1992. Covenants with and without a sword: Self-governance is possible. *American Political Science Review* 86: 404–17.

Pickett, S.T.A. and White, P.S., eds. 1985. *The Ecology of Natural Disturbance and Patch Dynamics*. New York: Academic Press.

Poole, P. 1995. Indigenous peoples, mapping and biodiversity conservation: A survey of current activities. *BSP Peoples and Forests Discussion Paper Series*, No.1. Washington DC: Biodiversity Support Program (BSP), World Wildlife Fund.

Porter, D., Allen, B. and Thompson, G. 1991. *Development in Practice: Paved With Good Intentions*, London: Routledge.

Powelson, J.P. and Stock, R. 1987. *The Peasant Betrayed: Agriculture and Land Reform in the Third World*. Boston: Lincoln Institute of Land Policy, Oelgeschlager, Gunn & Hain.

Ramamoorthy, T.P., Bye, R., Lot, A. and Fa, J., eds. 1993. *Biological Diversity of Mexico: Origins and Distribution*. New York: Oxford University Press.

Redfield, R. and Villa Rojas, A. 1934. *Chan Kom*. Carnegie Institute, Washington DC.

Rees, M.J. 1974. Law, land and religion: An analysis of *milpa* ownership. *Human Mosaic* 7: 21–30.

Rodas N., Flavio, O., Rodas, C. and Hawkins L.F 1940. *Chichicastenango: The Kiche Indians, Their Histotry and Culture*. Guatemala City: Unión Typográfica.

248 Janis B. Alcorn & Victor M. Toledo

Rzedowski, R. 1978. *Vegetación de México*. Mexico City: Editorial Limusa.

Sanderson, S.R.W. 1984. *Land Reform in Mexico: 1910–1980*. New York: Academic Press.

Sandoval, J.R. 1994. Paraíso ascoliado. *Revista Epoca No. 170* Septiembre 5, Suplemento: 4–11.

Schank, R. and Abelson, R. 1977. *Scripts, Plans, Goals and Understanding*. New York: John Wiley & Sons.

Sheridan, T.E. 1988. *Where the Dove Calls: The Political Ecology of a Peasant Corporate Community in Northwestern Mexico*. Tucson: University of Arizona Press.

Stresser-Péan, G. 1967. Problemes agraires de la Huasteca ou région de Tampico (Mexique). In *Les Problemes Agraires des Amériques Latines*, pp.201–14. Paris: Colloques Internationaux de Centre Nacional de la Recherche Scientifique.

Tapia Centeno, C. 1960. Paradigma Apologético (1725). In Aguinaga y Montejano, R. *Notas de Bibliografía Linguistica Huasteca* 6: 50–103

Toledo, V.M. 1992a. Biodiversidad y campesinado: La modernización en conflicto. *La Jornada del Campo* 10 November: 1–3.

Toledo, V.M. 1992b. Campesinos, modernización rural y ecología política: Una mirada al caso de México. In *La Tierra: Mitos, Ritos y Realidades*. pp.351–65, ed. J.A.Gonzalez Alcantud and M. Gonzalez de Molina. Granada: Anthropos.

Toledo, V.M. 1994a. *La Ecología*. Chiapas y el Artículo 27. Ediciones Quinto Sol, Mexico City.

Toledo, V.M. 1994b. Biodiversity and cultural diversity in Mexico. *Different Drummer* Summer 1994: 16–19.

Toledo, V.M. 1995. La ley agraria: Un obstaculo para la paz y el desarrollo sustentable. *La Jornada del Campo*, 7 March.

Toledo, V.M. and Barrera-Bassols, N. 1984. *Ecologia y Desarrollo Rural en Pátzcuaro*. Mexico City: Instituto de Biología, Universidad Nacional Autonoma de México.

Toledo, V.M., Batis, A.I.,. Becerra, R., Martínez, E. and Ramos, C.H. 1995a. La selva útil: Etnobotánica cuantitativa de los grupos indígenas del trópica húmedo de México. *Interciencia* 00: 00–00.

Toledo, V.M., Carabias, J., Mapes, C. and Toledo, C. 1985. *Ecología y autosuficiencia alimentaria*. Mexico City: Siglo Veintiuno Editores.

Toledo, V.M., Ortiz, B. and Medellin-Morales, S. 1994. Biodiversity islands in a sea of pastureland: Indigenous resource management in the humid tropics of Mexico. *Etnoecológica* 3: 37–50.

Ventocilla, J., Herrara, H. and Núñez, V. 1995a. *Plants and Animals in the Life of the Kuna*. Austin: University of Texas Press.

Ventocilla, J., Núñez, V., Herrara, H., Herrera, F. and Chapin, M. 1995b. Los indígenas kuna y la conservación ambiental. *Mesoamericana* 16(29): 95–125.

Vetancourt, A. de 1689. Teatro Mexicano. Mexico. 4 a. p., trat. 30., c. I, p.91–92.

Vogt, E.Z. 1976. *Tortillas for the Gods*. Cambridge, Mass.: Harvard University Press.

Warner, K. 1991. *Shifting Cultivators: Local Technical Knowledge and Natural Resource Management in the Humid Tropics*. Rome: Food and Agriculture Organization of the United Nations.

White, J.B. 1985. Law as rhetoric, rhetoric as law: The arts of cultural and communal life. *University of Chicago Law Review* 52: 684.

Wilken, G. 1987. *Good Farmers: Traditional Agricultural Resource Management in Mexico and Central America*. Berkeley: University of California Press.
Zizumbo Villarreal, D. and Colunga García-Marín. P. 1982. *Los Huaves: La Apropiación de los Recursos Naturales*. Chapingo, Mexico: Universidad Autonomo Chapingo, Departamento de Sociología Rural.

10
The resilience of pastoral herding in Sahelian Africa

MARYAM NIAMIR-FULLER

Introduction

Populations living in arid zones have developed many forms of safety nets and social mechanisms to help individual households buffer the effects of environmental variability. Examples include: reciprocal exchanges of livestock, re-stocking alliances, dowry and marriage price as a form of livestock redistribution and selective breeding, communal insurance schemes, labour exchanges, etc. Indications are that these safety nets are now gradually disappearing as communities loose their sense of unity, as their members become more settled and heterogeneous, as government administration and other external forces gradually weaken traditional authority, and as rules and regulations are gradually and increasingly ignored (Bovin and Manger, 1990; Niamir, 1990).

The average pastoral household has had to find different ways of adapting to adverse conditions, such as displacement, urban migration, settlement and cultivation, all of which have created serious environmental and socio-economic problems. The social–institutional linkages between the ecosystem and the production system have deteriorated, but have not been replaced with new linkages. The arid and semi-arid ecosystem in Africa may no longer be sustainable, given current trends and changes

There is a growing awareness among development practitioners and scientists that the time has come for a change in attitudes toward pastoralism, and a reversal of both 'benign neglect' and biased policies (Behnke, Scoones and Kerven, 1993; Scoones, 1995). This chapter will attempt to contribute to the debate by looking at local, traditional, pastoral systems and ecosystem processes and their resilience. The socio-cultural mechanisms that enable pastoralists to *buffer the effects* of ecosystem variability (e.g. re-stocking alliances, etc.) have been analysed elsewhere (e.g. ORSTOM 1986; Bovin and Manger, 1990) and will not be repeated here.

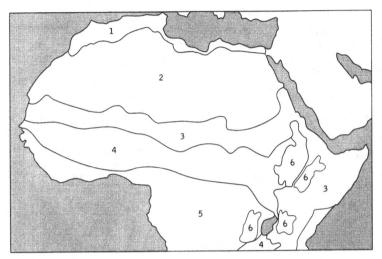

1 Mediterranean vegetation
2 Sahara Desert
3 Thorn (Acacia) woodland, wooded grassland and semi-desert vegetation
4 Broad-leaved woodland and wooded grassland
5 Lowland (Goineo-Congolian) rainforest
6 Afromontane vegetation

Figure 10.1.   Vegetation map of Africa and the Sahel.

Rather, the focus of this chapter is on how pastoralists *manage* ecosystem variability. What were the social–institutional mechanisms that allowed pastoralists to utilize environmental variability without deteriorating the ecosystem's functions – or even improving it – and can these mechanisms still be used today?

Background to Sahelian herders: challenges and changes

The Sahel region of Africa (the so-called inland 'coast' of the Sahara Desert) is a vast area stretching from west to east Africa. It is characterized by mainly loamy to sandy soils, and semi-arid to arid climate. Annual rainfall increases on a north–south gradient, and the resultant vegetation changes from annual grasslands and steppes in the north (with desertic elements on the transition to the Sahara) to savanna woodland and wooded grassland in the south (Figure 10.1).

This region is inhabited by about 160 million people, of whom 12–16% are pastoralists (Bonfiglioli and Watson, 1992; Figure 10.2). Pastoralism is defined as a production system in which 50% or more of household gross revenue (i.e. the total value of marketed production plus the estimated

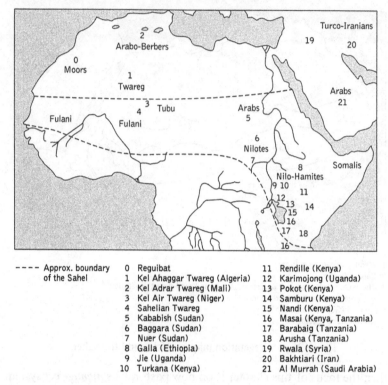

--- Approx. boundary   0 Reguibat   11 Rendille (Kenya)

--- Approx. boundary	0	Reguibat	11	Rendille (Kenya)
of the Sahel	1	Kel Ahaggar Twareg (Algeria)	12	Karimojong (Uganda)
	2	Kel Adrar Twareg (Mali)	13	Pokot (Kenya)
	3	Kel Air Twareg (Niger)	14	Samburu (Kenya)
	4	Sahelian Twareg	15	Nandi (Kenya)
	5	Kababish (Sudan)	16	Masai (Kenya, Tanzania)
	6	Baggara (Sudan)	17	Barabaig (Tanzania)
	7	Nuer (Sudan)	18	Arusha (Tanzania)
	8	Galla (Ethiopia)	19	Rwala (Syria)
	9	Jie (Uganda)	20	Bakhtiari (Iran)
	10	Turkana (Kenya)	21	Al Murrah (Saudi Arabia)

Figure 10.2.   Distribution of main pastoral groups in Africa and the Near East. (From Child *et al.*, 1984.)

value of subsistence production) comes from livestock or livestock-related activities.

An agropastoral production system is one in which more than 50% of household gross revenue comes from farming, and 10–50% from livestock (Swift, 1988). Of the 500–600 million people estimated to be living in arid and semi-arid lands of the tropics and semi-tropics, at least 6% (or 30–40 million people) are solely dependent on livestock (Child *et al.*, 1984), but the percentage of all pastoralists is not known. Figure 10.3 provides a measure of their extent in the world.

Pastoralism is predominantly an extensive production system; it includes nomadism, transhumance and semi-transhumance in order of decreasing mobility (Niamir, 1994a).

Nomadism is a production system that is highly mobile, but does not necessarily return to a 'base' every year, and does not include cultivation (e.g. nomads of the Sahara)

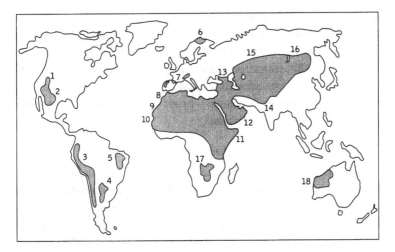

1 North American Indians (e.g. Navajo)
2 North and Central Mexican
3 Andean Pastoralists of Equador, Peru, Bolivia, and Chile
4 Pantanal and Chaco pastoralists
5 Northeast Brazil Vaqueiros
6 Laplanders
7 French, Italian, Spanish, and Portuguese herders
8 North African and Egyptian pastoralists
9 Sahara Desert nomads
10 Sahelian pastoralists
11 Pastoralists of Horn of Africa
12 Arabian penninsula and Middle East pastoralists
13 Turkic, Kurdish, and Iranian pastoralists
14 Afghani, Pakistani, Rajasthani, Nepali, and Tibetan pastoralists
15 Turkmen, Kazakh, Uzbek, and Tajik pastoralists
16 Mongolian and Chinese pastoralists
17 Zimbabwe and Botswanan pastoralists and hunter–gatherers
18 West Australian aborigine pastoralists and hunter–gatherers

Figure 10.3.   Schematic presentation of distribution of pastoral peoples in the world.

Transhumance is a production system that is highly mobile but moves between definite seasonal bases every year (e.g. Samburu of Kenya); it may include a non-sedentary form of cultivation (e.g. Zaghawa of Chad)

Semi-transhumance is a production system where only part of the family and/or livestock is seasonally mobile, and the rest is sedentary in one of the seasonal bases, and practising cultivation (e.g. Dinka of Sudan, and Karimojong of Uganda).

Pastoral populations in recent decades have been faced with major challenges. Among them are: desiccation of the environment due to recurrent droughts, encroachment of cultivation on rangelands, degradation of fallow lands and areas around permanent water sources, impoverishment of a major section of the population, breakdown of traditional institutions and systems for managing natural resources, separation of the ownership of land from its users, and inadequate delivery of services to mobile populations (health, education, production inputs). Most government policies towards pastoralists are either a result of 'benign neglect' (Swift, 1993) – forget them and they will change – or are directly aimed at changing the 'backward' lifestyles of pastoralists.

Rangeland encroachment

By misunderstanding and underestimating pastoral micro-economics, cultivation, tourism and wildlife conservation are thought to contribute more to the national economy than extensive livestock production. Therefore, government policies have encouraged both small and large-scale farms as well as large-scale gazetting of land for wildlife, at the expense of extensive livestock production. The result has been progressive and rapid encroachment onto rangelands, causing a shortage of range resources, particularly high-quality ones, and increasing land degradation, as farms are established on marginal lands and grazing pressure increases on remaining rangelands (Marchal, 1983; Little, Horowitz and Nyerges, 1987; Lane, 1991; Niamir, Lugando and Kundy, 1994).

Pastoral settlement

Settlement of pastoral people is still a common policy in most African countries, and is justified with the promise of socio-economic benefits (services, inputs, markets, infrastructure) as well as benefits from crop cultivation. Many pastoralists are also spontaneously settling in order to diversify their production base, in response to decreasing rangelands and livestock productivity, as well as to guarantee land tenure security.

However, not only do services not materialize, but there are also adverse ecological consequences: decreased mobility of animals and increased continuous grazing result in reduced vegetation diversity and increased soil degradation around the settlements, both of which reduce ecosystem resilience to disturbance. At the same time, lower grazing pressure on distant pastures often results in an invasion of the range by unpalatable plants. Finally, increased occurrence of deep cattle tracks leads to gully erosion (Brandstrom, Hultin and Lindstrom, 1979; Galaty, 1988; Warren and Rajasekaran, 1993).

Settlement results in loss of traditional knowledge of range management and ecosystems, and loss of traditional controls on range use, leading in turn to less efficient management of resources (Jacobs, 1980; Farah, 1993).

As a consequence, livestock productivity plummets, forcing pastoralists either to abandon livestock altogether or entrust the majority to mobile transhumants, thus depriving the family of milk and other benefits on a daily basis. Settled pastoralists are among the poorest of the poor; their production system is no longer in balance with the ecosystem and is no longer sustainable.

Land privatization

The theory of the tragedy of the commons (Hardin 1968) is still the current dogma among most governments officials in Africa (Roe, 1991). Land degradation in the semi-arid and arid regions is blamed on the seemingly chaotic movements of pastoralists, and on communal use of land. As a corollary, it is believed that land privatization will improve both productivity and the environment. This longstanding myth has been refuted by theoreticians (Runge, 1981; Feeny *et al.*, 1990; Moorehead, 1993), as well as by the consequences of privatization campaigns. Land privatization in Africa has led to accelerated social stratification, land grabbing, and land degradation (Peacock, 1987; Lane and Moorehead, 1993; Niamir, 1994).

Recent events in Africa – nationalization of land, sedenterization, land grabbing and expansion of cultivation – have resulted in a chaotic land tenure situation, where customary rights are fighting for supremacy over juxtaposed modern ones. There are now fresh attempts to re-introduce managed commons in the Sahel, as in Sudan, Senegal and Burkina Faso. Present populations and herds of livestock cannot be cost-effectively maintained in Africa by settled, intensive, or zero-grazing systems by the majority of pastoral households, who have little or no access to credit and other production inputs. Extensive production systems, relying on large tracts of common land, are the only viable short-term and medium-term strategies. But to avoid environmental degradation, the livestock need to be mobile, even if the people are not.

Public water for rangelands

Since the 1970s, water development has been the main strategy for the development of pastoral areas. Interventions were generally capital intensive, and were done by the government as a 'public' good on the understanding that pastoralists did not have the necessary capacity themselves. However, by being open access, these interventions have disrupted the traditional transhumance systems and schedules, and the territorial organizations and controls on range use, and have decreased mobility of livestock, leading to environmental problems. Short-term gains are quickly offset by medium-term losses, as the water points fall into disrepair and are not maintained by the government. The necessity of empowering local communities to take over the management and maintenance of water points is now being clearly felt in most countries.

De-stocking

One traditional drought-adaptation strategy of pastoralists is that of keeping large numbers of animals. This increases the probability of a quick regeneration of the herd after a drought. Pastoral herd size fluctuates in response to booms and busts in the ecosystem, but a large herd provides the buffer that protects the household from disaster, and is sustainable in the long run because of transhumance and mobility. However, land degradation, unfavourable terms of exchange between livestock and crops, and breakdown of regional livestock marketing structures, have resulted in economic marginalization of the majority of pastoral households, and concentration of wealth in the hands of a few, including urban folk (Little, 1985; Bovin, 1990; Ndagala, 1991). It is common to find that about 15% of the pastoral population control 80% of the livestock. On aggregate, livestock ownership per household is now less than in the last century, and the majority of households have fewer animals than the minimum required for long-term sustainability and survival. Some governments have experimented with forceful de-stocking (e.g. Tanzania), with little success. De-stocking, or even attempts to increase the offtake of animals through better marketing, will not be successful as long as they are targeted to the majority of households who are below the subsistence level.

Ranches and fencing

Policies that encourage private ranching in Africa have only benefitted the élite, and mostly non-pastoral entrepreneurs. Projects that implemented group ranches (where the ownership of the extensive ranch is given to more than one person) have worked only while the project has existed, and at considerable cost. Group ranches in Kenya, considered as a successful model in the 1970s, have resulted in expropriation of land by the rich, and loss of access rights by women and the younger generation (Galaty, 1993).

Fencing, as a tool for controlling livestock, is a major expenditure on ranches, or in establishing exclosures. However, it is too costly, and is usually torn down or falls into disrepair because its original purpose is misunderstood or unacceptable to all in the community. Experience has shown that livestock are controlled far better by herders, rather than fences, if the herders understand and accept range use controls and regulations.

Protection of reserves without fencing will work if the herder knows he will be able to use the reserve in the near future, that all others are also observing the same discipline, and that there is another range area that he can use instead (see, for example, experiences of various projects: PRODESO in Senegal, El Odaya/UNSO in Sudan).

The ecosystem: unpredictable, highly variable, but resilient

Arid and semi-arid ecosystems of Africa occur mainly in the Sahara–Sahel belt, but also in southern Africa's inland and coastal deserts (Kalahari, Namibia, etc.). They have long been the home of mobile populations – whether pastoralist or hunter–gatherer – because production systems have had to adapt to an especially harsh and unpredictable environment.

These ecosystems are characterized by low net primary productivity and high variability in ecosystem structure and productivity – both spatial and temporal. The most limiting factor is water in the arid and hyper-arid systems, but as the ecosystem becomes more humid, the most limiting factor gradually becomes soil nutrient content (Penning de Vries and Djitèye, 1982). As the ecosystem goes from arid to humid, climate – the most important driving factor in the arid zones – looses its important role to other factors: herbivory, human impact, etc. The discussion in this chapter is limited to the arid and semi-arid zones.

The failure of most range management projects in the last two decades to achieve sustainable growth of pastoral and agro-pastoral systems in arid and semi-arid Africa has prompted a major re-thinking of the applicability of their basic principles and concepts to the African context. At the same time, anomalies in our understanding of dryland ecosystems have developed as new research has highlighted facts and trends that cannot be explained by the old paradigm. A thorough review of principles and conceptual frameworks in ecosystem management is now underway.

The new approach applies the non-equilibrium ecological theory to arid lands (Ellis and Swift, 1988; Westoby, Walker and Noy-Meir, 1989; Friedel, 1991). It recognizes three characteristics of arid ecosystems: ecological variability, unpredictability, and high resilience. According to this view, arid ecosystems of Africa never achieve equilibrium because of the high degree of variability. Equilibrium implies that a system can be in a particular state for a significant amount of time (i.e. is stable). The arid ecosystem is constantly changing from one level or state to the other. The most determining factor in arid lands – climate – is rarely stable.

The principle of ecological equilibrium has long been used by classical range managers as the basis and justification for the concepts of *climax* and *ecological succession*. Researchers are questioning the usefulness of these principles in describing arid ecosystems, since equilibrium is rarely achieved, the climax is largely undefinable, and the pathway to it (succession) is not always constant, progressive or replicable.

The range management tools of *carrying capacity* and *stocking rate*, have been derived from the classical concepts of climax and succession. They are used to determine the number of animals that can be supported on rangelands. However, they are only useful in providing a generalized description of the ecosystem, averaged over space and time (de Leeuw and Tothill, 1993). They do not take into account yearly variability, nor spatial heterogeneity at the micro-level ('patches' or 'patchiness') (de Angelis and Waterhouse, 1987; Scoones, 1991).

The focus on rigid measurement and management tools has been counterproductive in Africa (Perrier, 1991). It is precisely the two factors of space and time, rather than a static average of primary productivity, that determine sustainable carrying capacity – a fact which traditional pastoral managers were well aware of and used through their tools of mobility and tracking.

Ecosystem resilience in this context, therefore, cannot be defined as the resistance to disturbance and speed of return to a stable equilibrium state (Pimm, 1984), but rather it is the magnitude of disturbance that can be absorbed before the system loses its capacity to respond to climatic variability, i.e. the capacity of the system to buffer disturbance (Perrings *et al.*, 1995). The new theory focuses attention on the long-term sustainability of the ecosystem, and away from yearly fluctuations. It focuses on ecosystem function rather than structure. Resilient ecosystems have a highly diverse range of ecological processes (there is little specialization) because of the need to adapt to an unpredictable environment, although they may show low structural (bio-)diversity. Arid ecosystems are more resilient than was previously thought.

Given the lack of proper long-term monitoring data, it is difficult to separate the effects of human/livestock impact from climatic fluctuations. In Sahelian rangelands the ecosystem does not lose its resilience until disturbances such as deforestation, over-cultivation and continuous grazing pressure simplify its structure and reduce its functional options (lower vegetation diversity, greater soil homogeneity leading to soil compaction and sand dune mobility). But vegetation change by itself is not a sensitive indicator of resilience since it is determined by climate variability and endogenous factors. Even if the landscape looks bleak one year, it will spring back to its full productivity *as long as* ecosystem resilience has not been damaged.

When management options are simplified, as when livestock are no longer mobile and continuously graze around a settlement, then the functional diversity of the ecosystem is reduced, with the consequence that the ecosystem loses its capacity to respond to varying climatic conditions. The resilience has contracted. This phenomenon has been observed in many different management situations in unpredictable environments (Perrings *et al.*, 1995).

The issue of land degradation in the Sahel is by no means settled, however. The debate continues between those who predict rapidly expanding deserts and degradation in the Sahel (e.g. Stiles, 1995), and those who take a less alarmist view. A ground-breaking study that has helped to redirect the debate was carried out in the Sudan in the late 1970s. It showed that land degradation and loss of ecosystem resilience are found around settlements, public water points and other concentration points, but not – as yet – in extensive rangelands where transhumance and mobility remain a strong component of the production system (Hellden, 1984). It can be argued that the great loss of old trees in the Sahel, often given as proof of extensive degradation, is not due to destruction of the ecosystem's resilience, but to increasing aridity of the climate since the last century, and therefore a manifestation of functional adaptability and resilience.

There is an urgent need for proper monitoring and ecological research of the arid zones of Africa. There is a need for the development of appropriate lists of indicators that are sensitive to ecosystem variability, productivity, resilience and functional diversity, while at the same time lending themselves to easy and rapid monitoring.

Resource users and their production system: adjustable, flexible, but marginalized

The arid and semi-arid zones of Africa are home to a wide range of production systems. A continuum can be drawn from pure pastoral or nomadic, through transhumant, to agropastoral, depending on the ecosystem's variability and productivity.

Nomadic systems are relatively rare nowadays, and confined mostly to the oases of the Sahara. They are entirely dependent on communal rangeland resources and have very long amplitudes of movement. Transhumant systems are similar, but their movements are clearly seasonal between well-defined territories. Semi-transhumants are semi-sedentary, i.e. they have a

260   *Maryam Niamir-Fuller*

dry or wet season base in which they cultivate or leave behind women, children and the elderly, while transhuming with the majority of the livestock to distant pastures on a seasonal basis. They rely on both communal and individual land. Agropastoralists, by definition, are fully sedentary, but still have considerable numbers of livestock, which are fed off nearby communal land and private fallow land, or entrusted to neighbouring transhumants.

The production continuum is characterized by gradually less dependence on communal land, as one moves from nomadic to agropastoral systems, and gradually greater dependence on private cultivated land. In addition, there is a reduction in mobility and in the heterogeneity of natural resources used by the production system, as distance or amplitude of movement decreases. Nomadic and transhumant systems are able to use far more different types of pastures, and different types of water sources, than fully or partially sedentary production systems (Niamir, 1990).

Social characteristics also change along the production continuum. From nomadic to agropastoral populations, one finds a greater ethnic heterogeneity in the local community, relatively greater population size, rate of population growth, and rate of out-migration.

Although, since the 1970s, nomadic and transhumant systems have been gradually changing to semi-transhumant and agropastoral systems, the grazing system of most groups is still dependent on transhumance (scheduled mobility between well-defined seasonal grazing areas). Transhumance is primarily defined by water availability, and secondly by the number and type of grazing areas. These 'macro' movements are choreographed in space and time, often complementing those of different social groups (Stenning, 1959; Allan, 1965; Sutter, 1978). In addition, most pastoral systems have 'micro-movements' or well-defined and scheduled rotation between range types within the grazing areas . The following discussion distinguishes between mechanisms and processes at the 'micro' level (i.e. households and groups of households, and their daily use and management of natural resources), and the 'macro' level (i.e. co-ordination between groups and tribes).

Micro-systems: indigenous technical knowledge, micro-mobility, tracking and key-site management

Indigenous technical knowledge

Indigenous technical knowledge (ITK) has long been the subject of anthropological research (e.g. Howes, 1980; Waren and Meehan, 1980).

Box 10.1 Some examples of traditional knowledge of plants among pastoralists

The Bambara of Mali have one of the most complete soil classification systems. They distinguish seven major soil types. The most detailed classification is for sandy soils, according to inundation potential, ease of cultivation and land-use potential (Aubert and Newsky, 1949). Of the many ethnobotanical studies, a few examples can be mentioned. The Lugbara of north-west Uganda collect a certain plant from whose ash they obtain salt. The seeds of waterlilies are used by the Dinka of southern Sudan to make beer. The Zaghawa of Chad and Sudan harvest many annual grasses for food and beer. The herders eat wild fruits on the job. Plants are also used for sugar, flour, nuts and medicine (Tubiana, 1969).

Since the mid-1980s, attempts have been made to harness it for the purposes of development. Local indigenous knowledge is increasingly being seen as a reservoir of ideas and solutions for development work by major donors and NGOs (e.g. OXFAM, CARE, ITDG). However, as yet there are very few examples where ITK has been directly incorporated into development projects. Development planners continue to function 'top-down' despite the rhetoric (Scoones and Thompson, 1994).

Indigenous technical knowledge varies according to localities, production types, age, sex, division of labour, education, specialization (e.g., marabouts and witch doctors), aptitude, economic class, etc. (Warren and Meehan, 1980). People organize their knowledge so that it can be of use in their lives. Knowledge is dynamic and adapted to various purposes through experimentation and generalizations to other situations. A reservoir of general knowledge accessible to all in the community is strengthened by in-depth knowledge held by a few specialists (such as leaders, scouts, witch doctors).

Indigenous knowledge of natural resources can be organized around several categories: nomenclature, descriptive knowledge, classification, and analytical knowledge (the latter including interactions and causality) (see Box 10.1.).

In the case of rangelands in arid and semi-arid Africa, spatial knowledge is organized around the classification of ecological patches (such as different pastures) and key-sites[1] (Niamir, 1994; Scoones, 1995). Temporal knowledge of each of the patches and key sites provides a dynamic outlook on how the productivity and value of the site change with time, and is incorporated into local systems of predicting the manner and amplitude of such temporal changes, particularly of droughts. Technical knowledge of patches and key-sites is translated into rules that control micro-mobility, for the purpose of allowing maximum benefits to the individual without impairing the benefits of the group as a whole, and without endangering the resilience of the ecosystem. Although the knowledge is extremely important by itself, it cannot simply be 'plugged' into development. Focus should be on the *process* by which the knowledge is translated into actions, rules and changes to the system.

Micro-mobility

Techniques of herd management and micro-mobility are designed to mimic the variability and unpredictability of the ecosystem. The schedule and distances of transhumance are usually prescribed within local customs, but are flexible to a certain degree, depending on rainfall, livestock disease epidemics, competition for land, and a host of other socio-economic factors such as political considerations, cultural events, market schedules, etc. Pastoralists diversify by having an appropriate mix of animal species in the herd to take advantage of different vegetation types and patchiness. For example, cattle and camels of the Rendille in Kenya are taken to distant pastures to take advantage of the wooded savannas, while their sheep and goats are kept close to the homestead to use the crop residues and riparian patches (Fratkin, 1986).

Other customary practices are designed to reduce risk: herd separation (sending the surplus animals on transhumance), which effectively reduces the risk of overgrazing in the home base area, takes advantage of better pastures elsewhere and maintains the nutritional standards of the home base residents; group herding pools resources and makes the most efficient use of available labour for other activities (such as scouting the terrain and seeking water points, taking care of sick animals, marketing, etc.); 'guest' herds (i.e. animals entrusted to a different herd) not only benefit the caretakers with increased milk, but also reduce the owner's risks of production since his animals are spread over a wider geographical base (Swift, 1977; Winter, 1984; Hussein, 1990).

A tool of primary importance to western range managers – stocking rate – is of little importance to pastoralists in Africa. Just as absolute carrying capacity is of no meaning in non-equilibrium ecosystems, stocking rate, when defined as control of numbers of animals per unit area per unit time, is of no meaning to pastoralists. Instead, there are other specific rules that track ecological dynamics, control micro-mobility, and result in manipulation of the actual stocking rate. These rules act through four main

---

Box 10.2 Some examples of rules governing micro-mobility in the Sahel

The Wodaabe Fulani follow the lunar cycle when moving to new pastures, which in effect results in one of the most highly mobile systems: the camp is moved every two to three days, and the herd moves out of an area every week (Stenning, 1959). By contrast, the Rufa'a al Hoi of Sudan move to a new pasture every 204 days (Ahmed, n.d.). The Fulani of northern Sierra Leone used to practise 'shifting pasturage', whereby they heavily stocked an area for two to three years, then moved elsewhere and rested the first area for 15–20 years (Allan, 1965). The Sukuma of Tanzania had a similar strategy but a longer rest period of 30–50 years (Brandstrom, *et al.,* 1979). The Barabaig of Tanzania allow 20 days of rest each season. Pastoralists generally avoid concentration points, such as water points and villages, at the crucial moments. The Somali and Maasai allow some concentration in the dry season, but insist on dispersal in the wet season (Western and Dunne, 1979; Behnke and Kerven, 1984), whereas, on the contrary, the Arabs of Chad and the Turkana disperse in the dry season because of lack of forage, and allow concentration in the wet season (Gilg, 1963; Barrow, 1988). The Twareg will enter an area only if they see no signs of trampling (Bourgeot, 1981). The Dinka of Sudan do not enter a pasture before the dung from the previous herd has disintegrated (Niamir, 1982). Both the Fulani and the Barabaig say that you must isolate yourself to feed to your livestock better (Ba, 1982; Niamir *et al.*, 1994).

---

variables: length of continuous grazing episode on a patch by a herd, frequency (rate at which the same pasture is visited and grazed), rest interval (time lapse between each visit), and minimum distance between grazed sites (dispersion).

Many examples can be found in the literature of rules that control the *length of continuous grazing on the same pasture* (Box 10.2). Most groups do not impose strict time-limits on the grazing of a particular pasture, but will move when specific indices show high grazing pressure (see Table 10.1), or when there are other needs (social gatherings, marketing, etc.). In semi-arid and arid Africa, the *rate at which the same pasture is re-grazed* is often not formally prescribed, but will be an indirect result of rules governing *dispersion* and *rest interval* (Box 10.2).

In addition to these strategies, pastoralists regulate grazing pressure through other rules as well. The Maasai of Kenya widen the radius of grazing around wells progressively as the wet season advances so as to leave enough forage around the wells for the dry season, and they delay going into the dry season area as long as possible. This strategy has been shown to increase total (average) carrying capacity by 50% (Western, 1982). The Twareg have a rule that those households with large herds must range further away from the water point than those with small herds (Gallais, 1975).

Range management projects in the past were typically ignorant of the presence of such traditional techniques and attempted to impose different, often inappropriate, kinds of reserves and rotation systems (Gilles, 1993). The greater the use of mobility as a management tool, the less chance of long-term degradation, and the more possibility of allowing short-term perturbations (e.g. periodic high grazing pressure) without endangering the functional performance of the ecosystem; therefore, the greater the use of mobility, the greater productivity off a unit land area.

By using the practices of mobility and daily monitoring or tracking of resources employing appropriate indicators, traditional African livestock production systems have been able to maintain a higher stocking rate on the range than individual ranchers. The latter have fixed stocking rates for different paddocks which are not fine-tuned to daily and spatial micro-variations. As a result, ranchers have to maintain a constant stocking rate that in the long term is below that of the traditional extensive system (Behnke and Scoones, 1991).

Livestock impact on wildlife and the complex interactions between the two have been studied extensively in Kenya, Tanzania and Zimbabwe. Pastoralists and wildlife have co-existed in the Sahel for over 2000 years. Grazing and burning by pastoralists have helped to shape the present highly valued landscape (Homewood and Rodgers, 1987). Livestock populations have shown no overall trend of increase, although populations of wildlife have. In the case of the Ngorongoro Conservation Area (NCA) in Tanzania, a study shows that disease interactions favour wildlife, not livestock, and that erosion rates are lower in the NCA than surrounding areas, despite its greater predisposition to erosion. Despite these facts, pastoralists are being expelled from the NCA (Homewood, Rodgers and Århem, 1987).

Tracking

Within the bounds prescribed by their rules on mobility, pastoralists depend upon tracking to determine daily movements. The ecosystem is monitored and grazing behaviour is fine-tuned on a daily basis[2] in order to maintain a sustainable balance. The use of pastures by livestock and their needs are tracked by herders and balanced with the productivity and potential of the ecosystem, in order to make optimal use of ecological heterogeneity.

Table 10.1 gives some examples of tracking indicators used by various groups in arid and semi-arid Africa. The role of scouts and herders is of paramount importance in these systems, because tracking requires continual feedback and evaluation of the ecosystem on a daily basis.

Tracking requires:

- freedom of movement
- the selection of monitoring indicators that are sensitive to ecological and biological changes,
- specialized labour and talent for tracking and evaluating ecological processes,
- free information exchange between different groups and tribes through scouts, herders, leaders, traders and other itinerants,
- an institutional structure at the local level that acts upon the information received.

Key-site management and improvement

Group ranches, grazing blocks and the first generation of herder's associations were all based on the management of large blocks of land. This has proven to be impractical, requiring large expenditures using remote techniques for resource surveys, and in some cases fencing. The resultant 'management plans' existed only on paper and did not leave Government Range Officers' desks. Spatial heterogeneity in ecosystem productivity implies that proper management and improvement of small, crucial, well-defined resources ('key site management') will have a beneficial 'spill-over' effect on the remainder of the rangelands, and will have a greater impact on the overall ecosystem, than spreading interventions thinly over large areas (Behnke, 1994). This is particularly relevant for evaluating the cost-effectiveness of range improvements.

There are several noteworthy traditional techniques for range improvement: bush fires on particular patches and at appropriate seasons to destroy old growth and take advantage of the new growth, to destroy parasites, etc. (Ware, 1977; McDermott and Ngor, 1983); reserves to allow spontaneous regeneration; and cutting away at bush infestation by hand or by goat browsing (Jacobs, 1980; Legesse, 1984; Riesman, 1984). Water points are located and constructed in such a way as to minimize concentration and grazing pressure, and intricate rules have been developed for their management and maintenance (e.g. Helland, 1982; Maliki *et al.*, 1984; Putman, 1984).

Table 10.1. *How pastoralists track ecosystem processes – some examples*

Tracking ecological processes	Examples of descriptive and trend variables	Examples of indicators	References
Climate	Local calendars more flexible than Western Wodaabe Fulani have 8 seasons/ year combining climatic variation with changes in plant phenology Turkana remember drought patterns: one in every 4 or 5 years is a good wet season	Behaviour of fauna Changes in plant phenology Changes in meteorological conditions (air temperature, lightning patterns, etc.)	Gulliver (1970); Knight (1974) Jackson (1982); Ba (1982); Maliki et al. (1984)
Soil agricultural potential	Soil described according to type, moisture content, geomorphology, mineral content, colour, topography Soil described according to potential for forage or crops or trees	Specific plant indicators Topography Shade Disease/parasites Soil colour and texture	Knight (1974); Tubiana and Tubiana (1977); Maliki et al. (1984); Winter (1984); Oba (1985); Stiles and Kassam (1986)
Ground water availability	Water table described according to water pressure, depth to water, soil profile Water location described according to forage availability	Specific plant indicators and vegetation community types Topography Specific wild fauna as indicators	Tubiana and Tubiana (1977); Ba (1982)
Forage availability	Detailed description of patches and pasture types, soils Description of catenas and topography	Landmarks that establish location of patches Soil type Specific plant indicators	Langley (1975); Adegboye et al. (1978); Ba (1982); Niamir (1982); Maliki et al. 1984

Category	Objectives	Criteria	References
Temporal environmental variability	Describe changes with drought and other rainfall variation in plant community Predict future changes Morphology and phenology of plants that allow resistance to stress and adaptation to drought	Forage quantity (density & height of grasses, soil cover) Tree cover Forage quality (leafiness, greenness, no trampling) Water quality and quantity Livestock behaviour (restlessness/stampedes, lustre of coat, faeces quality, number of cows in heat, animals acting satiated) Presence/absence of wild fauna indicator species Vegetation diversity Specific indicator plants Vegetation cover Livestock behaviour Preceding season's meteorological conditions	Knight (1974); Benoit (1978); Bernuse (1979)
Environmental degradation	Classify types of degradation Classify stages of degradation Causes	Specific indicator plants Plant composition Soil cover and compaction Grazing pressure (trampling, faeces) Livestock behaviour (especially milk yield)	Spencer (1965); Marchal (1983); Benoit (1978); Clyburn (1978); Niamir (1982); Western (1982); Homewood and Rodgers (1984)

Modern rangeland improvement technologies, such as reseeding, scar-ification, fertilizer application, bush clearing, controlled fire, etc., have been successful in developed countries, but their use in Africa has rarely extended beyond experimental stations and projects. Machinery, firelines, fertilizers, etc. are too costly and beyond the means of local communities. These technologies rarely involve local people in their design, experimenta-tion and application. Rangeland improvement can be within the reach of pastoral communities, if: it is based on indigenous knowledge and prac-tices; it responds to the production system needs; it requires low external inputs; the community has access to credit and other financial resources; modest and progressive objectives are set; and focus is placed on key-sites selected by the local community.

Macro-systems: property rights and macro-mobility

Common property regimes

Traditional pastoral systems have different territorial units, resource tenure regimes, and institutional set-ups. In general, four levels of nested resource tenure/institutions can be distinguished in pastoral systems:

1. Customary pastoral territory, belonging to the 'tribe' or sometimes to a federation of tribes;
2. 'Annual grazing area' within the larger tribal territory, whose boundaries are flexible (depending on rainfall and other factors), and where several clans, sections, sub-clans, etc. have priority rights of usage. The annual grazing area includes both the wet and dry season grazing areas.
3. Key-sites within each dry season base, where the primary user has prior-ity of usage as well as responsibility for management.
4. Group and individual resources/areas, where the household or a group of households is the primary owner or user (e.g. ownership of specific trees by individual households in Turkana, Kenya).

At each level, the corresponding institution creates and enforces infor-mal and formal rules for the use and management of resources (Niamir, 1990). As an example, Figure 10.4 provides the nested territories of the Ngok Dinka of the Sudan. The common principles underlying these rules are discussed in the next section.

Outside a tribal territory, there are buffer zones and overlapping territo-ries. Overlapping territories between two neighbouring groups are main-tained in order to avoid clashes, allow room for expansion, maintain

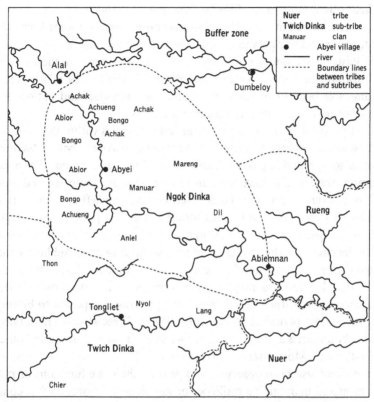

Figure 10.4.   Territorial rights of tribes, sub-tribes and clans in south-central Sudan. Dashed lines represent boundary lines between tribes and sub-tribes (From Niamir, 1982.)

ecosystem resilience, and provide a fall-back area in hard times. These over-lapping areas do not have very distinct boundaries, but are jointly managed by the neighbouring tribes, such as between the Samburu and Rendille of Kenya (Spencer, 1965).

Buffer zones (larger and more extensive than overlapping territories) were kept for the same reasons, but could be used by more than two groups. They often did not come under strict management by any group, but access to them was negotiated between parties concerned on an *ad hoc* basis. In recent years, overlapping and buffer zones have been taken over by immigrating communities (whether of the same ethnic groups or not), due to population expansion and destruction of local authority. The lack of such buffering areas puts added stress not only on the production system but also on environmental resilience.

Many of the traditional rules between two pastoral groups are also applied to tenurial relationships between pastoral and sedentary peoples. In almost all cases, the settled populations raise crops and some livestock, and therefore claim rights to rangelands immediately adjacent to their villages. Pastoralists, whether from near or far, establish rights to these rangelands for grazing, access to water, and passage. In most cases, relationships are amicable because of the specialization of each side, and options for trade/exchange (e.g. milk and manure for cereals).

Most pastoral groups establish large range reserves within their annual grazing areas in order to provide the production system with a 'savings bank' of forage, and to protect the resilience of the environment. The most common strategy is to have drought reserves that are protected and used only in the event of drought. These extensive lands are left ungrazed, often because they are distant or because forage is relatively poor, but constitute a major forage resource when drought hits. Not only are animals able to buffer the effects of the drought, but normal grazing areas are little used, so that their resilience is not damaged during these stressful periods (Odell, 1982). Another form of range reserve is protected and used only during large ceremonial gatherings when grazing pressure is expected to be very high for a short period of time (Schlee, 1987). These reserves have fairly distinct boundaries and are clearly recognizable by local herders and other habitual users. Most reserves are communal, but in a few cases, private reserves have also been observed, and provide the same functions of risk reduction and maintaining resilience of key-sites to a household as communal reserves do for a group (Ostberg, 1987; Ole kuney and Lendiy, 1994).

Researchers argue that the greater the variability of a natural resource, the more suited it is to being held and managed communally, since the relatively low returns from the arid resource do not warrant the costs of organizing and enforcing more exclusive forms of tenure (Behnke and Scoones, 1991). Communal forms of management have the added advantage of supporting a greater number of people on the same land area, without endangering ecological sustainability and resilience.

This conclusion is now being tested in a few development projects in the Sahel, and one result is that the management of common property is impossible unless the land is owned by a well-defined community. Group ranches in East Africa failed partly because the 'group' was an amalgam of individuals who had volunteered to join, lacking established socio-political links. Other factors of importance include: effective leadership and conflict-resolution mechanisms, tenure security, and effective linkages to government institutions (Shanmugaratnam *et al.*, 1992; FAO, 1992). Even private ranches in America continue to depend on public land (federal or state controlled) as a way of adjusting to environmental variability. The majority of pastoral peoples will continue to rely, in the short and medium terms at least, on communally held land. Ownership of the common land by established, legitimate, and representative local institutions provides the incentive for communal action on the management, maintenance and development of natural resources.

Macro-mobility

Macro-mobility of pastoralists is organized through several key principles: 'inclusion' rather than exclusion; some form of a 'fee' given to the owner of the land; varying degrees of priority allocated to multiple users of the same resource; flexibility in rules and agreements; incentives and sanctions for respecting the communal rules.

Transhumants follow the principle of *'inclusion'* rather than exclusion in the exercise of property rights. They recognize the right to use territories belonging to others, as well as the need imposed by environmental variability to allow flexibility and inclusivity in access rights to land. They distinguish between sojourn areas and passage routes, whether within or outside their own territory. Regular routes and sojourn areas are usually established through agreements negotiated through many generations. Examples are Sudan's Kababish (El-Arifi, 1979), the Twareg (Winter, 1984) and the Zaghawa of Chad (Tubiana and Tubiana, 1977). Figure 10.5 gives an example of transhumance routes of northwestern Sudanese tribes, and the buffer zones between them.

The *'fee'* exchanged for rights to use one's territory are usually non-material, reciprocal rights to demand access to the other's territory when needed. The principle of inclusivity has often been misinterpreted by outsiders to mean 'open access'. However, the very presence of a herder in another's territory puts him, and his herd and his tribe in a state of obligation towards the other tribe. Therefore, rights to another's territory are only exercised when both tribes know that in the future the favour can be reciprocated. The decision whether to exercise these rights is made by the tribe evaluating their ecosystem's potential, predicting future ecological variability, and judging current needs in relation to this potential, in addition to other socio-political considerations.

A feature of great importance is that although different people can use the same communal land, users are subject to regulations that determine their degree of *priority* of use. Any group has priority of use within the boundary of its 'home territory'. But this land can also be used by others

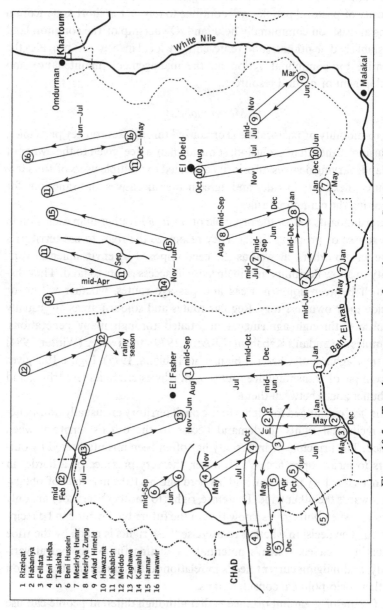

Figure 10.5. Transhumance routes of northwestern Sudanese tribes.

1 Rizeiqat
2 Habaniya
3 Fellata
4 Beni Helba
5 Taaisha
6 Beni Hussein
7 Mesiriya Humr
8 Mesiriya Zurug
9 Awlad Himeid
10 Hawazma
11 Kababish
12 Meidob
13 Zaghawa
14 Kawahla
15 Hamar
16 Hawawir

seasonally or infrequently (e.g. during major droughts). Transhumance between separate territories is negotiated and organized between the different priority users. Priority of rights to use land also determines the responsibility for management[3] of the resource. Priority users have the greatest or sole responsibility for the management of the resource, while other users have to abide by the rules developed by the former.

Access by other users is subject to *flexible negotiations*. The basic accords established through socio-political and ecological criteria can span a generation or more, but detailed agreements that determine actual yearly use are renegotiated frequently in order to adjust to environmental variability. Such negotiations are in the form of informal contracts agreed upon by leaders of both sides. A high degree of formal organization and the presence of laws and procedures have been observed, for example, in the Mecina empire of Mali (Wagenaar, Diallo and Sayers, 1986), Turkana of Kenya (Gulliver 1975), Tswana (Schapera, 1940), Rufa'a al Hoi of Sudan (Ahmed, n.d.), Somali (Rabeh, 1984), Berti of Sudan (Sandford, 1984) and Il Chamus of Kenya (Little *et. al.*, 1987).

Contrary to existing mythology, users of a common resource generally develop credible commitments to sustainable use without relying on external authorities (Ostrom, Walker and Gardner, 1992). Pastoral societies have *incentives* and *sanctions* for enforcing the communal rules. Enforcement of rules between tribes or distinct social units is based on the perceived interdependence of the tribes, and on future possibilities of reciprocated use of resources. Within a social unit (e.g. a clan), enforcement of rules is based on social ostracism, prestige from being model members, public sanctions and punishments, traditional accountability in political leadership, informal police force or observers, and observance of an informal 'fairness ethic' (Niamir, 1990; Swallow and Bromley, 1991).

Some of the intricate traditional range management rules and regulations still survive, but many have disintegrated due to external pressures (nationalization of land, expropriation of land by cultivators, rangeland shortages, desiccation, public water development, etc.). Without effective communal control on rangelands, the traditional grazing systems are no longer co-ordinated, certain key-sites are abused or destroyed, and there is no longer an incentive for protection of reserves or improvement of the resources (Homewood *et al.*, 1987; Barton, 1993).

However, customary rights were effective in managing tenure systems because of their flexibility and adaptability, based on local-level and *ad hoc* negotiation processes. It may still be possible to revive the rules in certain situations or, if not, to create new and adaptable rules based on old principles. Formal (government) recognition of such rules and rights, without creating parallel state institutions, may be necessary in order to provide political legitimacy to local authorities. But the very act of formalizing them may result in static and inappropriate laws. The greater the flexibility in property rights, the less the formal legal framework is able to cope with them (Vedeld, 1993; Behnke, 1994). There is a need for the creation of a two-tiered legal system: an overall legal framework at the national level that officially recognizes decentralized customary rules and common property and, at the local level, flexible rules and procedures developed by a decentralized institutional system that provides a forum for negotiation and conflict resolution.

Synthesizing design principles

The more arid the natural resource, the more suited it is to being held and managed communally, since it can maintain a greater number of people without endangering ecological sustainability and resilience. In arid ecosystems, there are three major ecological variables: spatial productivity, temporal productivity, and resilience over time. Traditional pastoral systems in Africa have adapted their socio-economic systems to these variables, using several main tools: indigenous systems of classification and evaluation of ecosystems, mobility, tracking, dispersion, diversity, reciprocity, flexibility, communal co-ordination, and key-site management and improvement. These mechanisms and variables allow pastoralists to take advantage of patches of productivity without destroying the resilience of the environment and endangering its sustainability. These mechanisms are translated into specific socio-economic structures, such as appropriate communal institutions, tenure regimes, rules and regulations, and conflict-resolution mechanisms.

Monitoring and tracking ecological variability, and predicting its sustainability and resilience, are appropriate technologies that herders have used with considerable success. In arid and semi-arid lands, because of extremes of variability, such technologies cannot be effective unless the production system has the flexibility to move its base in accordance with the variability. Mobility not only allows the household to minimize its production risks, but also subjects the ecosystem to short but intense pulses of grazing that allow it to maintain its long-term resilience and to function under as wide a range of climatic conditions as possible – even episodic devastating droughts (Perrings and Walker, 1995).

One viable design for property rights is to maintain their flexibility in accordance with the flexible nature of the production system. A 'nested' structure where the rights are tailored to the resource, and resource use is controlled inclusivity ('managed commons') is more appropriate to arid systems (Swift, 1995). For example, private rights can be assigned by the community-based authority to farms, water for domestic use, and specific resources such as rights to put bee-hives in certain trees. At the same time, communal rights can be assigned to grazing lands, forests, ponds, salt licks, etc.

Parallel to a nested property regime, there should also be a nesting of appropriate institutions designed to manage and maintain the resources, for example private individuals or households to manage private property; neighbourhood councils to manage specific key-sites; village councils to manage, establish and enforce rules on communal grazing; and tribal councils to negotiate access to other territories. Local governments should be incorporated in the design to provide an enabling environment and an 'even playing field' through establishment of standards, permits and sanctions and the enforcement of rules.

Production systems, adapted to their specific ecosystems, cannot be fitted into a common model. However, a few common strategies can be distilled from the foregoing discussion.

1. *Communal use and regulation* of arid and variable resources. Communal use and regulation of natural resources is feasible, in the long run being able to provide sustainable use of arid and highly variable resources, and able to sustain a higher number of people on the same unit of land than a comparable privatized production system. It is also more equitable since it allows poorer segments of the population to derive sustenance rather than be shunted into urban slums. The encouragement of planned communal use requires an enabling environment by the government, secure tenure rights, appropriate local institutions, and a decentralized process of decision making and authority.
2. *Spatial and temporal flexibility in resource use.* Mobility of animals as a resource management tool can be revived and encouraged by the authorities.[4] It requires supporting rules and regulations that are also flexible; pastoral institutions that have the authority to impose and enforce communal rules; temporary use rights to lands owned by others; establishment and development of transhumance routes; reciprocal arrangements for temporary use of land between neighbouring pastoral units; a revival of herding as a profession through training and equipment of professional herders; and elimination of the stigma attached to

mobility. In addition, and concurrently, appropriate technologies to intensify the production of home-based animals can be extended.

3. *Flexible drought-adapted strategies.* There is a need for the creation and revival of safety nets against the devastations of droughts. These safety nets can be based on traditional tools, such as insurance schemes, credit, restocking programmes, reciprocal exchanges, etc. Successful examples where such traditional mechanisms have been revived in the Sahel are due to a unified community, decentralized decision making, communal resources as collateral, and communal guarantees for repayment of loans (e.g. OXFAM's 'Habbanae' re-stocking schemes in Niger, UNSO's credit programme in northern Burkina Faso, and village-organized collection of dues for water point maintenance in Mandera, Northern Kenya).

4. *Multiple-uses and diversity.* The use of resources has to be fitted to the resource for optimum and sustainable production. Given the heterogeneity of patches and high functional diversity in arid ecosystems, the production system too should consist of multiple objectives and options. Multiple use, compared to mono use, means a lower production from each use, but higher production in the aggregate. It also means that all resource types will be used at a level below maximum, that no single one will be mined, and therefore the long-term resilience will not be hampered. Multiple use has the added benefit of acting as a cushion in times of drought. But even in normal times, multiple use builds resilience into production systems' outputs, and provides the resource manager with a diverse portfolio from which to adjust strategies to his household's requirements.

The diversity of social–ecological mechanisms used by pastoral cultures is a necessary adaptation to the unpredictable and highly variable environment (Figure 10.6). Wherever this diversity has been reduced, environmental degradation has followed. Wherever the flexibility offered by each mechanism has been curtailed, the production system has suffered and the community has lost its viability and ability to maintain environmentally sustainable production. Each mechanism is linked to and dependent on the others. Reducing dispersion in space (e.g. through concentration of herding units into settled villages, as in the case of villagization in Tanzania) can result in loss of mobility (and overgrazing around settlements), less negotiation over access, and fewer tiers of nested rights (increased conflicts and expropriation of land by outsiders), fewer incentives for monitoring (therefore less awareness of environmental feedbacks), more dependence on

Environment ←→	Social–ecological mechanisms	←→	Culture
productivity	mobility	decentralization	institutional
variability	dispersion	nested rights	complexibility
	diversity	flexible rules	heterogeneity
long-term	tracking	local enforcement	
condition and	multiple-use	monitoring	indigenous
sustainability	negotiation	scouting	knowledge
	reciprocity	drought adaptation	
	flexibility		

Figure 10.6.   Conceptualization of the social–ecological linkages and
mechanisms for greater resilience in pastoral systems.

centralized institutions, and loss of drought-adapted strategies such as group herding and reciprocal labour exchanges (Ndagala, 1991).

The 1990s are seeing the revival of transhumance in areas that for long have been privatized or under state control. In the Middle East, the traditional system of range reserves (ahmia) has been the subject of revival in development projects since the mid-1980s (Draz, 1981; Nesherwat, 1991). In Europe, Italian and Spanish communities have formed NGOs to reinstate the commons on pastures of the Pyrenées and the Alps. Restructuring in Mongolia has seen a spontaneous return to traditional forms of herd movement (Mearns, 1993).

In most countries of the Sahel, national policies directly or indirectly inhibit pastoral development; land tenure legislation and policies do not legally recognize customary tenure nor common land; centralized government bureaucracies place little or no priority on development in pastoral areas; agriculture and forestry policies discriminate against pastoralists; policies that promote government provision of services inhibit the emergence of local, private services; political representation by pastoralists within parliament or other national fora is weak; etc. A few priority areas can be seen as pre-requisites to the strengthening of pastoral resilience. One priority area is the redrafting of inappropriate national policies. In most countries of the Sahel, existing policies are coming under review due to the impetus provided by structural adjustment programmes and the Convention to Combat Desertification. However, most countries do not have a national-level institution capable of advocating the views of the pastoralists, and it is feared that they will once again be left out.

The 1990s have seen a mushrooming of 'natural resource management' projects and programmes, some of which are located in arid and semi-arid lands. Hardly any of these projects target mobile pastoralists, because the donor community is sometimes as much ignorant of pastoralism as are the governments. A second priority area for pastoral development is to convince governments and donors of the necessity for flexibility in resource use, and for regulated mobility in pastoral areas. Fortunately, most current-generation projects are more or less participatory, and the groundwork is there – although more field research is needed – to provide the right arguments and factual data to decision makers, and to allow the voice of pastoralists to be heard. However, more needs to be done to restructure development work to give pastoralists a real and effective voice.

Just as the health and resilience of an arid ecosystem are related to its functional diversity, the health and resilience of pastoral production are directly related to the diversity of its options, its adaptations and its objectives. Perhaps the strongest plea that can be made is for appropriate development programmes that avoid models that simplify the system, and emphasize diversity and complexity by focusing on the appropriate *process* of development.

Notes

1 Key-sites are resources of relatively limited geographical extent, but of great importance to the production system (e.g. water points and immediate area surrounding them), or critical seasonal factors (e.g. dry season reserves, salt licks, swamps, areas with special palatable plants, etc.).
2 Also known as 'opportunistic grazing strategy' (Sandford, 1983; Behnke and Scoones, 1991).
3 Here the distinction is made between 'utilization' (i.e. using a resource), and 'management' (i.e. devising rules and regulations about how the land should be used; planning improvements and maintenance regimes; in addition to using the land).
4 It should be recognized, however, that land degradation is not only due to high animal pressure, but also to deforestation and over-cultivation by settled populations. Therefore, keeping animals on the move is only part of the solution, and should be accompanied by afforestation, agroforestry and other resource-building measures in the settled areas.

References

Adegboye, R.O. *et al.* 1978. *A Socio-economic Study of Fulani Nomads in Kwara State*, Ibadan: Federal Livestock Department, Kaduna.
Ahmed, A.G.M. (n.d.) Nomadic competition in the Funj area. *Sudan Notes and Records*, Khartoum.
Ahmed, A.G.M. 1978. *Integrated Rural Development: Problems and Strategies. The Case of the Dinka and the Nuer of the Jonglei Project Area in the Sudan*. Executive Organ Development Projects in Jonglei Area, Report no.8, Republic of Sudan.
Allan, W. 1965. *The African Husbandman*, London: Oliver and Boyd.
Aubert, G. and Newsky, B. 1949. Note on the vernacular names of the soils of the Sudan and Senegal. In *Proceedings of 1st Commonwealth Conference on Tropical and Subtropical Soils*, CBSS Technical Communication no. 46, pp.107–9. Harpenden: Commonwealth Bureau of Soil Science.

Ba, A.S. 1982. *L'art veterinaire des pasteurs Saheliens.* ENDA serie Études et Recherches. No. 73–82. Dakar.

Barral, H. 1974. Mobilité et cloisonnement chez les eleveurs du Nord de la Haute-Volta: les zones dites 'd'endrodromie pastorale'. *Cahiers ORSTOM Serie Sciences Humaines* 11: 127–36.

Barrow, E. 1988. *Trees and Pastoralists: the Case of the Pokot and Turkana.* Social Forestry Network Paper No. 6b, London: ODI.

Barton, D. 1993. *Community Participation in Range Management and Rehabilitation in Kenya.* Workshop on New Directions in African Range Management and Policy, Woburn. London: ODI/IIED/Commonwealth Secretariat.

Baxter, P.T.W. and Butt, A. 1953. *The Azande and Related Peoples of the Anglo-Egyptian Sudan and Belgian Congo.* London: International African Institute.

Behnke, R. 1994. *Natural Resource Management in Pastoral Africa,* London: ODI IIED.

Behnke, R.H. and Kerven, C. 1984. *Herd Management Strategies among Agropastoralists in the Bay Region, Somalia.* Bay Region Socio-economic Baseline Study, Somalia.

Behnke, R.H. and Scoones, I. 1991. *Rethinking Range Ecology: Implications for Range Management in Africa.* London: ODI/IIED.

Behnke, R.H., Scoones, I. and Kerven, C., eds. 1993. *Range Ecology at Disequilibrium: New Models of Natural Variabilty and Pastoral Adaptation in African Savannas.* London: ODI IIED Commonwealth Secretariat.

Benoit, M. 1978. Pastoralisme et migration. Les Peuls de Barani et de Dokui (Haute Volta). *Études Rurales* 70: 9–50.

Bernuse, E. 1979. L'arbre et le nomade. *JATBA* 26(2): 103–28.

Bonfiglioli, A.M. and Watson, C.J. 1992. *Pastoralists at a Crossroads: Survival and Development of African Pastoralism,* NOPA Project, Nairobi: UNSO/UNICEF.

Bourgeot, A. 1981. Pasture in the Malian Gourma; habitation by humans and animals. In *The future of pastoral peoples: proceedings of the conference,* pp.165–82, ed. J.G. Galaty, D. Aronson, P.C. Saltsman and A. Chouinard. Ottawa: IDRC.

Bovin, M. 1990. Nomads of the drought: fulbe and wodabee nomads between power and marginalization in the Sahel of Burkina Faso and Niger Republic. In *Adaptive Strategies in African Arid Lands,* pp.29–58, ed. M. Bovin, and L. Manger. Uppsala: SIAS.

Bovin, M. and Manger, L., eds. 1990. *Adaptive Strategies in African Arid Lands,* SIAS, Uppsala.

Brandstrom, P., Hultin, J. and Lindstrom, J. 1979. *Aspects of Agropastoralism in East Africa.* Research Report no. 51. Uppsala: SIAS.

Child, R.D. Heady, H.F., Hickey, W.C., Peterson, R.A. and Pieper, R.D. 1984. *Arid and Semiarid Lands: Sustainable Use and Management in Developing Countries,* Washington DC: USAID.

Clyburn, L. 1978. The process of change in certain livestock owner and operating groups in West African Sahel. In *Proceedings of the 1st International Rangeland Congress,* pp.108–10, ed. D.N. Hyder. Denver: SRM.

de Angelis, D. and Waterhouse, J. 1987. Equilibrium and non-equilibrium concepts in ecological models. *Ecological Monographs* 57: 1–21.

de Leeuw, P.N. and Tothill, J.C. 1993. The concept of rangeland carrying capacity in sub-saharan Africa – myth or reality. In *Range Ecology at Disequilibrium,* pp.77–88. ed. R.H. Behnke, I. Scoones and C. Kerven. London: ODI/IIED.

Draz, O. 1981. *An Integrated Programme for Range Management and Sheep Feeding for the Near East and North African Countries.* Rome: FAO.

El-Arifi, S.A. 1979. Some aspects of local government and environmental management in the Sudan. In *Proceedings of the Khartoum Workshop on Arid Lands Management*, pp.36–9, ed. J.A. Mabbutt. Khartoum: University of Khartoum.

Ellis, J.E. and Swift, D.M. 1988. Stability of African pastoral ecosystems: alternate paradigms and implications for development. *Journal of Range Management* 41: 450–9.

FAO 1992. *International Workshop on Pastoral Associations and Livestock Cooperatives*, Njoro, Kenya. Rome: FAO.

Farah, M.I. 1993. *From Ethnic Response to Clan Identity: a Study of State penetration among the Somali Nomadic Pastoral society of Northeastern Kenya.* Stockholm: Almqvist & Wiksell.

Feeny, D., Berkes, F., McCay, B.J. and Acheson, J.A. 1990. The tragedy of the commons: 22 years later. *Human Ecology* 18: 1–19.

Fratkin, E. 1986. Stability and resilience in East African pastoralism: the rendille and the Ariaal of northern Kenya. *Human Ecology* 14: 269–86.

Friedel, M.H. 1991. Range condition assessment and the concepts of thresholds: a veiwpoint. *Journal of Range Management* 44: 422–6.

Galaty, J.G. 1988. Scale, politics and cooperation in organization for East African development. In *Who shares? Cooperatives and Rural Development*, pp.282–308, ed. D.W. Attwood and B.S. Baviskar. Oxford: Oxford University Press.

Galaty, J.G. 1993. *Individuating Common Resources: Sub-division of Group Ranches in Kenya Maasailand.* Workshop on New Directions in African Range Management and Policy, Woburn. London: ODI/IIED/ Commonwealth Secretariat.

Gallais, J. 1975. *Pasteurs et paysans du Gourma: la condition Sahelienne.* Paris: CEGET/CNRS.

Gilg, J-P. 1963. Mobilité pastorale au Tchad occidental et central. *Caheirs d'études Africaines* 3: 491–510.

Gilles, J.L. 1993. *New Directions for African Range Management: Observations and Reflections*, Workshop on New Directions in African Range Management and Policy, Woburn. London: ODI/IIED/Commonwealth Secretariat.

Gulliver, P.H. 1970. *The Family Herds: a Study of Two Pastoral Tribes in East Africa, the Jie and Turkana.* Westport: Negro University Press.

Gulliver, P.H. 1975. Nomadic movements: causes and implications. In *Pastoralism in Tropical Africa*, pp.369–86, ed. T. Monod. Oxford: Oxford University Press.

Hardin, G. 1968. The tragedy of the commons. *Science* 162: 1243–8.

Helland, J. 1982. Social organization and water control among the Borana. *Development and Change* 13: 239–58.

Hellden, U. 1984. *Drought Impact Monitoring: a Remote Sensing Study of Desertification in Kordofan, Sudan.* Institute for Natural Geography, no.61. Lund: Lund University.

Homewood, K. and Rodgers, W.A. 1984. Pastoralism and conservation. *Human Ecology* 12: 431–41.

Homewood, K. and Rodgers, W.A. 1987. Pastoralism, conservation and the overgrazing controversy. In *Conservation in Africa*, pp.111–28, ed. D. Anderson and R. Grove. Cambridge: Cambridge University Press.

Homewood, K., Rodgers, W.A. and Århem, K. 1987. Ecology of pastoralism in
    Ngorongoro Conservation Area, Tanzania. *Journal of Agricultural Sciences
    of Cambridge* 108: 47–72.
Howes, M. 1980. The uses of indigenous technical knowledge in development.
    In *Indigenous knowledge systems and development*, pp.335–51, ed. D.
    Brokensha, D.M. Warren and O. Werner. Washington DC: University Press
    of America.
Hussein, M.A. 1990. Traditional practices of camel husbandry and management
    in Somalia. In *The Multipurpose Camel: Interdisciplinary Studies on Pastoral
    Production in Somalia*, ed. A. Hjort af Ornås. Uppsala: University of
    Uppsala.
Jackson, I.J. 1982. Traditional forecasting of tropical rainy seasons. *Agricultural
    Meteorology* 26: 167–78.
Jacobs, A.H. 1980. Pastoral Maasai and tropical rural development. In
    Agricultural development in Africa: Issues of Public Policy, pp.275–300, ed.
    R.H. Bates and M.F. Lofchie. New York: Praeger.
Knight, C.G. 1974. *Ethnoscience: a Cognitive Approach to African Agriculture*,
    Paper prepared for SSRC Conference on Environmental and Spatial
    Cognition in Africa, May 1974.
Lane, C.R. 1991. *Alienation of Barabaig pastureland*. PhD thesis, University of
    Sussex.
Lane, C. and Moorehead, R. 1993. *New Directions in African Range Management,
    Natural Resource Tenure and Policy*. Workshop on New Directions in African
    Range Management and Policy, Woburn London: ODI/IIED/
    Commonwealth Secretariat.
Langley, P. 1975. The ethnolinguistic approach to the rural environment: its
    usefulness in rural planning in Africa. In Richards, P., ed. *African
    Environment: Problems and Perspectives*, pp.89–101. Special Report No.1.
    London: International African Institute.
Legesse, A. 1984. Boran-Gabra pastoralism in historical perspective. In
    *Rangelands, a Resource under Siege; Proceedings of 2nd International
    Rangeland Congress*, pp.481–2, ed. P.J. Joss, P.W. Lynch and O.B. Willians.
    Adelaide, Australia.
Little, P.D. 1985. Social differentiation and pastoral sedenterization in northern
    Kenya. *Africa* 55: 243–61.
Little, P.D. 1987. Risk aversion, economic diversification and goat production. In
    *Proceedings of the 3rd International Conference on Goat Production and
    Disease*, Tucson, USA, pp.428–30
Little, P.D., Horowitz, M.M. and Nyerges, A.E., eds. 1987. *Lands at Risk in the
    3rd World*. Boulder: Westview.
Maaliki, A. White, C., Loutan, L. and Swift, J.J. 1984. The wodaabe. In *Pastoral
    development in central Niger: Report of the Niger Range and Livestock
    Project*, pp.255–530, ed. J.J. Swift. Niamey: USAID, Ministry of Rural
    Development.
Marchal, J.Y. 1983. *Yatenga, Nord Haute-Volta: la Dynamique d'un Espace Rural
    Soudano-Sahelien*. Trav. et Doc. Paris: ORSTOM.
McDermott, J. and Ngor, M.D. 1983. *Grazing Management Strategies among the
    Tuic, Nyarrawng and Ghol Dinka of Kongor Rural Council: Prospects for
    Development*. Kongor Integrated rural development project. Rome: FAO.
Mearns, R. 1993. *Pastoral Institutions, Land Tenure and Land Policy Reform in
    Post-socialist Mongolia*. PALD Project Research Report No.3. Mongolia:
    IDS, Institute of Agricultural Economics.

282   Maryam Niamir-Fuller

Moorehead, R. 1993. Policy options for pastoral resource tenure in non-equilibrium environments. In *Proceedings of Workshop on Pastoral Natural Resource Management and Pastoral Policy in Africa*, pp.17–25. New York: UNSO.

Ndagala, D.K. 1991. *Pastoralism and Rural Development: the Ilparakuyo Experience*, New Delhi: Reliance Publishing House.

Nesherwat, K.S. 1991. *Socio-economic Aspects of the Traditional Hema System of Arid Land Management in Jordan*. Rome: FAO.

Niamir, M. 1982. *Report on Animal Husbandry among the Ngok Dinka of the Sudan*. Integrated Rural Development Project, Abyei, Sudan. HIID Rural Development Studies. Cambridge, Mass.: Harvard University.

Niamir, M. 1990. *Herder's Decision-making in Arid and Semi-arid Africa*. FAO Community Forestry Note No.4. Rome: FAO.

Niamir, M., ed. 1994. *Proceedings of Workshop on Pastoral Natural Resource Management and Pastoral Policy in Africa*, New York: UNSO.

Niamir, M., Lugando, S. and Kundy, T. 1994. *Barabaig displacement from Hanang District to the Usangu Plains: Changes in Natural Resource Management and Pastoral Production in Tanzania*. FTPP Working Paper. Rome: FAO, IIED, SUAS.

Oba, G. 1985. Local participation in guiding extension programs: a practical approach. *Nomadic Peoples* 18 :27–45.

Odell, M.J. 1982. *Local Institutions and Management of Communal Resources: Lessons from Africa and Asia*. Pastoral Network Paper No. 14e, London: ODI.

Ole kuney, R. and Lendiy, J. 1994. *Pastoral Institutions among the Maasai of Tanzania*. Paper presented at 6th PANET Workshop. Dar es Salaam: FAO.

ORSTOM 1986. *Nomadisme: mobilité et flexibilité?* Bulletin de Liaison No.8, Paris: Department H.

Ostberg, W. 1987. *Ramblings on Soil Conservation: an Essay from Kenya*. Vernamo: SIDA.

Ostrom, E., Walker, J. and Gardner, R. 1992. Covenants with and without a sword: self-governance is possible, *American Political Science* 86: 404–17.

Peacock, C.P. 1987. Herd movement on a Maasai group ranch in relation to traditional organization and livestock development, *Agricultural Administration and Extension* 27: 61–74.

Penning de Vries, F.W.T. and Djitèye, M.A., eds. 1982. *La productivité des paturages saheliens: une étude des sols, des végétations et de l'exploitation de cette ressource naturelles*. Report No. 918, Wageningen: Wageningen Centre for Agricultural Publication and Documentation.

Perrier, G.K. 1991. *The effects of policy development and organizational structure on the performance of range liestock development projects in Africa*. PhD dissertation, Utah State University, Logan.

Perrings, C.A., Mäler, K-G., Folke, C., Holling, C.S. and Jansson, B-O., eds. 1995. *Biodiversity Loss: Ecological and Economic Issues*. Cambridge: Cambridge University Press.

Perrings, C. and. Walker, B.H. 1995. Biodiversity loss and the economics of discontinuous change in semi-arid rangelands. In *Biodiversity Loss: Ecological and Economic Issues*, pp. 190–210, ed. C.A. Perrings, K.-G. Mäler, C. Folke, C.S. Holling and B.-O. Jansson. Cambridge: Cambridge University Press.

Pimm, S.L. 1984. The complexity and stability of ecosystems. *Nature* 307: 321–6.

Putman, D.B. 1984. Agro-pastoral production strategies and development in the Bay region. In *Proceedings of 2nd International Congress of Somali Studies*, pp.159–86, ed. T. Labahn. Hamburg: University of Hamburg Verlag.

Rabeh, O.O. 1984. The Somali nomad. In *Proceedings of 2nd International Congress of Somali Studies*, pp.57–69, ed. T. Labahn. Hamburg: University of Hamburg Verlag.

Riesman, P. 1984. The Fulani in a development context: the relevance of cultural traditions for coping with change and crisis. In *Life Before the Drought*, pp.171–91. ed. E.P. Scott. Boston: Allen & Unwin.

Roe, E. 1991. Analyzing sub-Saharan livestock rangeland development. *Rangelands* 15: 16–7.

Runge, C.F. 1981. Common property externalities: isolation, assurance and resource depletion in a traditional grazing context, *American Journal of Agricultural Economics*, Nov.1981: 595–606.

Sandford, S. 1983. *Management of Pastoral Development in the Third World*, Chichester: John Wiley & Sons/ODI.

Sandford, S. 1984. Traditional African range management systems. In Rangelands: a resource under siege, *Proceedings of the 2nd International Rangeland Congress*, pp.475–8, ed. P.J. Joss, P.W. Lynch and. O.B. Williams. Adelaide.

Schapera, I. 1940. The political organization of the Ngwato of Bechuanaland protectorate. In *African Political Systems*, pp.56–82, ed. M. Fortes and E. Evans-Pritchard. Oxford: Oxford University Press.

Schlee, G. 1987. *Holy Grounds.* Workshop on Changing Rights in Property and Problems of Pastoral Development in the Sahel, Manchester: Manchester University.

Scoones, I., ed. 1991. *Wetlands in Drylands: The Agroecology of Savanna Systems in Africa.* London: IIED.

Scoones, I., ed. 1995. *Living with Uncertainty: New Directions for Pastoral Development in Africa.* London: Intermediate Technology Publications.

Scoones, I. and Thompson, J. 1994. *Beyond Farmer First: Rural People's Knowledge, Agricultural Research and Extension Practice,* London: Intermediate Technology Publications.

Shanmugaratnam, N., Vedeld, T., Mossige A. and Bovin, M. 1992. *Resource Management and Pastoral Institution Building in the West African Sahel.* World Bank Discussion Paper no. 175. Washington DC: World Bank.

Spencer, P. 1965. *The Samburu: a Study of Gerontocracy in a Nomadic Tribe,* London: Routledge & Kegan Paul.

Stenning, D.J. 1959. *Savannah Nomads: a Study of the Wodaabe Pastoral Fulani of Western Bornu Province, Northern Region, Nigeria.* Oxford: Oxford University Press.

Stiles, D. 1995. Desertification is not a myth. *Desertification Control Bulletin 26*: 29–36.

Stiles, D. and Kassam, A. 1986. An ethno-botanical study of Gabra plant use, Marsabit District, Kenya. *Journal of the East Africa Natural History Society and National Museum* 76: 1–23.

Sutter, J.W. 1978. *Pastoral Herding in the Arrondissement of Tanout, Niger Range and Livestock Project.* Zinder: USAID.

284   *Maryam Niamir-Fuller*

Swallow, B.M. and Bromley, D.W. 1991. *Institutions, Governance and Incentives in Common Property Regimes for African Rangelands*. Workshop on New Directions in African Range Management and Policy, Matapos, Zimbabwe. London: ODI/IIED/Commonwealth Secretariat.

Swift, J.J. 1977. Pastoral development in Somalia: herding cooperatives as a strategy against desertification and famine. In *Desertification: Environmental Degradation in and around Arid Lands*, pp.275–305, ed. M.H. Glantz. Boulder: Westview.

Swift, J.J. 1988. *Major Issues in Pastoral Development with Special Emphasis on Selected African Countries*, Rome: FAO.

Swift, J.J. 1993. *Pastoral Policies*. Paper for Donor/Specialized Agency Consultation on Pastoral Development. New York: UNSO.

Swift, J.J. 1995. Dynamic ecological systems and the administration of pastoral development. In Scoones, I. ed., *Living with Uncertainty: New Directions for Pastoral Development in Africa*, pp.153–73. London: Intermediate Technology Publications.

Tubiana, M-J. 1969. La pratique actuelle de la cueillette chez les Zaghawa du Tchad. *JATBA* XVI(2–5): 4–83.

Tubiana, M-J. and Tubiana, J. 1977. *The Zaghawa from an Ecological Perspective*, Rotterdam: Balkema.

Vedeld, T. 1993. *Enabling Pastoral Institution Building in Dryland Sahel*. Paper for Donor/Specialized Agency Consultation on Pastoral Development, New York: UNSO.

Wagenaar, K.T., Diallo, A. and Sayers, A.R. 1986. *Productivity of Transhumant Fulani Cattle in the Inner Delta of Mali*. Research Report No. 13. Mali: ILCA.

Ware, H. 1977. Desertification and population: sub-saharan Africa. In *Desertification: Environmental Degradation in and around Arid Lands*, pp.166–202, ed. M.H. Glantz. Boulder: Westview.

Warren, D.M. and Meehan, P.M. 1980. Applied ethnoscience and dialogical communciation in rural development. In *Indigenous Knowledge Systems and Development*, pp.317–34, ed. D. Brokensha, D.M. Warren and O. Werner. Washington DC: University Press of America.

Warren, D.M. and Rajasekaran, B. 1993. *Using Indigenous Knowledge for Sustainable Dryland Management: a Global Perspective*, International Workshop on Listening to the People: Social Aspects of Dryland Management, Nairobi: UNEP.

Western, D. 1982. The environment and ecology of pastoralists in arid savannas. *Development and Change* 13: 183–211.

Western, D. and Dunne, T. 1979. Environmental aspects of settlement site decisions among pastoral Maasai. *Human Ecology* 7: 75–98.

Westoby, M., Walker, B. and Noy-Meir, I. 1989. Opportunistic management for rangelands not at equilibrium. *Journal of Range Management* 42: 265–74.

Winter, M. 1984. The Twareg. In *Pastoral Development in Central Niger: report of the Niger Range and Livestock Project*, pp.531–620, ed. J.J. Swift. Niamey: USAID.

11

Reviving the social system–ecosystem links in the Himalayas

NARPAT S. JODHA

Introduction

This chapter deals with the natural resource-friendly traditional patterns of resource use in the Hindukush–Himalaya (HK–H) region, their progressive decline under present-day circumstances, and possible approaches to their revival (Figure 11.1). This formulation of resource use, without getting into finer definitional issues, represents the operational dimensions of ecosystem–social system links in the context of fragile mountain areas. The chapter draws on the broad synthesis of inferences and understanding generated by more than 50 studies by different agencies in different parts of the region. It is often inferred that present-day society – particularly the

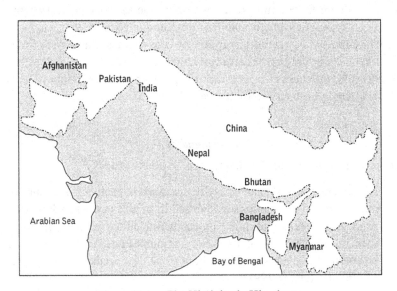

Figure 11.1.   The Hindukush–Himalayas.

285

policy makers, planners, and their technical advisors dealing with the mountain regions – is better equipped than the traditional communities in terms of knowledge of ecosystems and their functional dynamics. And yet, it is unable to design and implement a social framework (covering norms and mechanisms to govern a community's approach and interactions with nature) which could more appropriately respond to the imperatives of the natural resource base. Traditional communities, without the knowledge of formal principles of ecosystem (or natural resource system) dynamics, did understand the manifestations of these dynamics, largely in terms of the myriad opportunities and constraints for the community's survival. Consequently, they evolved norms and practices to regulate individual and collective behaviour (*vis-a-vis* nature) as well as technical and institutional mechanisms to support them. This in turn helped in shaping and sustaining positive ecosystem–social system links. The traces of such links, though under severe strain, can still be found in several parts of the HK–H region (or other relatively inaccessible ecosystems), where modern changes causing disruption of such linkages are yet to have their full impact.

Objectives

The chapter primarily focuses on the following aspects of human–nature interactions in the HK–K region.

(1) The circumstances and processes responsible for positive ecosystem– social system links, their manifestations and implications.
(2) The process of gradual disruption of ecosystem–social system links, and its underlying causes and consequences.
(3) Possible approaches to arrest the disruption and restore the above links by learning from (i) and (ii) above.

Basic premises

The central argument of the chapter can be briefly stated.

(i) Ecosystem–social system links are discussed in terms of dynamics of nature–society interactions manifested through a two-way adaptation process, i.e. society adapting its needs (including mechanisms to fit them) to the features of its natural resources base and adapting or amending the natural resource base to suit its needs (Jochim, 1981; Gadgil and Berkes, 1991).
(ii) In a given social–ecological context, the nature and composition of

adaptation measures, e.g. those directed towards rationing and diversifying the needs or those focused on manipulating and amending the resources to meet demands, represent a society's responses to the objective circumstances created by the specific features of natural resources, on the one hand, and the socio-economic driving forces, on the other. Factors such as the resource users' direct and crucial dependence on, as well as control over, local resources, and their close proximity to and functional knowledge of the resources (as found in the isolated or semi-closed indigenous/traditional communities such as those in remote mountain areas) provide an ecological basis for natural resource management systems (Ellen, 1981; Berkes, 1989; Davis, 1993). An absence of these features (as in the case of open and externally linked areas) usually generates natural resource-use systems, which are governed by social perspectives shaped less by ecological circumstances and more by man-made circumstances; they are more often insensitive to the limitation of the natural resources; and they are more extractive, potentially unsustainable and less conducive to two-way adaptations.

(iii) Restoration of the aforementioned features (i.e. those characterizing traditional systems) that facilitate effective ecosystem–social links and sustainable management of natural resources, seems almost impossible in today's world, with its complex socio-economic realities creating a variety of inter-system linkages, hierarchies and differences. These changes in turn create barriers and widen the distance between resource uses and decision making, and between natural processes and social processes influencing the resource base. Finally, they dilute or erode the local communities' interests and capabilities for sustainable resource use.

(iv) A closer examination of traditional and conventional resource-use systems and their underlying factors can help to identify some present-day functional equivalents of the aforementioned traditional features (i.e. close proximity, direct and crucial dependence), on which a strategy can be planned for restoring ecosystem–social system links.

Empirical base and context

The information base and geographical context

The formulation of the issues and the approaches presented in this chapter are based on field-level understanding gained through studies in selected areas of four countries, as a part of the work programme of the International Centre for Integrated Mountain Development during 1988–93 (Jodha, Banskota and Partap, 1992; Jodha, and Shrestha, 1994). The areas included West Sichuan and Tibet in China; Himachal Pradesh and the hill areas of Uttar Pradesh states in India; middle mountains of Nepal; and North – West Frontier Province of Pakistan.

Focus on mountain agriculture

The primary focus of studies and operational work was mountain (including hill) agriculture, covering all land-based activities such as cropping, horticulture, animal husbandry, forestry, and their support systems. In the very early stage of work, those involved in these studies were alerted, both by reviews of existing literature and by field investigations, to several negative trends characterizing agriculture and the overall natural resource situation in the region. These persistent negative changes manifested in the emerging prospects of unsustainability of resource use in most parts of the HK–H region.

These trends suggested that present patterns of resource use and production practices (in terms of choice of activities and resource-use intensity) were at odds with the imperatives of the features of natural resources in the region. The relevant key features of mountain areas (termed *mountain specificities*) included an incredibly high degree of inaccessibility, fragility, marginality, diversity, and unique production opportunities with comparative advantage to mountain areas compared to other regions (Jodha *et al.*, 1992). The integrated mountain specificities, providing both opportunities and constraints, represent the 'ecosystem' as people would understand it and respond (adapt) to it. Similarly, the patterns of human adaptations to the circumstances created by mountain specificities, as reflected through resource-use practices and the technological as well as the institutional arrangements supporting them, represent the 'social system' in the context of the present discussion.

Drawing on over 50 studies, including those by ICIMOD since 1987, inventories were prepared of past and present resource-use systems, production and consumption practices, and technological and institutional measures, including those relating to demand management and resource upgrading. The inventories were related to the imperatives of the aforementioned mountain characteristics to assess the degree of match or mismatch between the imperatives of resource features (representing ecosystem) and attributes of the resource-use systems, including their technological and institutional underpinnings (representing social systems). The understanding thus generated formed the empirical context of the issues dealt with in this chapter. Note that it is the 'mismatches' that gave rise to persistent negative changes. These changes, described as indicators of unsustainability (Jodha, 1990), relate to:

(i) 'health' of natural resources (e.g. increased land slides and other forms of land degradation indicate poor 'health');

(ii) productivity of land-based activities (e.g. persistent decline in crop yields and biomass yields from pastures); and

(iii) range and quality of resource management options (e.g. unfeasibility of resource regenerative agronomic practices, farming–forestry linkages).

Indicators under (i) and (ii) represent the disruption of ecosystems, while those under (iii) represent the disruption of social systems; and, finally, all three categories of indicators represent the process of one disruption reinforcing another.

Imperatives of mountain conditions and human responses

Under the traditional systems, people understood and responded to ecosystems in terms of constraints and opportunities generated by the specific features of their natural resource base. Thus the feature of biophysical resources or the circumstances created by them shaped the social perspective, i.e. values, norms and mechanisms for individual or collective interactions with nature (Jochim, 1981). In mountain areas, this was the context in which human responses were shaped by mountain specificities.

Based on the synthesis of descriptive or quantitative accounts of different mountain areas reported by different studies, Table 11.1 describes the situation in relatively broad terms.

The concrete practices incorporating the rationale behind the general features of the situation conveyed by Table 11.1 are too numerous and too varied to be dealt with meaningfully in this chapter. Even a quick glance at the traditional farming systems in HK–H and other mountain areas (Pant, 1935; Price, 1981; Bjonness, 1983; Guillet, 1983; Hewitt, 1988; Whiteman, 1988; Sanwal, 1989; Carson, 1992; Jodha *et al.*, 1992; Yanhua, 1992; Jodha and Partap, 1993; Jodha and Shrestha, 1994) will furnish a range of examples of the aspects highlighted by Table 11.1.

Table 11.1. *Mountain resource characteristics, their imperatives, objective circumstances and driving forces behind human response.*

Resource features and objective circumstances	Imperatives – driving forces	Responses, resource use practices*
Inaccessibility (caused by physical, terrain factors); imposing high degree of isolation, poor mobility, and limited external linkages, semi-closed	Survival strategies with direct and total dependence on local resources and high stake in their protection, regulated use and regeneration; local control of local resources, culture of self-management, evolution of systems from below based on closer proximity and knowledge of resource base	Ecology-driven resource management, using conservation and protection technologies, and institutional arrangements, evolved with closer feel of the resources and enforced through local autonomy and control of local resources; rationing of demand pressure on resources, and restricting extraction levels in keeping with subsistence needs
Fragility (caused by biophysical, topographic, edaphic characteristics); making resources highly vulnerable to irreversible degradation with small disturbance, restricting usage options, intensity levels	High risk of rapid resource depletion due to intensification inducing measures to balance ext raction and conservation of production base; narrow range of production options (only land extensive uses)	Technologies and usage practices combining intensive and extensive uses of natural resources; provision of institutional arrangements (e.g. common-property resources) against overextraction of fragile/ marginal resources, spatially and temporally differentiated resource use systems rationing; knowledge and capacity-based resource upgrading (e.g. by terracing, agroforestry, etc.)
Diversity (created by huge variations in biophysical features and elevations at shorter distances); creating opportunities for diversified interlinked production/ consumption activities	Local knowledge, skill and capacity-based diversification of resource use as a key element of survival strategies; sustainable productivity; health of natural resource base	Spatially and temporally diversified and interlinked activities with varying levels of intensification; diversification of demands to match the diversity of products and supplies, especially in a semi-closed situation

Niche (created by unique agro-climatic, biophysical situations), imparts comparative advantage to mountain areas in some activities and products (forests, horticulture, herbs, hydropower, etc.)	Potential for trade-based external linkages restricted by levels of knowledge, capacities to harness, etc.	A limited range of diversified activities directed to petty trading to supplement subsistence activities; local niche, demand and extraction facilities/capacities as key factors governing the exploitation of niche.
Implication	Adherence to two-way adaptation process	Ecology-driven systems of resource use conducive to sustainability (under low pressure of population and external demand)

Note:
* The table is based on a synthesis of accounts of concrete situations described in over 45 studies in mountain areas covering Nepal (18), China (15), India (7), Pakistan (3), Bhutan, Bangladesh and Myanmar (1 each) as synthesized by Jodha and Shrestha (1994).

The prevalence of traditional practices declines as one moves from relatively remote to more accessible villages. The consequences of these changes were quite visible, and in most cases people recognized them as part of their concerns and their vision of the future for their children (Jodha, 1995b). A detailed analysis of these issues is presented in the next section.

Dynamics of nature–society interactions: traditional systems

Based on the understanding provided by different accounts of situations in mountain areas, we can summarize the dynamics of ecosystem–social systems linkages, and try to address issues such as: what governs these linkages? how did they operate in the traditional context? and how are they disrupted in the present-day context? The issues involved are summarized in Table 11.2.

Although our context is mountain areas, the formulation and analysis presented below may have general applicability to many traditional or indigenous communities in semi-isolated situations. Accordingly, Table 11.2 indicates (i) the nature-dominated key objective circumstances under which the small and relatively isolated communities lived and managed their natural resources – through improved accessibility; (ii) the key driving forces and factors which shaped societal responses to the said objective circumstances; (iii) broad social responses in terms of concerns and adaptation strategies; (iv) technological and institutional mechanisms evolved and adopted for implementing the said strategies; and (v) consequences of (ii) to (iv) in terms of evolution of nature–society interactions and sustainability of resource use.

The above features of traditional systems are contrasted with the changes following the rapid population growth as well as the physical, administrative, and market integration of hitherto semi-closed/isolated systems or areas within the mainstream society. Integration, despite its various gains to mountain areas, has adversely affected the traditional resource management system (Banskota, 1989; Collier, 1990; Jodha, 1995a).

Table 11.2 is relatively self-explanatory, but its key points can be briefly summarized. The community's biophysical environment, characterized by a high degree of inaccessibility, imposed a certain degree of isolation and necessitated self-sufficiency. In the absence of effective outside linkages, their sustenance and welfare depended totally or crucially on the local resources. This forced them to adapt their requirements etc. by demand rationing as

Table 11.2. *Factors and processes associated with the nature–society interactions under traditional and present-day systems of resource use in mountain areas*

Situation under traditional systems	Situation under the present-day systems
A. Basic objective circumstances	
(i) Greater degree of inaccessibility, isolation and semi-closeness of systems; poor mobility and external linkages, etc. creating total and exclusive dependence on local resource base and high concern for its health and sustainable use	(i) Greater physical, administrative and market integration of traditionally isolated areas/communities with the dominant, mainstream systems, reducing critical dependence of the former on local resources and hence the degree of their stake in the conservation of local resources
B. Key driving forces/factors generated by (A)	
(i) Social survival/welfare strategies totally focused on local, diverse, fragile resources	(i) External linkages-based diversification of sources of sustenance, welfare and development, reducing the extent of critical stake in local resource maintenance
(ii) High collective stake in protection and regeneration of local natural resources	(ii) Role of functional resource knowledge marginalized due to imposition of generalized approaches from above for local resource management; wider gap between resource users and decision makers
(iii) Functional knowledge and closer understanding of limitations and potential of resources due to closer proximity and access to resources, little gap between resource user and resource itself	(iii) Erosion of local resource control, autonomy following the extension of mainstream, legal, administrative, fiscal arrangements to formerly isolated areas
(iv) Autonomy, local control over local resources (due to absence of external impositions)	(iv) Rapid demographic changes
(v) Low population pressure as permitted by biophysical constraints.	
C. Social Responses (concerns and adaptations) dictated or facilitated by (B)	
(i) Adoption and enforcement of production/extraction systems adapted to natural resource features through diversified usage, controlled usage-intensity; regenerating, upgrading, developing the resources, depending on capacities and needs	(i) Greater role of demand-driven measures leading to resource use intensification, overexploitation with greater extractive capacities and technologies
	(ii) Increased role of (unregulated) external demands, which are insensitive to local resource limitation

Table 11.2. (cont.)

Situation under traditional systems	Situation under the present-day systems
(ii) Controlling or rationing the demand pressure on resources through social and institutional sanctions, collective sharing, recycling, out-migration, etc.	(iii) Resource upgrading measures more generalized and less location specific
D. Mechanisms and means to execute social responses:	
(i) Collectively evolved site- and season-specific norms of resource use facilitated by direct access and proximity to resources and little gap between decision makers and resource users	(i) Largely externally evolved generalized rules guiding resource use, framed by legal and technical experts with little concern for local resource users' perspectives and limited knowledge of site-specific situations
(ii) Site, season, product and resource component-specific folk-technologies evolved over the generations facilitated by functional knowledge and close proximity to resource base	(ii) High science-based modern **R&D** as a source of technologies, ignoring rationale of traditional practices; ignore local resource perspectives
(iii) Formal/informal institutional arrangements guiding broad approach to resource management, access and usage regulation, facilitated by group action or community participation, and autonomy and local control over local resources	(iii) Institutional interventions evolved and designed for incomparable situations extended to these areas as a part of agricultural, rural development, etc.
E. Consequence: Ecology-driven natural resource management systems: (i) evolved by the communities having high stake in sustainability of the resources base; (ii) facilitated by functional knowledge of resources, close proximity to resources, and community control over the local resources.	Resource usage system driven by uncontrolled pressure of demand: (i) developed by experts without local participation (ii) enforced (rather un-enforced) by formal state machinery.

Note:
The table is based on a synthesis of accounts of concrete situations described in over 45 studies in mountain areas covering Nepal (18), China (15), India (7), Pakistan (3), Bhutan, Bangladesh and Myanmar (1 each) as synthesized by Jodha and Shrestha (1994).

well as their resource-use systems to the limitations and potential of local resources, rather than attempting to manipulate or overexploit resources to satisfy uncontrolled human needs. They had a high stake in the health and productivity of local natural resources. Close proximity to natural resources, local control of resources, intimate functional knowledge about them (again largely because of the closedness of the system), and lower pressure of population helped the communities to evolve folk-technologies and institutional arrangements, and to enforce them without external interference, for the protection, regeneration and regulated use of their resources. In the process, attitudes and norms of socio-economic behaviour, which had gradually evolved for the use of biophysical resources of the community, helped in linking social system with ecological systems to ensure sustainable use of resources in a subsistence context (Guillet, 1983; Hewitt 1988).

The present-day context

With the changed circumstances associated with the increased integration of hitherto isolated areas with the mainstream areas (as well as their population growth), ecology-driven social responses and resource management systems faced a rapid decline. While the integration of isolated/indigenous areas may be justified on several grounds, the process involved (using the norms and procedures characterizing the mainstream – i.e. prime land, industry and market-dominated areas) has marginalized the areas and communities in question. As a result, while the biophysical context remained largely unchanged, the socio-economic circumstances in these areas have changed rapidly (Bjonness, 1983; Ives and Messerli, 1989; Jodha *et al.*, 1992). Table 11.3 summarizes various changes, and their impact on the resource base, production flows and resource-use practices. The text will concentrate on one major change, that relates to population.

Population growth

The negative impacts of integration with external systems were accentuated by rapid population growth in these hitherto semi-closed mountain areas. While the population growth is largely an internal change, the external linkages have also played an important role in the process (Sharma and Banskota, 1992). Firstly, growth of life-saving-health facilities available from the outside, although much less than required, has reduced death rates even in the remote mountain areas. Second, the possibilities of external

Table 11.3. *Negative changes as indicators of emerging unsustainability in Hindu-Kush Himalayan region*

Visibility of change	Change related to*		
	Resource base	Production flows	Resource use practices/ management options
Directly visible changes	Increased landslides and other forms of land degradation; abandoned terraces; per capita reduced availability and fragmentation of land; changed botanical composition of forest/ pasture. Reduced water flows for irrigation, domestic uses, and grinding mills	Prolonged negative trend in yields of crop, livestock, etc.; increased input need per unit of production; increased time and distance involved in food, fodder, fuel gathering; reduced capacity and period of grinding/saw mills operated on water flow; lower per capita availability of agricultural products, etc.	Reduced extent of: fallowing, crop rotation, intercropping, diversified resources management practices; extension of plough to submarginal lands; replacement of social sanctions for resource use by legal measures; unbalanced and high intensity of input use, subsidization
Changes concealed by responses to changes	Substitution of: cattle by sheep/ goat; deep-rooted crops by shallow-rooted ones; shift to non-local inputs. Substitution of water flow by fossil fuel for grinding mills; manure by chemical fertilizers	Increased seasonal migration; introduction of externally supported public distribution systems (food, inputs): intensive cash cropping on limited areas	Shifts in cropping pattern and composition of livestock; reduced diversity; increased specialization in monocropping; promotion of policies/programmes with successful record outside, without evaluation
Development initiatives, etc., i.e. processes with potentially negative	New systems without linkages to other diversified activities and regenerative processes; generating excessive dependence on outside resource (fertilizer/pesticide-based	Agricultural measures directed to short-term quick results; primarily production (as against resource)-centred approaches to development; service-centred	Indifference of programme and policies to mountain specificities; focus on short term gains; high centralization; excessive, crucial dependence on external advice

| consequences† | technologies, subsidies); ignoring traditional adaptation experiences (new irrigation structure); programmes focused mainly on resource extraction. | activities (e.g. tourism) with negative side-effects | ignoring traditional wisdom; generating permanent dependence on subsidies |

* Most of the changes are interrelated and they could fit into more than one block.
† Changes under this category differ from the ones under the above two categories, in the sense that they are yet to take place, and their potential emergence could be understood by examining the involved resources-use practices in relation to specific mountain characteristics. Thus they represent the 'process' dimension rather than consequence dimension of unsustainability.

Source: Table adapted from Jodha (1990), Jodha and Shrestha (1994). It is based on data or description by over 45 studies from Nepal (18), China (15), India (7), Pakistan (3), Bhutan, Bangladesh and Myanmar (1 each) as synthesized by Jodha and Shrestha (1994).

relief support during crisis and scarcities, although less than adequate, have eroded internal demand-rationing measures (including traditional methods of population control such as the practice of the elder son in the family becoming a Buddhist monk, and not marrying – a practice still prevalent in parts of Bhutan). Third, and most important, the demographic impact of external linkages took place in terms of qualitative changes in the population. Growth of individualistic tendencies, and disregard or indifference towards social sanctions and collective action, which were key to traditional resource management, have emerged following the establishment of external links. The point here is not to deny the importance of external links, but to point out their side-effects in eroding traditional arrangements.

Thus the local-level ecosystem–social system linkages were disrupted by the emergence of a complex of internal (population) and external driving forces (e.g. interventions through market and state). The pressure on resources, encouraging their overextraction, increased. Unlike in the past, the total dependence on local resources ceased to be a key driving force to sustain people's stake in resource stability. The positive effects of local autonomy, control over local resources, close proximity to resources, functional knowledge of resources, and social cohesiveness, all of which in the past helped in the development of technologies and institutional responses, became weakened. Integration also meant the imposition of irrelevant technological and institutional measures from the outside (Banskota and Jodha, 1992).

Because of these rapid and major changes, the local communities were left without sufficient lead time or control over their resources and community affairs to amend their age-old coping strategies or to evolve new ones. Furthermore, they did not have the capacity or even incentives to resist the internal and external forces released by their integration with the stronger, external systems and the unprecedented growth of population. Their knowledge systems, social sanctions, collective sharing system, etc. became less effective or less feasible and less attractive (especially to the younger generations) in comparison with externally supported arrangements. In the final analysis, the whole complex (type and nature) of driving forces and patterns of responses to them changed (Jodha, 1995b). The net consequence was the emergence of what could be considered as indicators of unsustainability. Viewed in the context of the thematic framework of nature–society interactions mentioned earlier, these negative changes reflect a complex of disruptions, where social systems tend to behave independently of the imperatives of ecosystems. The two-way adaptation process is converted to a one-way adjustment, whereby resource manipulation and extraction are over-stretched to meet increasing human demands, rather than adjusting the latter to the limits of resource availability. This led to the breakdown of resource-regenerative, diversified production systems; indiscriminate resource-use intensification (often maintained through a high level of chemical, biophysical and economic subsidies); and the depletion of resources (Jodha, 1995a).

Restoring ecosystem–social system links: exploring the possibilities

By looking closely at the traditional systems, one can identify some key elements of the circumstances and processes which were responsible for positive ecosystem–social system links. These include the processes which generated community concerns, commitments, incentives and facilities for the protection and regulated use of natural resources; enhanced community capacities (both technical and institutional) to respond appropriately to biophysical circumstances through combining production and protection-centred measures; and motivated and facilitated enforcement of measures that helped in adapting community needs to resources rather than manipulating and overextracting the latter to meet unrestrained demands. Accordingly, one can identify three elements which individually or jointly strengthened the ecosystem–social system links and contributed to the natural resource-friendly traditional management systems: (a) a total dependence-driven stake in the protection of natural resources; (b) close proximity and a functional knowledge-driven approach to resource use; and (c) local control-determined sanctions and facilities governing resource use. Also, the smaller populations and greater social cohesiveness of traditional societies were the major facilitators of the above responses. The following discussion elaborates on these three points with a view to exploring some possibilities of reinstating these elements, or identifying their present-day functional equivalents, as parts of an incentive structure to facilitate positive social approaches essential for sustainable resource management in mountain areas. The main points are summarized in Table 11.4 and expanded upon in the text.

The key premise behind this exercise is that even though semi-isolation and other key objective circumstances characterizing traditional communities formed the basis of community stakes and sensitivity towards natural resources, the feasibility and viability of ecosystem-friendly attitudes and

Table 11.4. *Possibilities of reorienting current resource-usage systems in mountain areas by incorporating elements from traditional systems*

Circumstances, driving forces, response mechanisms characterizing traditional resource use systems	The element of traditional systems with scope for revival, reorientation and substitution in the present-day context
A. Total dependence-driven community stake in natural resource base (i) Key factor: almost total and exclusive dependence on local resource for survival (in a semi-closed, isolated subsistence-oriented context) inducing protection, regeneration and sustainable use of resources; the process was complemented by close proximity and functional knowledge of the resources which sharpened community's perception and diagnosis of resource situation (ii) Infeasibility (and undesirability) of (i) in the changed context of reduced isolation and access to external sources, etc.	**A. Rediscover areas of total/crucial dependence as sources of community stakes in natural resources** (i) Change of product/service – context of stake, ie, substituting (traditional) sustenance security by security of 'niche' ('high pay-off products/services with comparative advantage to the local communities) and use them as lead sector influencing overall natural resource management (eg, horticulture- or tourism-led initiatives in mountains)
B. Close proximity, direct access to resources, and their functional knowledge (i) Generated understanding and sensitivity to resource situation and its variability; helped in developing relevant folk-technologies; encouraged institutional arrangements for resource-use regulation (intensity, diversification, common-property regimes, etc.); reduced gap between resource user and decision maker, producer and produce consumer, and helped in regulating pressure on resources (ii) After integration with the mainstream (external systems), leading to distance between the resource and resource	**B. Functional substitutes for close proximity and first-hand knowledge of resource base** (i) Focus on sensitivity in place of proximity to resource situation, as the latter contributed to the evolution of resource management measures mainly by generating sensitivity towards the resources; re-orient, sensitize policy makers, development agencies (even market forces) to make them understand imperatives of mountain resources and act accordingly (ii) Evolve feedback mechanisms (about resource situation) by involving local communities as a substitute for instant

planners; multiplicity and diversity of resource users and pluralization of perception of stakes and marginalization of traditional knowledge systems, the physical proximity-dependent approaches are not feasible

C. Autonomy and community command over local resources

(i) A product of isolation or semi-closedness of areas/communities that facilitated effective community ownership of resources; helped in designing and enforcing resource-use regulations; reduced gaps between decision makers and resource users; encouraged community participation and group actions; and helped in resource-use rationing, collective sharing, resource recycling, and protection against pressure of (possible) external demand

(ii) The above features and functions have vanished or weakened with the integration of isolated areas with the mainstream economy and demographic changes. Revival of the system conflicts with 'centralization' and top-down approaches of the mainstream decision makers, governed more by the interests and perspectives of the mainstream

feedback provided by physical proximity in the past

(iii) Fill in knowledge gaps by collection, synthesis and application of resource-related information on using facilities offered by information technologies and communications

C. Restoring community management of local resources

(i) Build on the emerging trends toward decentralization, community-based development, participatory and bottom up approaches to development. Experiences of user group and NGO initiatives, etc. can help provide functional substitute arrangements for traditional group action through community control of resources

(ii) Take leads from successful experiences, community forestry, community irrigation systems and other grassroots level participatory initiatives, facilitate their replication and mainstreaming; sensitize decision makers to new possibilities

practices are not confined to small and isolated groups. By changing the forms of their manifestation and their operating mechanisms, these elements can be integrated into any resource management system and can prove effective in any context.

Dependence-driven community stake in resource health

In the relatively less accessible mountain areas, exclusive or total dependence for sustenance on local resources was the key incentive behind communities' concern and the follow-up actions which led to protection and regulated use of their natural resources. To reiterate, activities from the combining of production and conservation measures to the rationing of demand, as well as adherence to social sanctions regulating resource use, can be easily linked to the uniqueness of traditional societies and their incentive systems. The reinstating of such incentive or disincentive systems (i.e. by creating exclusive dependence for survival on local resources) is neither desirable nor feasible in the context of the changed situation in the mountains. However, one can explore other approaches to strengthen the dependence-driven community stake in the natural resource base. One such possibility involves change of the *product context* of dependence, which can be illustrated by indicating possibilities and actual cases where, by changing the product context of dependence, the community stakes in the natural resources can be strengthened.

By changing the product context of dependence we mean the following. At the local community level, the traditional security of sustenance can be substituted by the security of the biophysical 'niche', reflected through the comparative advantage (due to specific high-value options) potentially available to the community through physical and market integration of their area with the mainstream economy. There are cases of transformed areas in the HK–H region and other mountain regions where such niche-based gains have worked as new incentives for the protection and regulated use of the overall natural resources base by the communities.

For example, in areas such as Ningnan county (West Sichuan, China), where sericulture has recently become a lead activity with a high pay-off and comparative advantage to the area, communities attempt to manage and protect hill slopes, shrubs and waterflows on a priority basis because it helps in strengthening sericulture activities. Other examples include Himachal Pradesh (especially the apple zone) in India, or the Ilam district of Nepal, where multiple new activities are sustained through better management of natural resources in general. In the areas where mountain environment and landscape have become major tourist attractions (including the Alps and pockets of the Himalayas), the same logic of incentive-through-stake has helped improve the management of natural resources by the communities. There are many such examples where a stake in the lead sector/lead activity (due to biophysical and economic interlinkages) has induced and initiated a process of better management of overall resources by the communities (Jodha and Shrestha, 1994). The exception to the above includes cases where 'niche' is identified and harnessed (or extracted) without involving the local communities.

Yet further examples of the revival of the community stake in natural resources through common perception of needs, are the informal forest-protection committees (under joint forest management) in India and forest-user groups in Nepal. In most of such cases, having reached the threshold local level of scarcities of fodder, fuel, etc. due to degradation of forests resulting from external and internal demand pressures, communities (with and without the help of NGOs) have revived systems of collective protection and regulated use (Poffenberger and McGean, 1996). The same applies to the revival of community irrigation systems in the hills of Nepal, Pakistan and India.

The key factor in all the above success stories is the user groups' common perceptions of stakes and collective action. This represents a step towards the rehabilitation of communal integrity or the rebuilding of social capital. As a total picture involving natural resources and communities, this change manifests the revival of ecosystem–social systems linkages.

Physical proximity and functional knowledge of the natural resource base

We have argued that in traditional systems, the community's stake in its natural resources was an important driving force in shaping society's approach to ecosystem and inducing a conservation-oriented resource management system. An equally important role in this context was played by site- and season-specific functional knowledge of the resources. This in turn was gained through a close proximity and access to resources. Physical proximity and functional knowledge thus had a mutually reinforcing role in enhancing and sharpening people's perceptions of their stake in the better management of their natural resources.

In the following discussion, on the role of these features as a driving force or facilitator of traditional resource management systems is elaborated. These features were key factors in the strengthening of environment-friendly resource-use systems, the generation of folk-technologies, the creation of demand-rationing measures, and the institutional arrangements to facilitate their adoption and enforcement. The balancing of extensive and intensive types of land uses, various forms of resource-use diversification and flexibility, resource regenerating, recycling practices, methods of resource upgrading (i.e. by terracing), seasonal and periodic restrictions on product gathering from the village commons, are some of the concrete examples of instances where a better understanding of resource features and the availability of a longer lead time for informal experimentation helped communities.

Absence of any gap between decision making and the actual use of resources, as well as between the resource user (i.e. producer) and the product users (again facilitated by proximity and access), helped in encouraging flexible approaches to resource management to meet site- and season-specific differences and contingencies. They also helped in adjusting people's requirements (i.e. animal grazing intensity, seasonal collection and use of food, fuel and fodder) to the availability and potential of the resource base. Restriction on the collection of specific products during specific periods or from specific areas, and enforcement of grazing rotations by local communities in some villages, even today, are illustrations of such adjustments

Unlike in the past, the greatly modified present-day resource-use situation is characterized by: (i) a wider scatter of users of the resource products (due to market integration) and an equally wide gap between the producer (resource user) and the product consumer (e.g. the final user of herb and horticulture products and hydropower from mountain areas); (ii) the disassociation between usership and ownership of resources (due to growth of absentee landlordism in many areas); (iii) the disassociation of decisionmaking agencies (legal, fiscal and administrative authorities) and resource-using groups (i.e. the farmer or the community); and (iv) the distance and differences between technology developers and technology users. These circumstances restrict the scope for reinstating and strengthening resource management practices which are closely tied to physical proximity, direct involvement and accessibility to the resource base, and first-hand knowledge. However, to take fuller advantage of knowledge and understanding of the resource base, to re-orient an ecosystem-friendly social approach, and to design and implement relevant usage/management systems, it is not necessary to recreate the traditional situation characterized by semi-closedness and close physical proximity to resources. In the present-day context, the above goals can be achieved through better means of information acquisition, verification and synthesis, as well as communication and dissemination.

To benefit from the first-hand experience of the field situation and accumulated traditional knowledge about resources and resource-use systems, there are well-tested methods of involving local communities, e.g. through rapid rural appraisal/participatory rural appraisal (RRA/PRA), in the processes of collection, analysis and utilization of information. Such information can help create sensitivity toward natural resources among diverse stakeholders, from policy makers to urban consumers.

The key constraint in this respect is that the aforementioned means have been utilized only by experts from the outside. They have not been utilized for building sensitivity and understanding of mountain resources to develop technological and institutional measures relevant to local people's conditions. The focus of policy and programme interventions – be they agricultural research and development or integrated rural development – has lacked the mountain perspective, i.e. the understanding and incorporation of imperatives of mountain specificities (the high degree of inaccessibility, fragility, diversity and marginality), in the conception, design and implementation of development and welfare activities in mountain areas (Jodha, 1990). If these interventions are seen as part of broader social systems characterizing and influencing mountain areas, they once again reflect the rapidly vanishing links between social systems and ecosystems and their co-evolution in mountain areas. A first step towards filling in this gap can be made by initiating a process directed towards the following (Jodha *et al.*, 1992).

(i) Sensitization and re-orientation of the decision makers to create a policy environment friendlier to mountain conditions.
(ii) Involvement of the local communities in decision making and actions relating to local resources, to ensure relevance of interventions to the field situation.
(iii) Recognition and utilization of traditional knowledge systems by the formal research and development engaged in the development of technologies and policies for these areas.
(iv) Re-orientation of the whole process of project planning, designing and implementation by making it a bottom-up approach involving local communities and user groups.

The implementation of these and related measures, in our thematic context, implies an overhauling of the social systems to enhance their ability for linking with the ecosystems. In a less integrated form, these steps already form the mandate of several NGOs and donor-supported activities in many countries.

Conclusions: community control and rationing resource use

Regulation of resource use, rationing or limiting pressure on resources, mobilization of community and focused group action which helped in the sustainable use of natural resources in mountain areas in the past, were greatly facilitated by control of local resources by communities. This was a positive consequence of isolation and semi-closedness of communities, which prevented impositions from outside. This sort of autonomy available to communities was conducive to the evolution of both institutional and technological measures suited to local resources and survival needs. The close proximity and knowledge of local resource complemented the process.

With the integration of these previously isolated areas with mainstream society, the authority of the state, executed through different agencies, was extended to these areas. In the name of development, welfare, social and political integration, and even national security in many cases, the state usurped the resources and mandates that historically belonged to the people. With this process, both formal and informal control of the communities over local resources weakened or disappeared. The same thing happened to resource management arrangements supported by the autonomy and social sanctions of the community. Furthermore, following quantitative and qualitative changes on the demographic front, a 'disciplined' socio-economic behaviour, responsive to ecosystem needs, also disappeared.

In the contemporary world, restoring of traditional autonomy and control over resources enjoyed by isolated communities does not look possible. Its revival in some cases may conflict with the ruling culture and approach of the state, oriented toward greater centralization. Even the genuine efforts by some states towards decentralization and participatory development may not be sufficient to provide the traditional type of autonomy to village communities, and they may in the process disempower themselves or their bureaucracy. However, despite the above constraints, some limited form of autonomy and functional control of local resources by local communities, within the framework of overall legal control of the state, is possible.

Such possibilities are further strengthened by some emerging trends. Firstly, it is becoming increasingly clear that the management and protection of local-level resources through state agencies such as forest departments, are becoming progressively more difficult and costly. On the other hand, the involvement of local communities in local resource management has improved the situation in many areas (World Bank, 1995; Poffenberger and McGean, 1996). Secondly, the awareness and mobilization of local communities, for their rights and resources, enhanced through NGO activism, are emerging features of communities, even in less accessible areas. The successful negotiations of forest-user groups and community irrigation groups (helped by NGOs) to acquire control of resources in countries like Nepal, India and Pakistan in the HK–H region, are one case in point (Ostrom, 1990; Poffenberger and McGean, 1996).

However, these positive developments assume the building of social cohesion and the mobilization of communities based on shared perception and collective action. Such a social transformation may face several hurdles in the contemporary context. In addition to population increase, qualitative changes in mountain populations (Sharma and Banskota, 1992), reflected through rapid erosion of community cohesion, weakening of the culture conducive to group action and collective sharing, rapid growth of individualistic tendencies, and economic differentiation of communities, may obstruct the effective use of restored community authority over local resources for regulating resource use. Thus, the rebuilding of the social system is a prerequisite for developing or restoring effective linkages with the ecosystem.

Finally, the internal weakness of present-day village communities (constraining the community initiatives for resource-use regulation) may be complemented by external forces generated by market and political economy, as manifested through a range of fiscal and pricing arrangements. An overextraction of mountain resources (disregarding the imperatives of ecosystems), driven by the above forces, may continue despite increases in regulatory powers at the community level. This calls for a gradual process of change focused on the following steps.

(i) Making constant efforts for the greater involvement of communities and user-groups supported by NGOs for the planning and implementation of resource management initiatives, use of resource regenerative technologies, and regulation of resource use (Daly and Cobb, 1989; Cernea, 1991).
(ii) Taking the lead from the successful cases of participatory, decentralized resource management projects and focusing on their replication and mainstreaming (World Bank, 1995).
(iii) Helping build capacities and incentives for local communities to adapt to the changed circumstances and revive traditional practices for resource management in the changed contexts (Jodha *et al.*, 1992).

(iv) Introducing different norms for mountain product/resource pricing, reflecting their true worth or environmental cost by building on the conceptual leads provided by recent thinking in this area (Munasinghe, 1993).

(v) Introducing biophysical measures for compensation for resource extraction, e.g. reforestation or planting the same type of tree when a tree is cut for the timber market (Jodha and Shrestha, 1994).

The suggestions for creating present-day functional equivalents of traditional circumstances as presented in this chapter are indicative of the new possibilities. However, their design and implementation presume the fulfillment of several preconditions, including the commitment of decision makers and community-specific preparations.

References

Banskota, K. 1989. *Hill Agriculture and the Wider Market Economy: Transformation Processes and Experience of the Bagmati Zone in Nepal.* ICIMOD Occasional Paper No. 10. Kathmandu: ICIMOD.

Banskota, M. and Jodha, N.S. 1992. Mountain agricultural development strategies: Comparative Perspectives from the countries of the Hindu Kush–Himalayan region. In *Sustainable Mountain Agriculture*, pp.83–114. ed. N.S. Jodha, M. Banskota and T. Partap. New Delhi: Oxford and IBH Publishing Co. Pvt. Ltd.

Berkes, F., ed. 1989. *Common Property Resources: Ecology and Community-Based Sustainable Development.* London: Belhaven Press.

Bjonness, I.M. 1983. External economic dependency and changing human adjustment to marginal environments in High Himalaya, Nepal. *Mountain Research and Development* 3: 263–72.

Carson, B. 1992. *The Land, the Farmer, and the Future: A Soil Fertility Management Strategy for Nepal. ICIMOD Occasional Paper No. 21.* Kathmandu: ICIMOD.

Cernea, M. M., ed. 1991. *Putting People First: Sociological Variables in Rural Development.* New York, London: Oxford University Press.

Collier, G.A. 1990. *Seeking Food and Seeking Money: Changing Relations in Highland Mexico Community.* Discussion Paper No. 11. Geneva: United Nations Research Institute for Social Development (UNRISD).

Daly, H.E. and Cobb, J.B. Jr 1989. *For the Common Good: Redirecting the Economy Towards Community, the Environment and Sustainable Future.* London: Merlin Press.

Davis, S.H. 1993. *Indigenous Views of Land and the Environment. The World Bank Discussion Paper No. 188.* Washington DC: World Bank.

Ellen, R. 1981. *Environment, Subsistence and System: The Ecology of Small Scale Social Formations.* Cambridge: Cambridge University Press.

Gadgil, M. and Berkes, F. 1991. Traditional resource management systems. *Resource Management and Optimization* 18: 127–41.

Guillet, D.G. 1983. Toward a cultural ecology of mountains: the central Andes and the Himalayas compared. *Current Anthropology* 24: 561–74.

Hewitt, K. 1988. The study of mountain lands and peoples: a critical overview. In *Human Impacts on Mountains*, pp.6–23, ed. N.J.R. Allan, G.W. Knapp and C. Stadel. New Jersey: Rowman and Littlefield.

Ives, J.D. and Messerli, B. 1989. *The Himalayan Dilemma: Reconciling Development and Conservation.* London: Routledge.

Jochim, M.A. 1981. *Strategies for Survival: Cultural Behavior in an Ecological Context.* New York: Academic Press.

Jodha, N.S. 1990. Mountain agriculture: the search for sustainability. *Journal of Farming Systems Research Extension* 1 (1): 55–75.

Jodha, N.S. 1995a. *Sustainable Development in Fragile Environments.* Ahmedabad: Center for Environment Education.

Jodha, N.S. 1995b. *Hope and dismay in mountain regions.* In *Choosing our Future: Visions of Sustainable Future*, pp.50–58, ed. T. Nagpal, and C. Foltz, The 2050 Project, Washington DC: World Resource Institute.

Jodha, N.S., Banskota, M. and Partap, T., eds. 1992. *Sustainable Mountain Agriculture.* Vol. I *Perspectives and Issues;* Vol. II *Farmers' Strategies and Innovative Approaches.* New Delhi: Oxford and IBH Publishing Co. Pvt. Ltd.

Jodha, N.S. and Partap, T. 1993. Folk agronomy in the Himalayas: implications for agricultural research and extension. In *Rural People's Knowledge, Agricultural Research and Extension Practice*, pp.15–37. IIED Research Series, Vol. 1, No. 3. London: International Institute for Environment and Development.

Jodha, N.S. and Shrestha, S. 1994. Sustainable and more productive mountain agriculture: problems and prospects. In *Proceedings of the International Symposium on Mountain Environment and Development*, pp.1–66 (Part B). Kathmandu: ICIMOD.

Munasinghe, M. 1993. *Environmental Economics and Sustainable Development.* The World Bank Environment Paper No. 3. Washington DC: The World Bank.

Ostrom, E. 1990. *Governing the Commons: The Evolution of Institutions for Collective Action.* Cambridge: Cambridge University Press.

Pant, D.D. 1935. *The Social Economy of Himalayas: Based on a Survey in the Kumaon Himalayas.* London: George Allen and Unwin.

Poffenberger, M. and McGean, B. 1996. *Village Voices, Forest Choices: Indian Experiences in Joint Forest Management into the 21st Century.* New Delhi: Oxford University Press.

Price, L.W. 1981. *Mountain and Man: A Study of Process and Environment.* Berkeley: University of California.

Sanwal, M. 1989. What we know about mountain development: common property, investment priorities, and institutional arrangements. *Mountain Research and Development* 9: 3–14.

Sharma, P. and Banskota, M. 1992. Population dynamics and sustainable agricultural development in mountain areas. In *Sustainable Mountain Agriculture*, Vol.I pp.165–84, ed. N.S. Jodha, M. Banskota and T. Partap. New Delhi: Oxford and IBH Publishing Company Ltd.

Whiteman, P.T.S. 1988. Mountain agronomy in Ethiopia, Nepal and Pakistan. In *Human Impacts on Mountains*, pp.57–82, ed. N.J.R. Allan, G.W. Knapp and C. Stadel. New Jersey: Rowman and Littlefield.

World Bank, The 1995. *The World Bank Participation Source Book*. Washington DC: Environment Department, World Bank.
Yanhua, L. 1992. *Dynamics of Highland Agriculture. (A Study in Tibet, China)*. ICIMOD Occasional Paper No. 22. Kathmandu: ICIMOD.

12

Crossing the threshold of ecosystem resilience: the commercial extinction of northern cod

A. CHRISTOPHER FINLAYSON & BONNIE J. McCAY

Introduction: opportunities of systems crisis

C.S. Holling suggests that it is only at points of deep crisis in both the ecosystem and the social system that fundamental conceptual and structural change is possible (Holling, 1986). This idea can be usefully located within the theory of 'paradigm shifts' in science (Kuhn, 1962; 1970). A paradigm is a set of canonical assumptions, theoretical propositions, and methods. Broadened to science-based natural resource management, the theory may be stated as follows: when the accumulation of perceived failures significantly exceeds the perceived utility of management, the legitimacy and conceptual coherence of that management institution are weakened to the point where they are vulnerable to challenge and open to fundamental change.

The groundfish fisheries of the northwest Atlantic – cod, haddock, flounders and other species – are in such a crisis. Among the important causes are inadequacies and systematic errors in the science of fisheries stock assessment (Steele, Andersen and Green, 1992; Finlayson, 1994; Hutchings and Myers, 1994; Walters and Maguire, 1996). The question we pose, although cannot yet answer, is whether the current set of crises will open the door to change in the practice and theory of fisheries science and management.

To explore the problem, we turn to what has become the classic case of the failure of conventional science-based fisheries management: the collapse of the northern cod of Newfoundland and Labrador. The current crisis in the fishery of Newfoundland, a province of the east coast of Canada, has destabilized prevailing political and epistemological power relations, creating opportunity for renegotiation of those relationships. The relations most at stake have been those centred on scientists and science-based management.

311

The northern cod crisis

In July, 1992, the Canadian Minister of Fisheries declared a moratorium on all fishing for northern cod in Canadian waters. At that time, it appeared that the northern cod populations had declined to the point that they were on the verge of commercial extinction. The estimated biomass was about one third the average since 1962 (Coady, 1993: 74), and perhaps only 1% of the maximum estimate for which data are available (Myers *et al.*, 1996). Some 35 000 fishers and fish-plant workers were affected by this closure, not to mention the other businesses, families and community organizations dependent on the work of fishermen and plant workers. The government provided assistance for what was planned as a two-year closure to allow the stock to rebuild. Contrary to assumptions and expectations, the 1994 assessment indicated that the northern cod population had continued to decline, and to the point that it was possible to use the language of biological, not just commercial, extinction. As of 1996, the talk is of keeping the northern cod fishery closed for another 10 or 15 years. Fishery-dependent workers and their families have been promised assistance only until 1999, and for many assistance is being cut off in 1996.

The northern cod fishery and adaptations to periodic decline

Northern cod is a stock (or set of stocks) of *Gadus morhua* found in the waters off the southern half of Labrador to the northern half of the Grand Banks of Newfoundland, NAFO (Northwest Atlantic Fisheries Organization) management zones 2J, 3K, and 3L (Figure 12.1). The extra-ordinary abundance of cod was the major reason for early exploitation and settlement of Newfoundland by Europeans. In Newfoundland vernacular, the word 'fish' means 'cod'.

Northern cod have thrived on the productivity of an ecosystem strongly influenced by the mixing and upwelling caused by the Labrador Current, carrying cold water from the arctic, and the Gulf Stream, which brings warmer waters from the south (Gomes, 1994). The northern cod spawn on the offshore banks in the winter and spring months, particularly Hamilton Banks, off southern Labrador, Belle Isle and Funk Island Banks, off the north and northeast coasts of Newfoundland, and the northern part of the Grand Banks. On the Grand Banks, they are susceptible to an international fishery, beyond 200–miles, in areas such as the 'nose' and 'tail' of the Grand Banks and the Flemish Cap (Figure 12.1). Until recent decades, northern cod were rarely fished in the winter because of the storms and ice that

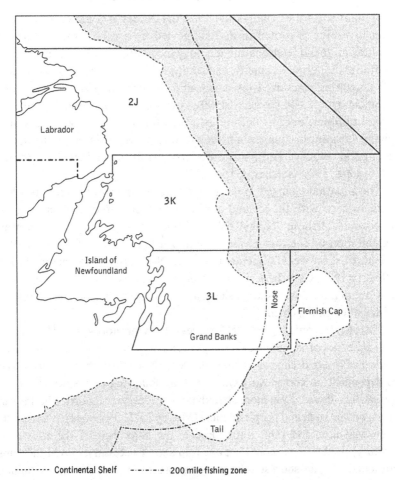

---------- Continental Shelf   ─·─·─·─·─ 200 mile fishing zone

Figure 12.1.   Newfoundland, Canada, within the 200-Mile Fishing Zone and Northwest Atlantic Fisheries Organization boundaries. (Adapted from Department of Fisheries and Oceans, Canada Communications Branch, Halifax, Nova Scotia, 1995)

dominate the north-west Atlantic Ocean. The principal fishery was during the summer months, when the cod migrated shoreward to feed. Relatively simple and inexpensive techniques, like traps, gill-nets, and hook-and-line gear, were more than adequate for a very productive if not always profitable fishery.

The northern cod fishery was the focal point of the distinctive culture of outport Newfoundland. The outports are small, once-isolated communities scattered around the coast adjacent to the fishing grounds. The

dominant pattern of small-scale, seasonal domestic commodity production in fishing is rooted in and supported by a robust folk culture of
egalitarian social relations and ideology mediating problems such as mer-
cantile exploitation, ethnicity, and competing gear-types, but also capable
of generating organized resistance and opposition (Faris, 1972; Sider,
1986; Martin, 1979; Cadigan, 1990). Although variable and dynamic, in
broad outlines 'outport culture' persisted for more than 300 years, sur-
viving substantial changes in fishing technologies, the larger political and
economic contexts, and natural fluctuations in the cod stocks (McCay,
1976; 1978; 1979; Sinclair, 1985; 1987).

The situation changed in the 1960s with the arrival of foreign trawlers
built to cope with the ice and storms of winter. Although there was an
important offshore 'banks' fishery from the beginning of European
exploitation of the region, the scale and fishing power of the post-1950
offshore fishery were far greater than ever. Northern cod catches peaked in
1968. In 1977, Canada declared a 200-mile exclusive fisheries zone, began
to Canadianize this offshore fishery, and developed a science-based system
of fisheries management.

Historical records suggest that for at least 100 years northern cod have experienced cycles of abundance (Harris, 1993). Every 30 years or so there was a
period of severe decline or 'failure', particularly for the inshore fishery which
is dependent on cod moving close to shore following prey such as capelin
(*Mallotus villosus*). Even more often there were localized 'failures'. But the fish
always came back and people adapted (McCay 1978). For example, the fishermen who used cod traps (introduced in the latter part of the nineteenth
century) tried to have more than one trap and, if necessary, moved the traps
around during the short summer season in order to reduce the risks of failure.
Most inshore fishermen also maintained a diversified fishery. If the trap season
was a failure, they were prepared for a late summer/early autumn fishing
season using handlines, long-lines, and gill-nets. Where possible, fishermen
also put out pots for lobsters, set nets for salmon, and jigged and seined for
capelin, herring and mackerel. In the spring they often hunted seals.

If local fishing proved poor for an extended period, some might go to
distant shores, such as the Labrador coast, to fish. Fishermen also coped
by engaging in what sociologists call 'occupational pluralism', finding work
in logging, highway maintenance or construction, even when that required
long-distance migration. From the 1960s, inshore fishermen began to invest
in vessels and gear that enabled a more diversified, mobile strategy and
reduced their dependence on the inshore migrations of cod. Central to this
was adoption of a larger vessel, called a 'longliner', the use of nylon and

then monofilament gill-nets, and the slow but steady development of markets for products other than salt cod.

Finally, the larger political economic system had its own buffers, some of which imposed severe constraints on fishermen. One of these was the 'truck' system whereby a season's catch was mortgaged to a local fish buyer against the costs of outfitting for the season plus, in many cases, the costs of feeding and clothing one's family. In this system, merchants – most of them 'factors' for companies based in the capital, St. John's, or in England – minimized their risks and maximized their profits by leaving the risky business of catching and curing fish to the fishing families, while dictating both the price of the fish they bought and the goods they sold.[1] This system of de facto indentured servitude to the local merchant and fish-buyer lasted in muted forms into the 1940s, when Newfoundland emerged from bankruptcy and royal receivership to become a province of Canada and participant in an industrialized welfare state.

Prior to 1949, the government provided modest economic buffers, including poor-relief and 'winter works projects', which became a perennial feature of Newfoundland outport adaptation. When Newfoundland, more-or-less independent since 1855, joined Canada in 1949, it gained a generous package of federal transfer payments providing social security to families and the elderly. More to the point, in 1954 the Canadian government enacted a special system of unemployment insurance, for nominally 'self-employed' fishermen, that helped mitigate the ups and downs of the inshore fishery. These buffers were extremely important to the survival and even growth of the outports of Newfoundland, but also contributed to their vulnerability in some ways, including the intensification of dependency, with its many costs (Hanrahan, 1988).

The current crisis may be just another one of those cyclic periods when 'fish is scarce, b'ye'. Yet the scale, scope and duration of the collapse of northern cod and the apparent truncation of the population to just two year-classes strongly suggest the possibility that it is not. Hence the heroic and costly measures imposed by the Canadian government, in the hope of giving the stock a better chance to recover. The question of whether heroes can learn from their mistakes is one to which we return at the end of this chapter. Our focus now is on the question of what caused the northern cod crisis.

Tragedies of the international commons

The idea that fishery problems in Newfoundland (and Canada more generally) might be 'tragedies of the commons' began to appear in official writings and speeches in the 1960s, 1970s and, especially, 1980s (Matthews, 1993), following the formulation of the economic theory of common property by Canadian economists H. Scott Gordon (1954) and Anthony Scott (1955) and its specific application to the situation in eastern Canada (e.g. Copes, 1982). Such a tragedy did take place, but not because Newfoundlanders enjoyed open access to their fishery; it came about through the introduction of powerful fishing technologies in the context of weak international controls.

Despite hundreds of years of a growing and essentially unregulated fishery, until the post-war advent of a large-scale, industrialized offshore fishery, the yield from northern cod can be pragmatically accepted as sustainable at an average of about 200 000 metric tons (mt). The historical data show considerable variability in the landings from place to place and year to year, but general stability within the bounds of 150 000 and 300 000 mt (Harris, 1990). The situation began to change after World War II. With much of the infrastructure of European agriculture in ruins, fish became a vital source of food. Under this impetus, and incorporating technologies developed during the war – inexpensive steel ship construction, powerful diesel engines, shipboard refrigeration and freezing, and electronics for precise navigation, long-distance communications, bottom imaging and fish-finding – the hungry nations of Europe, led by the Soviet Union and its Warsaw Pact member states, developed distant-water fishing capacity and combined these technologies into a new and devastatingly effective configuration: the factory freezer-trawler. With the size and strength to fish in ice-ridden waters and all but the worst storms, these ships could be directed by their corporate or state owners to wherever catch rates were highest. Supplied via motherships with food, fuel and fresh crews from their home ports, these vessels could fish the year round and stay at sea indefinitely. By the mid-1960s, their numbers were so great that the Newfoundland fishing banks at night were described as a 'city of lights' (see Warner, 1983).

Recalling that the northern cod stock had historically supported an evidently sustainable yield averaging 200 000 mt, notice in Table 12.1 how the total catches increased to triple or quadruple this figure in just 12 years, and then even more quickly collapsed. Also notice how completely the distant water fleets came to dominate the fishery even in a depleted state. Finally, notice the devastating effect on landings in the Newfoundland inshore fishery.

Constraints on the distant water fishery were minimal. The territorial sea was only 3, later 12, nautical miles from shore. As early as 1949, it had

Table 12.1. *Harvests of cod in ICNAF sub-areas 2J3KL, selected years 1956–75*

| Year | Total | Distant water nations | | Canada | | |
		Harvest	Share (%)	NF Inshore	NF Offshore	Other
1956	300.5	117.1	39.0	172.1	2.3	8.7
1960	393.6	228.9	58.2	157.3	2.5	4.9
1964	562.0	420.5	74.8	131.5	6.7	3.3
1968	783.2	659.8	84.2	101.0	20.2	2.2
1972	454.6	388.1	85.4	62.3	3.8	0.4
1973	354.5	310.0	87.4	42.7	1.4	–
1974	372.6	336.5	90.3	35.2	0.9	–
1975	287.5	245.0	85.2	41.1	0.9	0.4

Note: all catch figures given in thousands of metric tonnes.
Source: Munro (1980), based on ICNAF Statistical Bulletin 1975.

become evident to the nations sharing this fishery that some form of regulation would be necessary to protect the stocks from overfishing. This led to the formation of the International Commission for the Northwest Atlantic Fisheries (ICNAF) in 1951. For its first 20 years, ICNAF limited the regulation of its member fleets to a simple minimum mesh size. In 1971 it proposed the first overall quota for cod in the management zone. Through the early 1970s, ICNAF brought more species and stocks under nominal quota management, and the science of stock assessment improved. However, ICNAF was ineffective because of at least two classic problems of international environmental policy: quotas tended to be set at the 'least common denominator' level because states unhappy with them could defect; and enforcement could only be done by the flag-states of the fishing vessels (Haas, Keohane and Levy, 1993; Parsons, 1993; Peterson, 1993).

Meanwhile, the adaptive capacity of inshore fishermen and the larger system was stretched to the limit (McCay, 1978; 1979). Northern cod catches in the inshore fishery plummeted in the early 1970s. In the outports, many people put away their cod-traps and skiffs, thinking they would never use them again, and even the new longliner fishing operations were in trouble. Welfare rolls and out-migration increased. Reliable data on the numbers of fishers are scarce, but it is estimated that they declined to a very low point in 1974 (McCorquodale, 1983: 154).

Canada then took a tough stance in the ongoing United Nations Law of the Sea negotiations and, in 1977, unilaterally declared a 200-mile zone of exclusive fisheries jurisdiction. ICNAF was disbanded and replaced by the Northwest Atlantic Fisheries Organization (NAFO) to manage and regulate the high-seas fisheries in what remained of international waters. This included important nursery, feeding and spawning areas of the Grand Banks, the so-called 'nose and tail' of the shallow fishing banks that extend beyond the 200-nautical-mile limit. NAFO proved to be just as ineffective as its predecessor, ICNAF, and for the same reasons (Peterson, 1993). In addition, the Law of the Sea Convention failed to address the protection of fish stocks which straddled the 200-mile limit.

For northern cod, in particular, the question of the ability of NAFO to set and enforce conservative quotas was of great consequence. Up to 1986, NAFO worked reasonably well (Rowe, 1993: 14). Significant rebuilding of the 'straddling stocks' (cod, yellowtail flounder and other species found on both sides of the 200-mile limit) appeared to have occurred, contributing to the optimism, over-investment, and scientific hubris that will be discussed below. In 1986, Spain and Portugal joined the European Community (EC), and the EC dramatically shifted its policy 'from cooperation to conflict based on demands for vastly increased allocations of straddling stocks' (Rowe, 1993: 14). NAFO was very vulnerable, having a rule that any party objecting to what the majority agrees upon is not bound by that decision. The EC made use of this objection procedure and set unilateral quotas far higher than those of the NAFO Fisheries Commission (Rowe, 1993: 14). The EC failed to enforce even its own quotas, leading to massive overfishing, particularly by the Spanish and Portuguese vessels. In addition, vessels of non-member nations and member nations flying a flag of convenience fished these grounds unhindered by even fictitious restraints.

Thus, the small spawning biomass of northern cod that remained after the 1992 closure of the fishery inside the Canadian zone continued to be exposed to heavy fishing pressure at precisely the time and place where it was most vulnerable – the dense spawning aggregations on the slopes of the shelf outside of the 200-mile limit.[2]

The fishery for northern cod in international waters was open-access; yet it was also subject to a regime that had the potential for effective management of the open-access commons. However, the fundamental conditions of 'international anarchy' (Young, 1989), including the retention of sovereignty for enforcement of regulations by nation-states,[3] contributed to the ineffectiveness of NAFO (Peterson, 1993). Canada's move in the 'turbot war' of 1995,[4] using gun-boat diplomacy to enforce international quotas,

Table 12.2. *Quotas versus reported*
catches, NAFO/ICNAF sub.area 3NO
cod (nose & tail of Grand Banks),
1982–1991

Year	Quotas	Catches
	(metric tonnes)	
1982	17000	31605
1983	17000	28379
1984	26000	24394
1985	33000	36899
1986	33000	50645
1987	33000	41619
1988	40000	43150
1989	25000	33297
1990	18600	18384
1991	13600	21320*

*Estimated catch, included reported catch
by Canada
Source: Rowe 1983:15

may be viewed as part of the longer-term process of carving up this international commons into zones of extended jurisdiction and turning the
matter over to nation states.

Canada's heroic action in the 'turbot war' of 1995 may also be a de facto,
if unintended, way of diverting public attention from other causes of the
fisheries crisis – a crisis which had 'spread' to other species and to other
north-west Atlantic fishing grounds by the mid-1990s. We thus turn to the
question of what was happening within Canada to influence the development of and responses to the northern cod crisis.

Expanding beyond the limits of system resilience: post-1977

Vivid evidence of the depredations of a virtually unrestrained international fishery must have added to the reasons why the economic construct
of the 'tragedy of the commons' came to influence federal government fisheries policy in Canada (Matthews, 1993). However, the concept was well
enough entrenched in Canadian policy to support a programme, beginning
in 1975, of limited licensing for certain species (lobster, shrimp, crab,
salmon) and the means for distinguishing 'full-time' or 'bona fide' fishers
from others in the fisheries (Matthews, 1988). Consequently, when troubles

arose, the idea of 'too many people, too many boats' associated with 'common property' came to dominate official analyses and inform policy. The metaphor of the 'tragedy of the commons' is based on established economic theory and thus eventually won out over competing but less well-articulated paradigms rooted in appreciation of the quality of life (Matthews, 1993).

Between 1974 and 1980, the number of licensed fishers in Newfoundland increased three-fold, from 12792 to 35080 (McCorquodale 1983: 154). Subsequently the number declined and fluctuated but did not go below 20,000 until 1991 (DFO, 1993: 96). Fish-processing operations increased as dramatically. In 1974 there were 89 licensed fish plants in Newfoundland; in 1980 there were 138, and in 1992, 173 despite continuous decline in landings of groundfish from 1986 on (Kingsley, 1993: 127). Much of this expansion of capacity came from public funding, fuelled by optimistic resource projections (Kingsley, 1993: 127). A major expansion of the fish-processing workforce also took place, particularly among women and young men (15–24 years) (Carter, 1993: 141).

Thus there was the kind of expansion of capacity, in relation to the post-1977 improvement in cod stocks, that might be expected in an open-access commons. However, as is the case elsewhere (Marchak, 1988–89), government policy played a strong role in shaping this tragedy of the commons. Subsidies for new vessels and upgrading also fuelled a major increase in inshore and nearshore fishing capacity, particularly after 1985/1986 (DFO, 1993: Figure 23, p.117). The number of active offshore vessels, stern trawlers over 100 feet in length, declined, particularly in the early 1980s and then again at the end of the 1980s, when the signs of serious trouble in groundfish stocks became apparent (DFO, 1993: Figures 13 and 14, p.105). These vessels were owned by vertically integrated companies.

The five major companies in the offshore fishery were consolidated into two in 1983. Despite belief that the cod stocks had rebounded from their depleted state in 1977 (reflected in increases in the quota from 135000 mt in 1978 to 215000 mt in 1982), by the early 1980s the offshore companies were on or over the brink of financial collapse. The federal government assessed the economic, social and political costs of allowing these firms to go bankrupt and deemed them too high. The solution was a complicated and expensive restructuring scheme which combined the five companies into two – National Sea Products (NatSea) and Fisheries Products International (FPI) – with the federal government as the majority shareholder and the Bank of Nova Scotia (the primary creditor) and the Atlantic provinces as minority shareholders. The plan was to restore the two corporations to operating profitability as quickly as possible and then return them to private sector ownership. Ironically, they only became consistently profitable in the early 1990s when, with the demise of Atlantic Canada's groundfish, they began processing cod and other species caught elsewhere and functioned as international seafood brokers.

The crisis of the early 1980s led to a task force analysis of the Newfoundland fishery (Kirby Task Force on Atlantic Fisheries 1982), which again emphasized the problem of 'common property' as the major factor leading to the overdevelopment of harvesting capacity. Several commentators have emphasized the extent to which this perspective on the problem in Canada ignored information about community-based systems of informal and formal property rights and resource management, and obscured the structural differences between the inshore and offshore sectors of the fishery (e.g. McCay, 1978; Barrett and Davis, 1984; Matthews, 1993). For example, in the Kirby report, the poverty of Newfoundland inshore fishermen is traced to the open-access nature of the ground fish fishery, ignoring issues such as differential access to resources and market power (Barrett and Davis, 1984: 129).

Science and its centrality

Another factor left out of the official analyses, and overlooked until very recently in discussion of Newfoundland's fishery problems, is the problem of the science of the commons and how science relates to policy (Neis, 1992; Steele *et al.*, 1992). The problem may very well be the *centrality of science* in the process of deciding such critical questions as the total allowable catch (TAC) (Finlayson, 1994). The TACs, and projected TACs, were central to the optimism that fuelled boat-building, fish plant licensing, and decisions to go fishing or work in a fish plant instead of finishing school. They were also often based on erroneous assumptions, inaccurate data and faulty science.

Shortly after claiming a 200-mile limit, Canada created the Department of Fisheries and Oceans (DFO), with the foremost goal being the rebuilding and rational scientific management of the northern cod. Generously supplied with material, human and financial resources, the DFO's programme was a superb example of scientific fisheries management done the way that scientists had long advocated (Parsons, 1993).

The goal of objectivity in science was emphasized in a later restructuring of the DFO to separate science from management completely, creating the Science Branch. The Science Branch is responsible for research leading to stock assessments, which are the basis of advice given to the Minister of Fisheries. Scientific advice was finalized at an annual meeting of the Canadian Atlantic Fisheries Scientific Advisory Committee (CAFSAC). By 1982 the DFO was claiming that the rebuilding process was well under way and was predicting a 1987 northern cod quota of 400 000 mt and a long-term sustainable yield of 550 000 mt (Kirby Task Force, 1982).

Problems in the science of northern cod assessment

Despite optimism expressed in the 1982 Kirby Task Force Report and subsequent TACs for northern cod, many inshore fishermen worried as the weight of their catches and the sizes of the fish they caught became smaller. The Newfoundland Inshore Fisheries Association, a coalition of inshore fishers working with university scientists and lawyers, forced a reassessment of the data and methodologies behind those TACS (the 'Keats Report', see Steele *et al.*, 1992), based on an internal DFO document, presented to CAFSAC in 1986, which described the mounting inaccuracies of the method of stock assessment being used (CAFSAC, 1986):

... from 1977 to 1985, DFO's calculations of the size of the fishable biomass [and thus the F0.1-based Total Allowable Catch (TAC)] had been out on average by 107%, and because this retrospective analysis tends to become more accurate the farther back one goes, and the more information one gathers about a given year-class through fishing mortality, even more alarming were the 117% overcalculation in 1979, 154% overcalculation in 1978, and 220% overcalculation in 1977.

Martin, 1995: 7.

Both the DFO document and the Keats Report were ignored in the quota-setting process, but mounting concern within and without the fisheries agency finally led to two government-sponsored studies: the Alverson Report of 1987 and the Harris Report of 1990, both of which identified serious problems in data and methodologies, and one of which (Harris, 1990) prompted action. The Harris Report followed alarming reports that the 1989 DFO research vessel survey showed that previous estimates of northern cod abundance were very wrong.

In the 1989 reassessment, the relatively crude 'bulk-biomass' model used for northern cod stock assessment since 1977 was replaced with a more sophisticated age-structured model. This, in concert with a change in the relative weighting of the research vessel survey and offshore catch-per-unit-of-effort indices of abundance used to 'tune' the model, led to a drastically revised estimate of stock size and trajectory. The 1989 assessment concluded that the exploitable biomass (fish aged four years and older) had not grown five-fold since 1978 as previously believed. It had grown only about three-fold and was then static or possibly in decline. Further, if the stock size had been seriously overestimated, then the quotas, set to achieve a target fishing mortality (expressed numerically as some value of 'F' such as $F_{0.1}$ or roughly 20% of the exploitable biomass), had, in fact, resulted in annual removals by fishing of one-third or more of the available population (Task Group on Newfoundland Inshore Fisheries, 1987). If this was true, then not only was the stock much smaller than had been thought, but its ability to reproduce itself may have been weakened, perhaps dangerously so (Harris, 1990).

Thus the fisheries scientists concluded that the earlier claims had overestimated the northern cod stock's abundance by as much as 100% and underestimated the level of fishing mortality by roughly 50%. The 1990 quota was slashed to 197000 mt, with considerable concern expressed that this was still too high and that the quota should be set at no more than 125000 mt to prevent further depletion (Harris, 1990).

Why the huge difference between the 1989 and earlier stock assessments? We briefly review some of the problems that surfaced in the *post mortem* of the northern cod crisis, referring the reader to other sources for fuller treatment and, in some cases, different interpretations (Neis, 1992; Steele *et al.*, 1992; Lear and Parsons, 1993; Finlayson, 1994; Martin, 1995; Myers *et al.*, 1996; Walters and Maguire, 1996).

First, northern cod was treated as a unit stock, when it was known to be comprised of distinct populations with different migratory patterns, contributing in discrete ways to inshore and coastal fisheries.[5] Failure to incorporate this variability into the models may have contributed to overestimation of biomass (deYoung and Rose, 1993; Lear, personal communication in Finlayson, 1994) as may failure to appreciate the extent to which 'recruitment overfishing' was taking place (Walters and Maguire, 1996). Heavy distant water fleet fishing in the 1960s and 1970s probably depleted some spawning units to the point of virtually ending inshore fisheries in some areas, notably Division 2J in Labrador (see Figure 12.1; Walters and Maguire, 1996).

There is also tremendous variability in the survival of year-classes of northern cod,[6] but limitations of data and modelling require that it be handled as a constant in the stock assessment models, based on the calculated average recruitment of past years. However, the average of an extreme stochastic variable is not likely to approximate the true current value. Moreover, the time series of recruitment data used to derive the average is now thought to have been a period of unusually strong recruitment, thus adding yet another source of overestimation to the model (Hutchings and Myers, 1994). The same problems apply to highly variable growth rates (Templeman, 1966)[7] and estimates of natural mortality.

Natural mortality (M) includes disease, starvation, cannibalism, old age, and predation by other species of fish and marine mammals. With no realistic way of measuring M, it is incorporated as an assumed constant with a value of 20%.[8] Given that M must also vary due to the same complex interactive factors driving the extreme observed variability in growth and recruitment, the choice of any constant value to represent M in the model must be considered to be highly arbitrary and speculative.

Just how unrealistic and unreasonable this practice is has been driven forcibly home by the apparent continued decline of the stock into 1995, three-and-a-half years since the imposition of the moratorium. In the absence of all but incidental fishing mortality, one would expect the biomass to show moderate to strong rebuilding or, at worst, stability. Instead, the post-moratorium assessments have concluded that the age four-plus biomass has collapsed further. If true, this violates the most basic assumption of fisheries management: that *fishing* mortality is the dominant variable determining population size and its trajectory.[9]

Inshore and offshore fisheries

Another factor contributing to the northern cod stock assessment error was a bias toward the use of offshore, as opposed to inshore, landings as an index of abundance to 'tune' the stock assessment model (which was primarily based on survey data rather than landings).[10] Inshore fisheries landings were not systematically used in those assessments, and thus evidence of decline in the abundance and size of inshore-migrating fish was apparently missed (Neis, 1992; Steele *et al.*, 1992). Catch-per-unit-of-effort data were only collected from the offshore fleet. This contributed to an overestimate of biomass because of the special conditions of that fishery. The fleet included vessels able to fish around and through pack ice, exposing the cod to a kind of fishing mortality they had not encountered in earlier times. Moreover, northern cod spawn in dense aggregations on the offshore slopes of the banks. Offshore trawlers targeted cod during these pre-spawning and spawning concentrations, which enabled them to maintain high catch rates on a declining stock.[11]

The original and ultimately successful challenge to the Science Branch's claims about the robust health and vigorous rebuilding of northern cod came from the inshore sector of the fishery – a group with no meaningful input to the stock assessments and on the extreme margins of the general policy-formation process. Despite these apparent handicaps – but consistent with the Kuhnian theory of the origins of paradigm shifts – this sector was able to mobilize and sustain sufficient cultural and political resources to force a genuine and substantive internal re-evaluation of scientific stock assessment.

For almost a decade, inshore fishermen had been questioning conclusions used to support the quotas for northern cod. The day-to-day operational reality of the inshore fishing community increasingly diverged from that of the DFO's science-based construction of reality: the inshore sector was landing progressively fewer – and smaller – fish while the offshore trawlers' catches were continuing to increase. Fishers adjusted by relying more on gears suited for catching the smaller fish, such as a modernized small-mesh cod trap, and processors adjusted their cutting machines and found markets for smaller cod. However, fishers also worried. Individually, through their political representatives, and through NIFA, inshore fishers began to claim that the stock was in danger – that the scientific description of a healthy, growing stock must be wrong – and that the northern cod quotas, particularly those for the corporate offshore fleet, should be immediately and significantly reduced.

A remarkable metaphysical change had occurred in their thinking. There had always been failures in the fishery in the past, but they were seen as natural and transient. When the European fishery escalated, people began to see that fishery failures could be caused by fishing itself, and they began to voice a new possibility:

Now a few mistakes and a few bad decisions could cause a failure that was not natural but man-made. Now there could be a failure and the fish wouldn't come back. Now there was someone to blame. And this was utterly different than anything they had known before.

Interview with Bernard Brown conducted in St John's on August 3, 1990; cited in Finlayson, 1994: 108.

This change in perception fuelled the movement that led to the 1986 Keats Report and the late 1980s reassessments of the status of northern cod, culminating in the Harris Report (1990). The offshore sector did not join in the criticism of the DFO's construction of reality until the critical reassessment of 1989 precipitated drastically reduced quotas for the 1990 offshore fishing year. Even then, the offshore perspective was diametrically opposed to that of the inshore sector. The offshore companies argued that the DFO had been right in its earlier assessment: there were plenty of fish out there, and this was supported by the operational reality of their skippers. The local newspaper carried a letter from one of the deep-sea trawler skippers that began: 'I've been fishing northern cod for eight years and I tell you there are more fish there now then [sic] there were eight years ago. . . .' (William Cox quoted in the St John's *Evening Telegram*, Feb. 24, 1990, p. A1; cited in Finlayson, 1994: 108).

The official response from the DFO, until the 1989 stock assessment, was to dismiss the inshore sector's perception of the stock's status as an artifact of resource availability: the stock was healthy and continuing to rebuild but, for reasons probably related to changes in the ocean climate, the cod were simply not migrating inshore in their usual numbers (DFO, 1983; Alverson, 1987). Why the DFO's response changed is discussed at greater length in Finlayson (1994).

Reasons for errors in stock assessment

There are several directions to take in understanding why the northern cod stock assessment was so poor, and why those who saw or suspected its problems did not speak up or were not heard. First we must acknowledge that the scientists involved used assessment methods and population dynamics assumptions that were widely believed to be sound and appropriate but turned out not to be so, and that the blame must be shared widely. As Carl Walters and Jean-Jacques Maguire recently wrote:

The fault, if there is one, lies with *all* fisheries assessment scientists for not questioning assumptions more rigorously, with fisheries managers and politicians for pretending to place confidence in the assessment and preventing scientists from openly discussing uncertainties in the assessments, and with fishermen and fish processors for misreporting, dumping, and discarding.   *Walters and Maguire,* 1996, p. 3.

However, it is important to be more precise about the social, political and ideological factors at play. A more precise hypothesis is that scientists knew the truth but were not heard or not allowed to speak because those charged with making fisheries policy had reasons to favour more generous assessments. The facts that offshore fishery companies were heavily in debt and that many jobs were at stake in their processing plants may have played a role in closing eyes and ears until it was too late. The 1986 warning about inaccuracies in the stock assessment method coincided with the privatization of the two large companies (c.f. Martin, 1995). However, it must be noted that scientists themselves played a role in the creation of a climate of over-optimism in the late 1970s and early 1980s, at the key moment of extended fisheries jurisdiction (Walters and Maguire, 1996), and thus helped create a situation in which their more pessimistic advice might not be welcome.

A second more sociological analysis is that once the northern cod situation became a public issue, the agency and its Science Branch behaved as most bureaucracies would, taking a defensive stance and trying to manage public relations. There are other bureaucratic forces, including a promotion system which gives little credit for contributions to organizational function or relations with client groups in comparison with work on issues deemed of scientific importance (Finlayson, 1994).

Third, the idea that fisheries management is primarily a scientific/technical undertaking may be part of the problem. Wild fish populations are not managed directly (with the partial exception of salmon) but through management of the socio-economic enterprise of fishing. Managers do not manage fish, they manage a nation's fishing industry; a complex amalgamation of individual social actors who, in turn, comprise a dynamic and interactive collection of subcultures and interest groups. However, with the partial and recent exception of economics, the social sciences and social scientists have had no discernable influence on the development and implementation of management policy and process. This serious omission may help account for the fact that few fishers in Newfoundland seem to believe that the DFO understands them or their industry. Many of the DFO's policies and implementing regulations are regarded as naive, ignorant, or irrational.

Another possible reason concerns relationships between scientists and policy-makers. The organization of the fisheries agency may have contributed to the problem, paradoxically, by isolating scientists from policy (c.f. Susskind (1994), who makes a similar argument with respect to international environmental policy).[12] One result was that most scientists lost touch with the fishing industry, which may have further marginalized and delegitimized information from the inshore fishery. Moreover, when scientists had to present their assessments to officials in other branches and to the scientific advisory body (CAFSAC), they felt pressured to give the simpler answers or at least to be consistent, omitting questions they had about reliability, degrees of uncertainty, or the very models and data-sets they were using (see interview in Finlayson, 1994: 79). It is at least arguable that greater involvement of scientists in the policy process, and greater openness of scientific debate, would improve the environment for truly critical reflection about what is happening to fish stocks and what should be done to protect them.

We would go further to suggest the de-centring of science in that process, so that the information, knowledge and concerns of fishermen and community members can play more direct roles. In response to criticism in the wake of the bad news of 1989, the DFO had developed innovative programmes in which scientists worked more closely with fishermen in their research and where fishermen voluntarily collected logbook data of use to scientists. However, momentum was lost with the 1992 closure of the northern cod fishery. In addition, there were signs that this had devolved to a one-way communication process, i.e. taking a few leading fishermen out on the research boats to show them how scientific data-gathering was done.

Conclusion: can heroes learn?

We began this chapter with a reference to Holling's theory of systems crisis and the necessary conditions for fundamental change in fisheries management regimes. We turned to the case of northern cod, both as a textbook example of science-based management and as an empirical test of the systems crisis theory. The stock has collapsed to the point where biological extinction is not out of the question. The fishery dependent upon this stock has been closed indefinitely at enormous social and economic cost. Tens of thousands of men and women in hundreds of communities in Newfoundland and Labrador are facing the loss of both their livelihoods and their way of life. The federal government has already spent billions of dollars on a suite of crisis-response programmes and, with no end to the crisis in sight, will probably need to spend billions more in the years to come. These should be necessary and sufficient conditions to precipitate a full-blown systems crisis, or at least significant changes in the scientific bases for fisheries management.

Canada's actions in closing the domestic northern cod fishery and incurring international wrath by trying to enforce quota cuts in the straddling stocks fishery were heroic indeed. A question asked earlier in this chapter was 'can heroes learn?' The northern cod crisis (and its earlier manifestations) has the potential of opening the door to significant structural, ideological, and political changes in fisheries management and policy.

In the classical Kuhnian model of paradigm shifts, the successful challenge comes from the periphery of the established body of scientists. We should look, then, to individuals and groups currently marginal to or minorities within established management institutions for the ideas and leadership necessary to revolutionize fisheries management. At this point in time, in various 'epistemic communities' focused on fisheries conservation in North America, there are signs of such ideas and individuals in debates about 'adaptive management', or the importance of dealing with uncertainty by designing intervention to facilitate learning (Holling, 1978; Walters, 1986; Lee, 1993; Walters, Goruk and Radford, 1993); from the same direction come proposals to rely more on closed areas or 'refugia' than on finely tuned TACs, accepting the difficulty of obtaining precision in assessments (e.g. Clark, 1996; Walters and Maguire, 1996). Another source of change is found in discussions of the need for 'ecosystem management' rather than species-based management, as well as related attempts to deal explicitly with chaotic or stochastic phenomena in marine ecosystems and fisheries (Wilson *et al.*, 1990). A third source of potential change is found in the notion of 'co-management', or power sharing between government agencies and non-governmental groups (Jentoft, 1989; Pinkerton, 1989), as well as in 'participatory research', whereby scientists and fishermen and other community members collaborate in various dimensions of fisheries research and where the knowledge of fishers is accorded value alongside the knowledge of scientists.

Serious consideration of these or other alternative perspectives has not yet happened on a major scale in Newfoundland. The linkages between the social and the ecological systems are very weak, and the resiliency of some aspects of the social system, but not others, is very strong. What is collapsing along with the stock is the social structure of the fishery to the extent that it was dependent on the northern cod. To be sure, there have been some changes in management attributable to the stock crisis, but they are evolutionary rather than revolutionary. The explanation for this is properly the subject of additional focused research, but we will offer some initial speculations.

First, consensus is lacking about what the lessons are. For example, within the DFO, scientists and administrators have been engaged in a debate about whether overfishing or environmental changes are the principal causes of the collapse of the northern cod stock and other groundfish stocks of the region (e.g. Hutchings and Myers, 1994; Myers, Hutchings and Barrowman, 1996). The role of DFO scientists in recommending quotas that were, in retrospect, too high to protect the fish stock, is also a matter of debate; the interpretation we give here, based on Finlayson's research, including extensive interviews with the scientists involved (1994), is controversial, as is the notion that it will be necessary to incorporate 'traditional ecological knowledge' of the fishers as well as data from inshore fisheries in the stock assessments of the future (Neis, 1992).

Some institutional experimentation and perhaps learning are evident. Early in the crisis, before the moratorium, two DFO scientists, George Rose and Ben Davis, had begun working more closely with inshore northern cod stocks and inshore fishers (Davis, Lundrigan and Ripley, 1994). A multidisciplinary team from Memorial University is engaged in a government-funded analysis that includes collaboration between social scientists and fisheries biologists in interviewing fishers about their ecological knowledge and understanding (Neis, personal communication, 1995).

Moreover, the value of a more participatory, inclusive decision-making process is reflected in a major change. In 1993, the Minister of Fisheries and Oceans created the Fisheries Resources Conservation Council (FRCC), comprised of scientists, academics, leaders of industry groups, and other experts outside the DFO. Final authority for resource assessments and recommending quotas and other conservation strategies to the minister was removed from the Science Branch of the DFO and vested in the FRCC. In existence only two years, it has nonetheless moved far more quickly and decisively than did the DFO under the old system. Its recommendations have led to radical reductions in the quotas for many Atlantic Canadian groundfish stocks and the complete closures of fishing on others. While certainly not popular, these measures have been generally accepted by the industry as necessary in a way that they would not if they had originated from within the DFO. The FRCC's findings and decisions have a legitimacy that derives foremost from its broadly inclusive and representative membership and its open and democratic process. This legitimacy is further enhanced by the council's perceived independence from the political bureaucracy of the state and the perceived competence of its members.

However, for the most part, public and private attention to the question of what caused the collapse of the northern cod and how such a disaster could be prevented from recurring – a key link between ecological and social systems – has diminished. One reason is that confrontations with foreign fishing vessels over straddling stocks management and Canada's subsequent role in UN deliberations diverted attention away from the critical question of how and whether scientists, working within the present political, social and bureaucratic structures and with the ideological underpinnings of fisheries science, can avoid the tragic hubris they showed in the past.

Another reason may be the rather quick and substantial intervention by the federal government with relief packages and the perceived role of the union representing the majority of Newfoundland fishery workers in securing that compensation. In accepting payments of up to $400 Canadian per week, the people directly affected by the failure of the state's management of the resource also implicitly accepted its continued central authority. It is surely significant that Richard Cashin, the head of the fisheries union during the negotiations of the compensation package, resigned his post to accept the chair of a federal task force on the future structure of the fishing industry. The task force's final report (DFO, 1993) largely endorsed the federal call for a policy of 'rationalization' of the fishing industry in Newfoundland. In this case, rationalization is understood to mean divorcing fishery management policy from social policy to create a new industry that is economically self-sufficient. What this will mean in detail is a matter of considerable debate, but in general terms it is agreed that at least half of the currently employed people, vessels and processing plants must leave the fishery permanently.

In a way this can be seen as a left-handed confirmation of the Holling theory. Rather than a revolution in the basic structure and process of management, management has exploited the chaos and fear of the crisis to impose a revolutionary restructuring policy on the fishing industry, one that it had been advocating in various forms since the 1982 Kirby Task Force report but which had been successfully resisted by the industry until the northern cod moratorium.

Science and its application to quota setting have become marginal to the broader task of coping with the immediate social and economic consequences of the collapse of Newfoundland's groundfish fisheries. Most money and attention has gone to the allocation of financial aid and retraining of fishermen and fish-plant workers, focusing on the issue of 'downsizing the industry'. Downsizing policies, including early retirement and 'buy-out' programmes requiring the vessel or gear licence holder to quit fishing forever, are justified in part by humanitarian concerns for people unable to fish or pay off their mortgages during the moratorium on the cod fishery. However, like the retraining programmes, they are predicated on a wide consensus, voiced by representatives of the major fishing companies, the unions, and government agencies, that the fishery of the future should not support as many people as the fishery of the recent past did.

The tragedy-of-the-commons model plays a strong if implicit role in this diversion of effort, leading to an overly simplified account of both problem and solution. If overfishing was the major cause of the collapse of the northern cod, then overfishing must be prevented. So far so good; but what does an attempt to reduce the number of people licensed to fish have to do with overfishing? The collapse of the northern cod is indeed a tragedy, and it is indeed a tragedy of the 'commons' broadly construed. However, the northern cod fishery 'commons' was neither unregulated nor open-access, certainly not since 1978. Nonetheless, at this point the thinking represented in the economic model of the open-access fisheries intrudes, leading too quickly to the 'downsizing' solution to what are complex problems caused in some potentially measurable part by government policies, corporate interests, international situations, and the errors and uncertainties of science.[13] Although the numbers of people whose jobs and lives are at stake may be fewer in the future, the crisis has not led to the kinds of institutional changes in science and policy that will help prevent future assaults on the ecological resilience of fish stocks.

Acknowledgements

This research was sponsored in part by the Beijer International Institute of Ecological Economics, The Royal Swedish Academy of Sciences, Stockholm, Sweden, with support from the World Environment and Resources Program of the John D. and Catherine T. MacArthur Foundation and the World Bank. The research was conducted as part of the research programme Property Rights and the Performance of Natural Resource Systems. It was also sponsored by the Rutgers University Grants Program, the National Science Foundation, the National Sea Grant College Program, and the New Jersey Agricultural Experiment Station. The chapter was prepared for the project Linking Social and Ecological Systems; Institutional Learning for Resilience, The Property Rights Programme, The Beijer International Institute of Ecological Economics, The Royal Swedish Academy of Sciences; it is based in part on a paper presented at the Annual Meetings of the American Anthropological Association, Washington, D.C., November 15–19, 1995.

Notes

1 As older Newfoundlanders recall, a quintal of fish (112 pounds) was worth the same as a barrel of flour, no matter how many fish were caught.
2 In 1992, at the time of the closure of the northern cod fishery, the Minister of Fisheries allowed a small-scale handline fishery for personal and recreational use only; this was widely abused for a black market in cod and hence closed in the winter of 1993–4. In the later summer of 1994, the minister opened a very short-term recreational fishery, which was closed very quickly soon thereafter. By-catches remain a problem in some fisheries as well. However, it may be safely assumed that the Canadian fishery has indeed been 'incidental' compared with earlier fisheries and with the international fishery.
3 Another factor contributing to the ineffectiveness of NAFO was a decision rule that called for majority votes but allowed objectors to create their own rules.

4 In 1994, NAFO agreed to a closure of the directed fishery for cod but allowed
   fishing for certain other species, including turbot (*Reinhardtius
   hippoglossoides*, 'Greenland halibut'). A similar process of objection and
   overfishing occurred in early 1995, prompting Canada to arrest a Spanish
   fishing trawler in international waters, a unilateral action that challenged
   international law at its foundations. The outcome was a path-breaking
   agreement for independent observers on fleets in international waters, a
   change in the allocation between the EC and Canada, and new momentum for
   an international 'straddling stocks' agreement at the United Nations which
   was completed in the summer of 1995.

5 Tagging studies (Templeman, 1979; Lear, 1984) and acoustic tracking (Rose,
   1993) have shown that components of this stock retrace distinctly different
   routes in their annual inshore/offshore migrations and contribute in discrete
   ways to the inshore and coastal fisheries.

6 The eggs and larval stages of cod are pelagic, drifting with the currents until
   settling to the bottom some months later as post-larval juveniles. Little is
   known about the intervening four to six years before they 'recruit' to the adult
   stock, but variation in the suitability of the settlement habitat and nursery
   areas, food supply, and predation are assumed to be the critical factors
   driving the extreme observed variability (several orders of magnitude) in
   recruitment of succeeding year-classes (de Young and Rose, 1993; Myers *et
   al.*, 1993).

7 Once recruited to the adult stock, northern cod feed on shrimp and benthic
   invertebrates during the months in the deep offshore waters, and almost
   exclusively on capelin during the spring and summer. Growth rates are also
   highly variable, both geographically and inter-annually, depending most
   strongly on relative food availability and water temperature. On average, post-
   recruit fish of a given year-class will be three-quarters of the length and half
   the weight at the northern end of their range as at the southern end. In years
   of unusually cold water or low abundance of capelin, growth may slow to
   nearly nil (Templeman, 1966).

8 Like the average value representing recruitment, M is derived from inferential
   post hoc calculations. It is taken to be the difference between consecutive
   years' assessments after subtracting total fishing mortality and adding growth
   and recruitment.

9 On this point there is considerable disagreement. Leading DFO scientists
   continue to argue that the inability to control fishing mortality is the primary
   cause for the collapse of the northern cod stock (Hutchings and Myers, 1994).

10 More generally, the attempt to 'tune' the virtual population assessment (VPA)
   model used to estimate fish abundance relied on a faulty assumption about the
   relationship between catch per unit effort and abundance, as if this were linear
   when it was more likely non-linear, as fishers became better at finding fish and
   as technology was used to cope with ice (c.f. Hutchings and Myers, 1994;
   Walters and Maguire, 1996).

11 A recent analysis suggests that the fleet was also increasing its discard rate as
   the fish stocks declined; that is, fishing mortality was much higher than
   evident in the landings data because proportionally more of the actual catch
   was thrown back into the sea, as the fleets had greater trouble finding the
   more marketable larger fish (Myers *et al.*, 1996).

12 As is not unusual, the science of fisheries management was separated from the
   policy and implementation branches of the agency, ostensibly to help provide
   more objective scientific appraisals.

334   A. Christopher Finlayson & Bonnie J. McCay

13 That the illogical leap to downsizing has happened is not, of course, simply
    the result of the dominance of a way of thinking. There are many people and
    organizations with interests in reducing the involvement of rural
    Newfoundlanders in the inshore fishery, ranging from corporations interested
    in trimming their labour and facilities costs to taxpayers concerned about the
    costs of bail-outs from fisheries collapses.

References

Barrett, G. and Davis, A. 1984. Floundering in troubled waters: the political
    economy of the Atlantic fishery and the Task Force on Atlantic Fisheries.
    Journal of Canadian Studies 19: 125–37.
CAFSAC (Canadian Atlantic Fisheries Scientific Advisory Committee) 1986.
    *Advice on the Status and Management of the Cod Stock in NAFO Divisions
    2J, 3K and 3L*. CAFSAC Advisory Dec. 86/25.
Carter, B.A. 1993. Employment in the Newfoundland and Labrador fishery.
    In *The Newfoundland Groundfish Fisheries: Defining the Reality;
    Conference Proceedings,* pp.132–75, ed. K. Storey. St John's:
    Institute of Social and Economic Research, Memorial University of
    Newfoundland.
Clark, C.W. 1996. *Refugia*. Paper presented to the National Academy of Sciences
    International Conference on Ecosystem Management for Sustainable Marine
    Fisheries, February 19–24, 1996, Monterey, California.
Coady, L.W. 1993. The groundfish resource crisis: ecological and other
    perspectives on the Newfoundland fishery. In *The Newfoundland Groundfish
    Fisheries: Defining the Reality*, pp.56–77, ed. K. Storey. Conference
    Proceedings. St John's: Institute of Social and Economic Research, Memorial
    University of Newfoundland.
Cadigan, S.T. 1990. Battle Harbour in transition: merchants, families and the
    state in the struggle for relief in a Labrador community during the 1930s.
    Labour/Le/Travail 26: 25–50.
Copes, P.1982. Implementing Canada's marine fisheries policy: objectives, hazards
    and constraints. *Marine Policy* 6: 219–35.
DFO [Department of Fisheries and Oceans. Canada] 1983. *Trap Cod: Some Facts
    About Unpredictable Catches and Small Fish*. Ottawa: Department of
    Fisheries and Oceans.
DFO. 1993. *Charting a New Course: Towards the Fishery of the Future.
    Report of the Task Force on Incomes and Adjustments in the Atlantic
    Fishery*. Ottawa: *Communications Directorate*, Department of Fisheries and
    Oceans.
Davis, M.B., Lundrigan, P. and Ripley, P. 1994. *A Description of the Cod Stock
    Structure in Placentia Bay, NAFO Subdivision 3Ps*. Department of Fisheries
    and Oceans, Atlantic Fisheries Research Document 94/32.
deYoung, B. and Rose, G.A. 1993. On recruitment and distribution of Atlantic
    cod (*Gadus morhua*) off Newfoundland. *Canadian Journal of Fisheries and
    Aquatic Sciences* 50: 2729–41.
Faris, J.C. 1972. *Cat Harbour: A Newfoundland Fishing Settlement*. St. Johns:
    Memorial University of Newfoundland.
Finlayson, A.C. 1994. *Fishing for Truth; A Sociological Analysis of Northern
    Cod Stock Assessments from 1977–1990*. St. John's: Institute of
    Social and Economic Research, Memorial University of Newfoundland.

Gomes, M.do C. 1994. *Predictions Under Uncertainty; Fish Assemblages and Food Webs on the Grand Banks of Newfoundland.* St. John's: Institute for Social and Economic Research, Memorial University of Newfoundland.

Gordon, H.S. 1954. The economic theory of a common property resource: the fishery. *Journal of Political Economy* 62: 124–42.

Haas, P.M., Keohane, R.O. and Levy, M.A., eds. 1993. *Institutions for the Earth; Sources of Effective International Environmental Protection.* Cambridge, Mass.: MIT Press.

Hanrahan, M. 1988. *Living on the Dead: Fishermen's Licensing and Unemployment Insurance Programs in Newfoundland. ISER Research and Policy Papers No. 8.* St. John's: Institute for Social and Economic Research, Memorial University of Newfoundland.

Harris, L. 1990. *Independent Review of the State of the Northern Cod Stock.* Prepared for The Honourable Thomas Siddon, Minister of Fisheries. Ottawa: Communications Directorate, Department of Fisheries and Oceans.

Harris, L. 1993. Seeking equilibrium: an historical glance at aspects of the Newfoundland fisheries. In *The Newfoundland Groundfish Fisheries: Defining the Reality; Conference Proceedings,* pp.1–8, ed. K. Storey. St. John's: Institute of Social and Economic Research, Memorial University of Newfoundland.

Holling, C.S., ed. 1978. *Adaptive Environmental Assessment and Management.* New York: John Wiley & Sons.

Holling, C.S. 1986. Resilience of ecosystem: local surprise and global change. In *Sustainable Development of the Biosphere,* pp.292–317, ed. E.C. Clark and R.E. Munn. Cambridge: Cambridge University Press.

Hutchings, J.A. 1996. Spatial and temporal variation in the density of northern cod and a review of hypotheses for the stock's collapse. *Canadian Journal of Fisheries and Aquatic Sciences* 53: 943–62.

Hutchings, J.A. and Myers, R.A. 1994. What can be learned from the collapse of a renewable resource? *Canadian Journal of Fisheries and Aquatic Sciences* 51: 2126–46.

Jentoft, S. 1989. Fisheries co-management: delegating government responsibility to fishermen's organizations. *Marine Policy* April: 137–54.

Kingsley, R.G. 1993. Overview of the Newfoundland and Labrador groundfish processing industry. In *The Newfoundland Groundfish Fisheries: Defining the Reality;* pp.125–75, ed. K. Storey. Conference Proceedings, St. John's: Institute of Social and Economic Research, Memorial University of Newfoundland.

Kirby Task Force on Atlantic Fisheries 1982. *Navigating Troubled Waters: A New Policy for the Atlantic Fisheries,* December 1982. Ottawa: Department of Fisheries and Oceans, Communications Branch.

Kuhn, T. 1962. *The Structure of Scientific Revolutions.* Chicago: University of Chicago Press.

Kuhn, T. 1970. *The Structure of Scientific Revolutions,* 2nd edn. Chicago: University of Chicago Press.

Lear, W.H. 1984. Discrimination of the stock complex of Atlantic cod (*Gadus morhua*) off southern Labrador and eastern Newfoundland, as inferred from tagging studies. *Journal of Northwest Atlantic Fisheries Science* 5: 143–59.

Lear, W.H. and Parsons L.S. 1993. History and management of the fishery for northern cod in NAFO divisions 2J, 3K and 3L. In *Perspectives on Canadian Marine Fisheries Management,* ed. L.S. Parsons and W.H. Lear. *Canadian Bulletin of Fisheries and Aquatic Sciences* 226: 55–89

Lee, K.N. 1993. *Compass and Gyroscope; Integrating Science and Politics for the Environment*. Washington DC: Island Press.

Marchak, M.P. 1988–89. What happens when common property becomes uncommon? *BC Studies* 80 3–23.

Martin, C. 1995. The collapse of the northern cod stocks. *Fisheries* 20: 6–8.

Martin, K.O. 1979. Play by the rules or don't play at all: space division and resource allocation in a rural Newfoundland fishing community. In *North Atlantic Maritime Adaptations*, ed. R. Andersen. The Hague: Mouton.

Matthews, D.R. 1988. Federal licensing policies for the Atlantic inshore fishery and their implementation in Newfoundland, 1973–1981. *Acadiensis* 17: 83–108.

Matthews, D.R. 1993. *Controlling Common Property; Regulating Canada's East Coast Fishery*. Toronto: University of Toronto Press.

McCay, B.J. 1976. *Appropriate technology and coastal fishermen of Newfoundland*. Unpublished doctoral dissertation, Columbia University, University Microfilms, Ann Arbor.

McCay, B.J. 1978. Systems ecology, people ecology, and the anthropology of fishing communities. *Human Ecology* 6: 397–422.

McCay, B.J. 1979. 'Fish is scarce': fisheries modernization on Fogo Island, Newfoundland. In *North Atlantic Maritime Cultures*. pp.155–89, ed. R. Andersen. The Hague: Mouton.

McCorquodale, S. 1983. The management of a common property resource: fisheries policy in Atlantic Canada. In *The Politics of Canadian Public Policy*, pp.151–71, ed. M.M. Atkinson and M.A. Chandler. Toronto: University of Toronto Press.

Munro, G. 1980. *Atlantic Report, A Promise of Abundance: Extended Fisheries Jurisdiction and the Newfoundland Economy*. Ottawa: Minister of Supply and Services.

Myers, R.A., Hutchings, J.A. and Barrowman, N.J. 1996. Hypotheses for the decline of cod in the North Atlantic. *Marine Ecology Progress Series* 138: 293–308.

Myers, R.A., Mertz, G. and Bishop, C.A. 1993. Cod spawning in relation to physical and biological cycles of the northern Northwest Atlantic. *Fisheries Oceanography* 2: 154–65.

Neis, B. 1992. Fishers' ecological knowledge and stock assessment in Newfoundland. *Newfoundland Studies* 8: 155–78.

Parsons, L.S. 1993. Management of marine fisheries in Canada. *Canadian Bulletin of Fisheries and Aquatic Sciences* 225.

Peterson, M.J. 1993. International fisheries management. In *Institutions for the Earth; Sources of Effective International Environmental Protection*, pp.249–305, ed. P.M. Haas, R.O. Keohane and M.A. Levy. Cambridge: Mass.: MIT Press.

Pinkerton, E., ed. 1989. *Cooperative Management of Local Fisheries; New Directions for Improved Management and Community Development*. Vancouver: University of British Columbia Press.

Rose, G.A. 1993. Cod spawning on a migration highway in the Northwest Atlantic. *Nature (London)* 366: 458–61.

Rowe, M. 1993. An overview of straddling stocks in the Northwest Atlantic – 'peace in our time?' In *The Newfoundland Groundfish Fisheries: Defining the Reality; Conference Proceedings*, pp.9–20, ed. K. Storey. St John's: Institute of Social and Economic Research, Memorial University of Newfoundland.

Scott, A. 1955. The fishery: the objectives of sole ownership. *Journal of Political Economy* 63: 116–24.

Sider, G.M. 1986. *Culture and Class in Anthropology and History; A Newfoundland Illustration.* Cambridge: Cambridge University Press.

Sinclair, P.R. 1985. *From Traps to Draggers: Domestic Commodity Production in Northwest Newfoundland.* St John's: Institute of Social and Economic Research.

Sinclair, P.R. 1987. *State Intervention and the Newfoundland Fisheries: Essays on Fisheries Policy and Social Structure.* Aldershot: Avebury.

Steele, D.H., Andersen, R. and Green, J.M. 1992. The managed commercial annihilation of northern cod. *Newfoundland Studies* 8: 34–68.

Susskind, L.E. 1994. *Environmental Diplomacy; Negotiating More Effective Global Agreements.* New York: Oxford University Press.

Task Force on Atlantic Fisheries. 1982. *Navigating Troubled Waters; A New Policy for the Atlantic Fisheries. Highlights and Recommendations, Report of the Task Force on Atlantic Fisheries,* Michael J.L. Kirby, Chairman, Ottawa: Minister of Supply and Services Canada.

Task Group on Newfoundland Inshore Fisheries 1987. *A Study of Trends of Cod Stocks off Newfoundland and Labrador Influencing Their Abundance and Availability to the Inshore Fishery.* A Report to the Honourable Tom Siddon, Minister of Fisheries, Canada, Submitted by the Task Group on Newfoundland Inshore Fisheries, November 1987. 101 pp. plus appendices.

Templeman, W. 1966. Marine resources of Newfoundland. *Bulletin of the Fisheries Research Board of Canada* 154.

Templeman, W. 1979. Migration and intermingling of stocks of Atlantic cod, *Gadus morhua,* of the Newfoundland and adjacent areas from tagging in 1962–66. *International Commission of the Northwest Atlantic Fisheries Research Bulletin* 14: 5–50.

Walters, C.J. 1986. *Adaptive Management of Renewable Resources.* New York: Macmillan.

Walters, C.J., Goruk, R.D. and Radford, D. 1993. Rivers inlet sockeye salmon: an experiment in adaptive management. *North American Journal of Fisheries Management* 13: 253–62.

Walters, C. and Maguire J-J. 1996. Lessons for stock assessment from the northern cod collapse, *Reviews in Fish Biology and Fisheries,* 6(2): 125–37.

Warner, W.W. 1983. *Distant Water: The Fate of the North Atlantic Fisherman.* Boston: Little, Brown and Company.

Wilson, J.A., Townsend, R., Kleban, P., McKay S. and French J. 1990. Managing unpredictable resources: traditional policies applied to chaotic populations. *Ocean & Shoreline Management* 13(3&4): 179–97.

Young, O.R. 1989. *International Cooperation: Building Regimes for Natural Resources and the Environment.* Ithaca: Cornell University Press.

Part IV
Designing new approaches to management

Introduction

Each of the ecosystems dealt with in Part III (tropical forests, semi-arid lands, mountain environments, temperate marine fisheries) has suffered from ecological 'surprises'. Chapter 12 showed in some detail how human use of resources can cause unexpected, precipitous and possibly irreversible changes in the natural environment. Natural sciences have made considerable progress in understanding how ecological surprises and system flips come about. By contrast, relatively little progress has been made in understanding surprises in social–ecological systems and in defining management practices that might help avoid system flips, together with the social mechanisms behind these practices. The four chapters in Part IV attempt to understand these issues.

Chapter 13, by Holling, Berkes and Folke, begins by bringing together two strands of resource management thought that have been discussed separately in the past. The first involves rethinking resource management in a world of uncertainty and surprise, using systems approach and adaptive management. The second involves rethinking resource management social science by focusing on cultural capital and on property rights. The authors argue that conventional resource management science does not even seem to be able to explain resource collapses, and apply alternative approaches to explore problems of sustainability. They find that resource management problems typically tend to be systems problems, where aspects of systems behaviour are complex and unpredictable, and where causes are always multiple. Characteristically, the problems are non-linear in nature, cross-scale in time and space, and have an evolutionary character. The analysis is applied to combined social–ecological systems which constitute, in effect, integrated systems with critical feedbacks across temporal and spatial scales.

Chapter 14, by Pinkerton, deals with conventional resource management science in forestry and its alternatives, illustrating the issues raised in Chapter 13. The author traces the development of a number of interrelated alternative approaches to improve the use of the North American northwestern temperate rainforest. At issue is 'industrial forestry', a single-purpose, short-sighted approach benefiting only one sector. Further, as Pinkerton argues, 'timber rights are often held by powerful interests who have captured the regulatory agency, the scientific discourse, and the political ideology. . . '. Not only do the politics of forestry prevent a paradigm shift towards ecosystem management, but the current scientific paradigm itself is an impediment. Against this background, Pinkerton outlines various movements of wholistic (holistic) forestry, and concentrates on an aboriginal group, the Gitksan of British Columbia. The focus is on a clan chief of the Gitksan who provided leadership in the development of a local wholistic forestry plan with its own 'neo-traditional' characteristics that included planning at the small watershed level, using traditional knowledge, incorporating personal and spiritual identification of people with their land, and planning for multiple forest uses. Pinkerton identifies the Gitksan effort as a creative model of co-management in which scientific and traditional modes of taking care of a forest are combined, and in which the recovery of forest health parallels the recovery of human health, socially and individually.

Chapter 15, by Acheson, Wilson and Steneck, also provides a critique of conventional resource management, in this case in fisheries, and its alternatives. The authors argue that resource management by the numerical approach, through single-minded emphasis on quota management to exercise control over the numbers of each species killed, is not working well in the world's fisheries. A review of 30 peasant and tribal societies reveals three fundamental differences between contemporary fishery science and traditional wisdom. First, management rules in these societies are integral to local culture and ideologies which have no analogues in Western science, and they are enforced by social sanctions. Second, resources tend to be managed by local communities that have coastal rights over areas they know intimately. Third, these societies tend to use diverse combinations of various techniques, among which quota management is conspicuous by its absence. The authors argue for a new approach that combines: (a) 'parametric management' or the ecosystem approach; (b) the development of fishing restraints that affect *how* fish are caught instead of *how many* fish are caught; and (c) a decentralized management strategy that depends primarily on local-level management, consistent with the temporal and spatial scale of the resource to be managed.

The concluding chapter, by Folke, Berkes and Colding, focuses on management practices based on ecological knowledge, and social mechanisms behind these practices for resilience and sustainability. The synthesis is based mainly on the findings of the chapters in this volume. The list of social–ecological practices includes the rotation of resource areas and the management of succession and landscape patchiness. Practices identified perhaps for the first time in the literature on traditional societies include the management of ecological processes at multiple scales, responding to and managing pulses and surprises, and nurturing sources of renewal. The social mechanisms behind these management practices are clustered into four groups: those dealing with the generation, accumulation and transmission of ecological knowledge; structure and dynamics of institutions; cultural internalization; and the expression of worldview and ethics. The chapter then returns to the hypotheses posed in Chapter 1 to generate seven principles for building resilience, and to identify promising avenues for further inquiry. The authors propose that management practices, social mechanisms and principles may also be of use in the designing of more sustainable resource management systems in the industrial world.

13

Science, sustainability and resource management

C.S. HOLLING, FIKRET BERKES & CARL FOLKE

Introduction

It is probably no exaggeration to say that there is a worldwide crisis in resource management. There is certainly a crisis in resource management science. It has been argued that sustainability is neither a realistic goal nor a useful concept; some authors have questioned if the sustainable use of living resources would ever be possible. The concept of sustainable development has been called an oxymoron, and the utility of scientific research in designing policies and managing resource exploitation for sustainability has been questioned (Ludwig, Hilborn and Walters, 1993). Judging by the volume of largely sympathetic responses in the *Ecological Applications* special issue (Vol. 3, No. 4, 1993) devoted to this subject, these views appear to be widely shared among ecologists.

Such views are probably correct if sustainable use is defined as the first phase of exploitation of a single species, natural science as small-scale reductionist biology, and social science as essentially traditional neo-classical economics. This chapter argues that the above assumptions are not correct, and that the issue is not usefully defined in that way. Rather, there are other, more promising kinds of natural science and social science to provide policy directions towards sustainable use of natural resources.

Our objective in this chapter is to bring together two resource management approaches that have so far been discussed separately in the literature. The first involves rethinking resource management science in a world of uncertainty and surprise, using systems approach and adaptive management (Holling, 1978; 1986; Walters, 1986; Lee, 1993). The second involves rethinking resource management social science by focusing on cultural capital (as an integral part of a triad with economic capital and natural capital), and on property-rights systems (Berkes and Folke, 1994a; 1994b).

342

These promising strands are by no means part of the current conventional wisdom. Their acceptance requires major changes in resource management policy and practice, and forces a re-examination of the science on which current policy and practice are based.

The history of renewable resource use is full of examples of overexploitation, and there is a long and honourable tradition among ecologists and other environmental scientists of deeply held and pessimistic personal beliefs about future prospects with regard to population, food production, pollution and 'growth and progress' in general. What the 'tragedy of the commons' of Hardin (1968) and the views of Ludwig *et al.* (1993) have in common is a belief that human greed and shortsightedness almost always lead to overexploitation and often to the collapse of the resource.

As a generalization, this belief is not supported by the evidence. Rather, the analysis of a large number of case studies indicates that the resource management outcome is shaped by a whole array of decision-making arrangements, including property-rights regimes, incentive structures, cultural factors and institutions (Feeny *et al.*, 1990). However, the works of both Hardin (1968) and Ludwig *et al.* (1993) contain arguments for which there is good evidence. This includes Hardin's (1968) argument for the necessity of 'mutual coercion, mutually agreed upon', and the observation by Ludwig *et al.* (1993) that resource management using scientific data is confounded by the large spatial scale of natural processes, by the high degree of natural variability, and by the inherently unpredictable behaviours generated by the interactions between ecological systems and human behaviour. The result is a complex system best seen as a co-evolutionary one, with changing functional controls in the ecosystem, in the economy and in the society.

At the minimum, such views serve as an antidote to the technological optimism and cornucopian beliefs of such authors as Simon and Kahn (1984). In contrast to the environmentalist view, elements of this second view are deeply imbedded in the Western worldview that developed after the Middle Ages. The belief in unlimited material progress, that is achievable through economic growth and technological progress, is one of a suite of beliefs that characterize Western culture. Many who hold these beliefs argue that humans have infinite capacity to learn, economies have limitless ability to adapt and substitute, and problems of the world unfold fast enough to be perceived, yet slowly enough for the response to be effective and timely. This view generally assumes that the time stream of natural and human events is smooth and continuous, and that the behaviour of natural systems is narrow in range of variability and predictable.

Both views are in the realm of 'beliefs', and since both have partial elements of truth within them, their proponents have the ability to mobilize compelling examples and causal arguments that convincingly support these diametrically opposite views. This makes for great intellectual debate but hardly provides the sort of analysis that the public, the media and policy makers can use as a basis for action.

However, more constructive scientific analyses are possible. The pessimistic views of environmental and resource scientists can be tempered with the results of extensive research and practical experience in systems and mathematical modelling, population dynamics, large-scale experimental and management designs, and adaptive methods of management that focus on the interaction among renewable resources, management agencies, developers and resource users. Once the analysis is based on this rigorous body of knowledge and experience, the message is entirely different. One can turn from the conclusion that management is impossible, to a set of tested prescriptions for management that could be made sustainable.

The prescriptions arise from several generalizations that can be made from the body of literature. The argument that sustainability is neither a realistic goal nor a useful concept is only made convincing by assuming there is only one kind of science and one exploitive definition of sustainability. We shall deal with each in turn.

One kind of science?

The conventional worldview that has come to dominate Western culture is based on Newtonian physics as the model for science. For two and a half centuries, physicists have used a mechanistic view of the natural world to develop a conceptual framework based, among others, on Newton's mathematics, Descartes' philosophy, and Bacon's scientific method. The general conception of reality from the seventeenth century onward saw the natural world as a multitude of separate material objects assembled into a huge machine. It was believed that complex phenomena could be studied and controlled by reducing them to their basic building blocks and identifying the mechanisms by which parts of the machine interacted. This approach, called reductionism, has become an essential part of Western culture. Other sciences, including biology and economics, came to adopt the mechanistic and reductionistic views of classical physics as the model of description of reality. However, just as physics moved in other directions in the meantime (e.g. Capra, 1982), so too has biology (e.g. Kauffman,

1993) and economics (e.g. Anderson, Arrow and Pines, 1988). In brief, scholars at the leading edge of traditional disciplines have been creating an understanding and theory of the behaviour and evolution of complex social, ecological and economic systems, a theory that can begin to inform practice. Moreover, those creations have moved beyond each of their disciplinary roots into syntheses that integrate them. The critiques of reductionist biology or, for example, neoclassical economics, are now becoming dated. Those bodies of scholarship are being superseded by true innovative integration of economics and ecology (e.g. Arrow *et al.*, 1995; Dasgupta and Mäler, 1995; Perrings, Turner and Folke, 1995) and ecology and social sciences (e.g. Berkes and Folke, 1994a; Holling and Sanderson, 1996).

The nature and limitations of Western science are studied by historians and philosophers of science but, paradoxically, they remain ill-understood by many of the practitioners themselves. Scientific disciplines are often cocooned; scientific purity is assured by the assiduous avoidance of societal issues. For example, many scientific articles in ecology journals are based on field experiments in which human influence is thought not to exist. Science is often thought to be objective and value free, quantitative and hence precise.

In the field of resource management, science developed under the conventional reductionistic and mechanistic worldview and, further, it was shaped by the utilitarian premises of the early industrial era. Nature was viewed merely as a storehouse of raw materials; resources were thought to be valuable only to the extent that they could be used to create wealth. The Linnean model of ecology had more to say about the human mission to extract rather than to conserve (Worster, 1977: 53). This new way of regarding nature also involved a transformation in ethical attitudes. The drive for the domination of nature resulted in the desacralization of the world. Land and natural resources, as well as people's own labour, came to be viewed as commodities (Polanyi, 1980).

Many concepts and practices of resource management science can be seen in the light of this process of commodification of nature. For example, the concept of 'maximum sustained yield' is still widely used in fisheries management even though it ignores ecosystem interactions and environmental variability (Larkin, 1977). The concept is geared for the efficient utilization of fishery and other resources as if stocks were discrete commodities in space and time, but evidently it is not conducive to sustainability in any larger context (Gadgil and Berkes, 1991).

Many critics have highlighted the inadequacy of conventional science, science that is disciplinary, reductionist, mechanistic, and detached from people, policies, and politics. But there is another stream of science that differs considerably from the first.

This other stream is represented by systems approaches and parts of evolutionary biology that extend the analysis of populations, ecosystems, landscape structures and dynamics, to include the interactions of social systems with natural systems. The applied form of this stream has emerged regionally in new forms of resource and environmental management in which uncertainty and surprises become an integral part of an anticipated set of adaptive responses (Holling, 1978; Walters, 1986; Lee, 1993; Gunderson, Holling and Light, 1995a). The new stream is fundamentally interdisciplinary and combines historical, comparative and experimental approaches at scales appropriate to the issues. It is a stream of inquiry that is fundamentally concerned with integrative modes of inquiry and multiple sources of evidence.

It is this stream that has the most natural connection to related ones in the social sciences that are historical, analytical and integrative. It is also the stream that is most relevant for the needs of policy and politics.

The first stream can be characterized as a science of the parts. It emerges from traditions of experimental science in which a narrow focus enables the formulation of hypotheses, the collection of data, and the design of critical experiments in order to test hypotheses. The goal is to narrow uncertainty to the point where the acceptance of an argument among scientific peers can become essentially unanimous. It is appropriately conservative and unambiguous, but it achieves that by being incomplete and fragmentary.

The second stream can be characterized as a science of the integration of parts. It uses the results and technologies of the first, but identifies gaps, develops alternative hypotheses and multivariate models, and evaluates the integrated consequence of each alternative by using information from planned and unplanned interventions in the whole system that occur, or are designed to be implemented, in nature. Typically, the goal is to reveal the simple causation that often underlies the complexity of time and space behaviour of complex systems. Often, there is more concern that a useful hypothesis might be rejected than a false one accepted – 'don't throw out the baby with the bath water'. Since uncertainty is high, the analysis of uncertainty becomes a topic in itself.

The premise of this second stream is that knowledge of the system we deal with is always incomplete. Surprise is inevitable. We use the term surprise to denote the condition when perceived reality departs *qualitatively* from expectation. Surprises occur when causes turn out to be sharply different from what was conceived, when behaviours are profoundly unexpected, and when action produces a result opposite to that intended (Holling, 1986). Not only is the science incomplete, the system itself is a moving target, changing because of the impacts of management (Gunderson *et al.*, 1995a) and the progressive expansion of the scale of human influences (Daly, 1994; Folke, Holling and Perrings, 1996).

In principle, therefore, there is an inherent unknowability, as well as unpredictability, concerning these evolving, managed ecosystems and the societies with which they are linked. The essential point is that evolving systems require policies and actions that not only satisfy social objectives but, at the same time, also achieve continually modified understanding of the evolving conditions and provide flexibility for adaptation to surprises. Science, policy and management then become inextricably linked.

One kind of sustainability?

In resource management science, the concept of sustainability goes back a great many years; a German scientist, Faustmann, used the concept in 1849 to calculate the forest rotation period that would maximize returns (Ludwig, 1993). At least since the 1930s, sustained yield has been a goal in all areas of living resources management. The biological concept of sustained yields has been closely linked with the economic concept of sustained yields, as in 'maximum sustained yield' and the 'maximum economic yield' in fisheries, although the priority has been a matter of debate between biologists and economists.

The concept of sustained yield developed in the service of a utilitarian worldview in which nature was seen as a storehouse of raw materials, and resources as merely commodities, and humans had 'dominion over nature'. The mathematical treatment of the supply and demand of these resources/commodities employed a reductionist approach, and essentially ignored the fact that these 'products' were embedded in ecosystems.

As already mentioned, the MSY concept, for example, treats stocks as discrete elements in space and time, predictable in isolation from other elements in the ecosystem, and assumes away natural variability (Larkin, 1977). This kind of resource management, a science of the parts, may be suitable for conventional exploitive development but not for sustainable use – if sustainable use is defined more broadly to include a wider range of ecological, social and economic objectives.

The idea of 'sustainable development', as promoted in the *World Conservation Strategy,* referred to such a broader range of objectives:

meeting basic human needs while maintaining essential ecological processes and life-support systems, preserving genetic diversity, and ensuring sustainable utilization of species and ecosystems (IUCN/UNEP/WWF, 1980). The popularization of the term came with the World Commission on Environment and Development. Its report was long on problem descriptions and short on policy prescriptions, but it provided the standard definition of sustainable development as 'development that meets the needs of the present without compromising the ability of future generations to meet their needs' (WCED, 1987; 8).

The two kinds of sustainability are in sharp contrast. The maximum sustainable yield focuses on the resource-as-commodity, with a prescription of the ways in which it can be efficiently utilized. The time horizon of sustainability is 'in perpetuity' as far as the ecologist is concerned; but commodities obey the laws of the market and there is no guarantee that the law of the market favours conservation. Clark (1973; 1976) has shown why a slowly reproducing living resource will be treated by the market like a non-renewable resource (because the exploiter will make more money depleting the resource quickly and re-investing elsewhere), and why long-term ecological values cannot be protected by the market (because the future value of any resource will be discounted to nearly zero in 15–20 years from the point of view of the economic decision maker).

In practice, the history of living resource use in the Industrial Era is characterized by a general pattern of *sequential exploitation* of stocks, from the more accessible to increasingly less accessible areas, and from the most marketable to the less and less marketable species (Berkes, 1989). This pattern is made possible in part by the development of transportation networks and markets, and is subsidized by the use of auxiliary energies, such as fossil fuels (Odum, 1971; Hall, Cleveland and Kaufmann, 1986). Sustained yields have often been, in effect, the first phase of exploitation of a single species, followed by a sequence of species and of geographic frontiers. The reality of the market place is that discounting of resource values rewards 'frontier economics' behaviour; under certain conditions, it is economically optimal for the exploiter to drive the stock to extinction and to move on.

By contrast, the second kind of sustainability, i.e. sustainable development, does not discount the future, and embodies three imperatives: (1) the environmental imperative of living within ecological means; (2) the economic imperative of meeting basic material needs; and (3) the social imperative of meeting basic social needs and cultural sustainability. Thus, sustainable development is concerned with much more than single stocks.

It tries to cover the broad range of environmental concerns as well as economic and social concerns.

Also, it is concerned with cultural capital, factors that provide human societies with the means and adaptations to deal with the natural environment and actively to modify it (Berkes and Folke, 1994a). Natural capital, economic (or human-made) capital and cultural capital are all interrelated. For example, attitudes towards the environment and resource use are shaped by culture and evolve through history. Attitudes towards nature are reflected in the way societies are organized to use their resources. There is a great diversity of institutions for resource use (Berkes, 1989). These institutions, the locally evolved environmental knowledge base used by institutional organization, and the worldviews and ethics that underpin the societies in question are all parts of the cultural capital of a society.

The practical problem with sustainable development is that we simply do not have the body of theory and practice to live up to these imperatives. It is little wonder that many scholars detest the use of the label 'sustainable development' as a veneer of environmental respectability on the process of continuing unsustainable practice.

Yet, this does not mean that we should reject the concept. As Lee (1993) puts it, '. . . sustainable development is not a goal, not a condition likely to be attained on earth as we know it. Rather it is more like freedom and justice, a direction in which we strive'. It is a process, not a final state (Robinson *et al.*, 1990). This is good advice for analysing cases of living resources management, as many of the short-term 'successes' ultimately clash with long-term sustainability for which we strive.

It has been profoundly disturbing to analysts that in many cases of renewable resource management, the success in managing a target variable (food, fibre) for sustained production has led to an ultimate pathology of: (1) more brittle and vulnerable ecosystems; (2) more rigid and unresponsive management agencies; and (3) more dependent societies (Holling, 1986). Examples include the initial decades of chemical control of spruce budworm in Canadian forests – more and more control effort seem to result in larger and larger infestations when they do occur; and forest fire suppression in Yellowstone National Park in the USA – almost half of the park burned down in one major fire in 1988, following a century of fire suppression.

There are many examples of apparently successful management, later leading to environmental backlash or surprise; examples range from pesticide use, to the damming of major African rivers (Farvar and Milton, 1972). The very success (i.e. profitability) of a well-managed fishery tends to trigger its own demise by attracting additional capitalization and fishing effort until all resource rents are dissipated, a well-known phenomenon in fishery economics (Clark, 1985). These pathologies seem to support the opinion that sustainable development is an oxymoron. Moreover, they occur not only in the area of renewable resources, but also in other areas of environmental management. Examples include the failures of the rigid policies of regulation of toxic materials by the US Environmental Protection Agency, and the narrow implementation of the Endangered Species Act by the US Fish and Wildlife Service.

However, if we examine these pathological examples over a longer span, we find that another kind of surprise may occur. There are cases in which external and internal crises, amplified by the pathology, trigger a sudden lurch in understanding, a redesign and expansion of policy, and a return of flexibility and innovation. There have been enough cases to prompt a book-length exploration of examples of pathological exploitation, followed by crises and learning. Examples range from the Everglades of Florida, the forests of New Brunswick, the estuary of Chesapeake Bay, the Great Lakes, and the Baltic Sea (Gunderson *et al.*, 1995a). In New Brunswick, for example, the intensifying deadlock in forest management, combined with slowly accumulated and communicated scientific, economic and social understanding, led to an abrupt transformation of forest policy that became freed from local constraints, and set in an adaptive framework designed to achieve both ecological and economic benefits (Baskerville, 1995). It is a policy that functions for a whole region by transforming and monitoring the smaller-scale stand architecture of the landscape and by releasing the productive and innovative capacities of industry.

Viewed this way, the message from pathological examples becomes entirely different. The examples do not prove that sustainability is impossible; rather, they indicate that pathology itself may trigger learning and innovative redesign towards sustainability. Indeed, breakdown may be a necessary condition to provide the understanding for system change.

The ecological analogy of this phenomenon may be found in the cycle of ecosystem development. In a forest ecosystem, a four-stage cycle of renewal – conservation, creative destruction, reorganization – repeats itself again and again. The first two phases, renewal (the establishment of pioneering species) and conservation (the consolidation of nutrients and biomass into a climax stage) lead to a system which is so stable, so dependent on conditions remaining constant, that it becomes 'brittle'. Such brittleness invites environmental surprises such as fire, insect pest outbreak or disease. When surprise happens, accumulated capital is suddenly released for other kinds of opportunity (creative destruction). This very rapid stage is followed by reorganization, in which, for example, nutrients released from the trees by fire will be fixed in other parts of the ecosystem as the renewal stage starts again (Holling, 1986; Holling *et al.*, 1995). The pattern of non-sustainable development or pathological exploitation followed by crisis and learning might still be rare in parts of the world in which resource development is in the frontier stage. British Columbia, the Canadian province in which the analysis of Ludwig *et al.* (1993) was shaped, might be one example of frontier economics. But the crisis-and-learning pattern is not at all uncommon where there has been a longer history of resource use, with the parallel development of local resource management systems. Examples abound in the common-property literature (McCay and Acheson, 1987; Berkes, 1989; Ostrom, 1990; Bromley, 1992).

'For most of recorded history', writes Ophuls (1992: 190), 'societies have existed at the ecological margin', and had to adapt to living within their ecological means. The Industrial Era has been a relatively short-lived anomaly of affluence in human history. In the last 400 or so years, the carrying capacity of the globe markedly expanded. Western European societies treated much of the world as a vast resource frontier. Although there are some views to the contrary (e.g. Simon and Kahn, 1984), it appears that this frontier is running out.

The West does not have a well-developed science for 'living at the ecological margin'. However, there is some information to help reconstruct human–nature relationships that were characteristic of human societies over the vast majority of human history, that is, with the exception of the 400-year period of industrial expansion. Most of this evidence comes from societies that were never part of the dominant Western industrial world but were 'marginal' themselves.

Historically, a pattern of co-evolutionary adaptations between social systems and natural systems must have been the norm, with the adaptations in many cases driven by crises, learning and redesign (see Chapter 5). There are places and societies that practised sustainable resource use, not merely of species but entire ecosystems, and some of their adaptations survive to date (Gadgil, Berkes and Folke, 1993; Norgaard, 1994). For this reason, ancient cultural practices of resource use are more than anthropological curiosities; they are part of humanity's wealth of adaptations that can serve the contemporary world as well (Gadgil *et al.*, 1993).

To conclude, it is not useful, except as a debating position, to define sustainable development as exploitation of a single species. It begins to be useful as a guide to science, policy and action when the definition of sustainability focuses on the processes of social and economic development that invest in the maintenance and restoration of critical ecosystem functions, that synthesize and make accessible knowledge and understanding of the combined social–ecological system, and that build trust among citizens based on this understanding.

Generic features of resource management issues

There is a worldwide crisis in resource management because the existing science that deals with the issue seems unable to prescribe sustainable outcomes. Existing resource management science does not even seem to be able to explain resource collapses. Resource managers, decision makers and citizens are frustrated because they are not hearing clear and consistent answers to key questions concerning environmental and renewable resource issues. The answers are not simple because we have just begun to develop the concepts, technology and methods that can address the generic nature of the problems. Characteristically, these problems tend to be systems problems, where aspects of behaviour are complex and unpredictable and where causes, while at times simple (when finally understood), are always multiple. They are non-linear in nature, cross-scale in time and in space, and have an evolutionary character. This is true for both natural and social systems. In fact, they are one system, with critical feedbacks across temporal and spatial scales. Therefore interdisciplinary and integrated modes of inquiry are needed for understanding. Furthermore, understanding (but not necessarily complete explanation) of the combined system of humans and nature is needed to formulate policies.

Non-linear nature of the problem

A general characteristic of resource management problems is that they are fundamentally non-linear in causation. They demonstrate multi-stable states and discontinuous behaviour in both time and space. Non-linearity results in unpredictable behaviour, either because periodically small changes can propagate dramatically and flip the system into another development path, as in chaos theory, or because stability regions collapse as slow processes accumulate and move the system from one set of controlling mechanisms and processes to another, as in catastrophe theory. It is the non-linear property that generates the four-stage cycle of exploitation, conservation, renewal, reorganization (Holling, 1986).

Resilience is a crucial concept in this respect. Resilience is the capacity of the system to absorb disturbance. It reflects the ability of the system to stay on the same branch of development when it is going through the four-stage cycle. Loss of resilience will move the system closer to thresholds, and ultimately cause it to flip from one equilibrium state to another. Such threshold effects occur in large-scale ecosystems, for example when loss of resilience irreversibly flips a tropical forest into a grassland ecosystem, or a savanna ecosystem into a bush-shrub landscape (Folke *et al.*, 1996). Collapse of fisheries and stock markets may be explained in a similar way. In ecosystems that have lost resilience, such as the Great Lakes, stocks may collapse and not come back despite conservation measures. In such systems, several kinds of changes are possible, but it is not possible *a priori* to predict even the kind of changes that will actually occur (Holling, 1973), let alone their probability. Ecosystems are often characterized by changes that could not, on looking back, have been anticipated; these are 'surprises' in which reality departs qualitatively from expectation.

Loss of resilience and threshold effects also occurs in social systems. The hunting and trapping territory system of the Cree Indians in James Bay, eastern subarctic Canada, came under pressure in the 1920s when a railway was constructed into the area. As the outsiders started to deplete the animal resources of the area, the territorial system and common-property institutions of the Cree also broke down, and Cree hunters themselves joined in the frontier behaviour that saw the near-extirpation of beaver over the vast eastern subarctic by 1930 (Berkes *et al.*, 1989).

The inherent unpredictability of ecosystems plays havoc with conventional resource management science, which starts with the assumption of a clockwork, predictable world in which sustained yields can be predicted, given enough information. For example, effective protection and enhancement of salmon spawning through the use of fish hatcheries on the west coast of North America quickly led to more predictable and larger catches by both sport and commercial fishermen. That triggered increasing fishing and investment pressure in both sectors, pressure that caused an increasing number of the less productive natural stocks to become locally extinct. That left the fishing industry precariously dependent on a few artificially enhanced stocks whose productivity began declining in a system where larger-scale physical oceanic changes began to contribute to unexpected impacts on distribution and abundance of fish. A similar pattern is being repeated in the management of Baltic salmon stocks. As a consequence of initially 'successful' management, the ecosystem and its natural resources became more vulnerable to surprise and crisis as resilience decreased. At the same time, the management institutions became more rigid and less responsive to feedbacks.

Many such empirical cases challenge the belief in a mechanistic, predictable world that can be understood through reductionism. Policies that assume smoothly changing and reversible conditions, and limitless ability of the economy to adapt and substitute, lead to reduced options, limited potential and perpetual surprise. The political window that drives 'quick fixes' for quick solutions simply leads to more unforgiving conditions for decisions, more fragile natural systems, and more dependent and distrustful citizens.

The linear, equilibrium-centred view of nature no longer fits the evidence, and is being replaced by a non-linear, multi-equilibrium view. This trend in ecology and natural resources parallels trends in economics and other fields in which equilibrium-centred thinking is losing ground to theories of self-organization (Perrings et al., 1995).

The non-linear nature of changes in resource systems requires that resource management science find new and appropriate tools to deal with the issue. Concepts that are useful come from non-linear dynamics and theories of complex systems (Costanza et al., 1993). A comprehensive understanding of linked social systems and natural systems requires the synthesis of several conceptual frameworks. A major step in this direction has been the integration of ecological and economic thinking (e.g. Jansson et al., 1994). Another major step involves the integration of relevant social science concepts; focus on property-rights and institutions, for example, cuts across a number of natural science and social science fields (Berkes and Folke, 1994a).

Property-rights systems do not change smoothly but show discontinuous behaviour. Like the depletion of natural capital, the degradation of institutions and loss of cultural capital result in sudden changes, as in the collapse of a society in civil war. Where the 'tragedy of the commons' does occur, it does not do so as a result of 'human greed', as Ludwig et al. (1993) would have it, but as a result of institutional failure. Cree hunters of James Bay, knowledgeable and careful with their resources one moment, may become active participants in the destruction of their own resources the next moment, as the incursion of outsiders creates open-access conditions in place of the indigenous communal management system.

Cross-scale nature of the problem

Social and ecological systems are nested in time and space from the cell to the ecosphere, with numerous non-linear feedbacks. Recently, Gunderson, Holling and Light (1995b) used the term panarchies to capture the evolutionary nature of adaptive cycles that are nested one within the other across space and time scales. They suggest that each level goes through the cycle of growth, maturation, destruction and renewal, and that all living systems, ecological as well as social, exhibit properties of the adaptive cycle and of panarchical relationships across scales. This view emphasizes that periods of gradual change and periods of rapid transformation coexist and complement one another. Characteristics of such nested systems have also been discussed by Günther and Folke (1993).

Problems of cross scale in nested systems are increasingly caused by slow changes reflecting decadal accumulations of human influences on air and oceans, and decadal to centuries-long transformations of landscapes. Those slow changes cause sudden changes in fast environmental variables that directly affect the health of people, productivity of ecosystems, and vitality of societies. The evolution of new diseases caused by large-scale land-use changes, human population growth and expanding economic activities are examples of such phenomena. The span of connections is intensifying so that the problems of a proper relationship and management of linked social and ecological systems are now fundamentally cross-scale in space as well as time. Biogeochemical and hydrological flows are being transformed on the global level. National environmental problems more and more frequently have their source not only at home but also half a world away, witness greenhouse gas accumulations, ozone hole, AIDS, deterioration of biodiversity. Natural planetary processes mediating these issues are coupling with the human, economic and trade linkages that have evolved exponentially among nations since Would War II.

Therefore, the science needed is not only interdisciplinary, it is cross-scale. And yet, the very best of environmental and ecological research and models have achieved their success by being either scale independent or constrained to a narrow range of scales. Analysis should focus on the inter-actions between slow phenomena and fast ones, and monitoring should focus on long-term, slow changes in structural variables. Hierarchical theory, spatial dynamics, event models, satellite imagery, and parallel pro-cessing perhaps open new ways to violate, successfully, the hard-won expe-rience of the best ecosystem modellers, i.e. never include more than two orders of magnitude or the models will be smothered by detail.

Western resource management, as discussed above, tends to cut off the feed-backs between the periods of gradual change and the periods of rapid trans-formation which coexist and complement one another. Periods of gradual change are supported, and rapid transformations are seen as disturbances which should be eliminated. As previously emphasized, this strategy leads to loss of resilience, reduction in variability, and more brittle systems. Such a strategy has dominated the development of modern, industrial society during the last decades, and still does, not recognizing that it exacerbates cross-scale interactions that challenge ecological thresholds on regional and even global levels.

In contrast, many traditional local communities have recognized the necessity of the coexistence of gradual and rapid change. In their institutions they have accumulated a knowledge base for how to respond to environmental feedbacks. They have developed social mechanisms that interpret the signals of creative destruction and renewal of ecosystems and cope with them before they accumulate and challenge the existence of the whole local community. Disturbance has been allowed to enter at smaller scales in the panarchy of nested sets of adaptive cycles. There is a culturally evolved 'monitoring' system that reads the signals, the disturbances, and thereby is more successful in avoiding the build-up of an internal structure that will become brittle and invite large-scale collapse. The local institution has evolved so that renewal occurs internally while overall structure is maintained. The accumulation and transfer of this knowledge between generations have made it possible to be alert to changes and continuously adapt to them in an active way. They have been a means of survival.

In contrast to Western-oriented management, these local institutions do not undermine their existence by degrading their ecological life-support system, thereby losing ecological and institutional resilience. Instead, they have maintained the resilience of the linked social–ecological system, by responding to renewal and opportunity. The local community has become a part of the dynamics of its surrounding ecosystem.

Locally linked social–ecological systems are very fragile to external disturbance caused by, for example, trade opportunities, new resource exploiters and governmental programmes based on Western value systems. However, the brittleness to such disturbance is not a reflection of rigid and unflexible organization of the local institution. It is caused by external cultural transformation. The irony is that most of these transformations for development generate the institutional pathologies of resource management founded on a reductionistic and mechanistic worldview, which often exacerbate cross-scale problems, problems which the linked traditional social–ecological system was successful in avoiding.

Evolutionary character of social and natural systems

Both the ecological and social components of resource management problems have an evolutionary character, as in self-organization and adaptive management. Self-organization is a key characteristic of all living systems. The study of evolution requires the concept of function as well as the concept of organization, the way components are connected within systems and the way systems are embedded in larger systems. The relationship between organisms and their environment is regulated by feedbacks. The physical environment sets limits on the growth of populations; the organisms, in turn, actively modify their environment to improve their chance of survival. Evolutionary perspective draws attention to processes by which organisms adapt to and co-evolve with their environment.

Evolution and self-organization are also found in social systems, as evidenced by the incredible diversity of common-property institutions that have been documented for a variety of resource types from a variety of geographic areas (McCay and Acheson, 1987; Berkes, 1989; Bromley, 1992). Yet, assuming that no such self-organization could occur, policy recommendations were typically made to impose externally designed solutions (Hardin, 1968). The legacy of non-evolutionary thinking is that a monolithic set of centralized resource management prescriptions has been spreading to all corners of the globe, often at the expense of locally evolved systems (Gadgil and Berkes, 1991).

A telling case comes from the field of fishery management. Wilson *et al.* (1994) have shown, for example, that traditional fishery systems are characterized by rules and practices limiting 'how' people fish, rather than by attempting to regulate 'how much' should be taken. The fact that traditional strategies are found so widely and have a track record over a long period suggests that they are highly adaptive. The evolutionary character reflects the fact that ecological and social systems can change qualitatively to generate and implement innovations that are truly creative, in the sense of opportunities for novel co-operation and feedback management. The same cannot be said for quota management under the current wisdom of fishery science, given the inherent unpredictability of yields and difficulty of enforcement.

Some of the most sophisticated common-property institutions are found in areas in which these systems have developed over a long period of time, in the order of hundreds of years. Examples include Spanish *huertas* for irrigation, Swiss grazing commons (Ostrom, 1990), and marine resource tenure systems in Oceania (Johannes, 1978). In other areas, common-property institutions have evolved over a short period of time (in the order of one decade) in response to a management crisis. An example is the Turkish Mediterranean coastal fishery in Alanya (Berkes, 1992). Yet other systems have collapsed and recovered over a period of time, sometimes more than once (Berkes *et al.*, 1989).

Concluding remarks

Integrative resource management can proceed by a design that simultaneously allows for tests of different management policies and emphasizes
learning by doing. Called adaptive management, this approach treats policies as hypotheses, and management as experiments from which managers
can learn (Holling, 1978; Walters, 1986). Adaptive management effectively
breaks down the barrier between research and management, and provides
one solution to a fundamental problem in resource management: the
difficulty of conducting controlled experiments. As it proceeds in a stepwise
fashion, responding to changes and guided by feedback from the resource,
adaptive management allows for institutional learning.

Indigenous knowledge and management systems are similar in some
ways to adaptive management. These traditional systems, which may be
characterized as knowledge–practice–belief complexes, also proceed in an
adaptive fashion by mutual feedback mechanisms and co-evolution
(Gadgil et al., 1993). Traditional systems rely on the accumulation of
knowledge over many generations, and knowledge is transmitted culturally.
They parallel adaptive management in their reliance on learning-by-doing,
and the use of feedback from the environment to provide corrections for
management practice. They differ from science-based systems generally by
the absence of testable hypotheses and generalizable theories, and by the
integration of moral and religious belief systems with management (Gadgil
et al., 1993).

The parallels between adaptive management and indigenous management systems are probably not accidental. Flexible social systems that
proceed by learning-by-doing are better adapted for long-term survival
than are rigid social systems that have set prescriptions for resource use. In
light of this, adaptive management in modern society could be seen as a
replication of traditional ecological knowledge systems in the framework
of contemporary science. It is a sort of rediscovery of principles applied in
traditional social–ecological systems. It is a search for a sustainable relationship with life-supporting ecosystems, a social and institutional
response to resource scarcity and management failure. By responding to
and managing feedbacks from the ecosystem instead of blocking them out,
adaptive management has the potential to avoid ecological thresholds at
scales that threaten the existence of social and economic activites, just as
the adaptive behaviour of many traditional social–ecological systems.

However, learning-by-doing requires time, but there is not as much time
and flexibility left in the present world, due to the magnitude of human

influences on the planet. Learning-by-doing experiments may become very costly since local change may trigger regional and global change (Holling, 1994). Therefore, drawing on traditional ecological knowledge and understanding the mechanisms behind the development, evolution and sustainability of successful social–ecological systems may speed up the process of adaptive management. Surprises – the gaps between expectation and perceived reality – were presumably much smaller in the historical adaptive management systems. Therefore, phenomena that are regarded as surprises in contemporary society may not have been surprises in traditional social – ecological systems, simply because of institutional learning for resilience of the combined social–ecological system. Resource crises may have a constructive role to play – by triggering the opportunity for renewal and redesign in systems capable of learning and adapting. Such systems should be designed to allow for internal renewal while overall structure is maintained.

Thus, at the heart of sustainable development is renewal and the release of opportunity, both social and ecological, and at relevant temporal and spatial scale in the panarchy of nested adaptive cycles. That is why the phrase sustainable development is not an oxymoron. The problems are not amenable to solutions based on knowledge of small parts of the whole, nor on assumptions of constancy or stability of fundamental ecological, economic and social relationships. Such assumptions produce policies and science that contribute to a pathology of rigid and unseeing institutions and increasingly brittle natural systems. Learning from traditional social–ecological systems and combining insights gained in adaptive management may counteract such pathologies and enhance institutional learning and understanding for resilience of the linked social–ecological system in any society.

References

Anderson, P., Arrow, K. and Pines, D. 1988. *The Economy as an Evolving Complex System*. Redwood City, Calif.: The Santa Fe Institute Studies in the Sciences of complexity. Addison Wesley.

Arrow, K., Bolin, B., Costanza, R., Dasgupta, P., Folke, C., Holling, C.S., Jansson, B.-O., Levin, S., Mäler, K.-G., Perrings, C. and Pimentel, D. 1995. Economic growth, carrying capacity, and the environment. *Science* 268: 520–21.

Baskerville, G. 1995. The forestry problem: Adaptive lurches of renewal. In *Barriers and Bridges to the Renewal of Ecosystems and Institutions*, pp.37–102, ed. Gunderson, C.S. Holling and S. Light. New York: Columbia University Press.

Berkes, F., ed. 1989. *Common Property Resources: Ecology and Community-Based Sustainable Development*. London: Belhaven.

Berkes, F. 1992. Success and failure in marine coastal fisheries of Turkey. In *Making the Commons Work*, pp.161–82, ed. D.W. Bromley. San Francisco: Institute for Contemporary Studies.

Berkes, F., Feeny, D., McCay, B.J. and Acheson, J.M. 1989. The benefits of the commons. *Nature* 340: 91–3.

Berkes, F. and Folke, C. 1994a. Investing in cultural capital for the sustainable use of natural capital. In *Investing in Natural Capital*, pp.128–49, ed. A.M. Jansson, M. Hammer, C. Folke, and R. Costanza. Washington DC.: Island Press.

Berkes, F. and Folke, C. 1994b. *Linking Social and Ecological systems for Resilience and Sustainability.* Beijer Discussion Paper No. 52. Stockholm: The Beijer Institute.

Bromley, D.W., ed. 1992. *Making the Commons Work: Theory, Practice and Policy.* San Francisco: Institute for Contemporary Studies.

Capra, F. 1982. *The Turning Point. Science, Society and the Rising Culture.* New York: Simon and Schuster.

Clark, C.W. 1973. The economics of overexploitation. *Science* 181: 630–34.

Clark, C.W. 1976. *Mathematical Bioeconomics: The Optimal Management of Renewable Resources.* New York: Wiley-Interscience.

Clark, C.W. 1985. *Bioeconomic Modelling and Fisheries Management.* New York: Wiley.

Costanza, R., Wainger, L., Folke, C. and Mäler, K.-G. 1993. Modeling complex ecological economic systems. *BioScience* 43: 545–55.

Daly, H.E. 1994. Operationalizing sustainable development by investing in natural capital. In *Investing in Natural Capital*, pp.22–37, ed. A.M. Jansson, M. Hammer, C. Folke and R. Costanza. Washington DC.: Island Press.

Dasgupta, P. and Mäler, K.-G. 1995. Poverty, institutions, and the environmental resource-base. In *Handbook of Development Economics,* Volume III, pp.2371–463, ed. J. Behrman and T.N. Srinivasan. Amsterdam: Elsevier.

Farvar, M.T. and Milton, J.P., eds. 1972. *Careless Technology: Ecology and International Development.* Garden City, New York: Natural History Press.

Feeny, D., Berkes, F., McCay, B.J. and Acheson, J.M. 1990. The tragedy of the commons: Twenty-two years later. *Human Ecology* 18: 1–19.

Folke, C., Holling, C.S. and Perrings, C. 1996. Biological diversity, ecosystems and the human scale. *Ecological Applications* 6: 1018–24.

Gadgil, M. and Berkes, F. 1991. Traditional resource management systems. *Resource Management and Optimization 8*, 127–41.

Gadgil, M., Berkes, F. and Folke, C. 1993. Indigenous knowledge for biodiversity conservation. *Ambio* 22: 151–6.

Gunderson, L., Holling, C.S. and Light, S., eds. 1995a. *Barriers and Bridges to the Renewal of Ecosystems and Institutions.* New York: Columbia University Press.

Gunderson, L., Holling, C.S. and Light, S. 1995b. Breaking barriers and building bridges: a synthesis. In *Barriers and Bridges to the Renewal of Ecosystems and Institutions,* ed. L. Gunderson, C.S. Holling and S. Light, New York: Columbia University Press.

Günther, F. and Folke, C. 1993. Characteristics of nested living systems. *Journal of Biological Systems* 1: 257–74.

Hall, C.A.S., Cleveland, C.J. and Kaufmann, R. 1986. *Energy and Resource Quality.* New York: Wiley-Interscience.

Hardin, G. 1968. The tragedy of the commons. *Science* 162: 1243–8.

Holling, C.S. 1973. Resilience and stability of ecological systems. *Annual Review of Ecology and Systematics* 4: 1–23.

Holling, C.S., ed. 1978. *Adaptive Environmental Assessment and Management.* London: Wiley.

Holling, C.S. 1986. Resilience of ecosystem: Local surprise and global change. In *Sustainable Development of the Biosphere*, pp.292–317, ed. W.C. Clark, and R.E. Munn. Cambridge: Cambridge University Press.

Holling, C.S. 1994. An ecologist view of the Malthusian conflict. In *Population, Economic Development and the Environment*, pp.79–103, ed. K. Lindahl-Kiessling, and H. Landberg. Oxford: Oxford University Press.

Holling, C.S. and Sanderson, S. 1996. Dynamics of (dis)harmony in ecological and social systems. In *Rights to Nature*, pp.57–85, ed. S. Hanna, C. Folke, and K.-G. Mäler. Washington DC.: Island Press.

Holling, C.S., Schindler, D.W., Walker, B.W. and Roughgarden, J. 1995. Biodiversity in the functioning of ecosystems: An ecological synthesis. In *Biodiversity Loss*, pp.44–83, ed. C. Perrings, K.-G. Mäler, C. Folke, C.S. Holling, and B.-O. Jansson, New York: Cambridge University Press.

IUCN/UNEP/WWF 1980. *World Conservation Strategy: Living Resource Conservation for Sustainable Development.* Gland, Switzerland: IUCN.

Jansson, A.M., Hammer, M., Folke, C. and Costanza, R., eds. 1994. *Investing in Natural Capital: The Ecological Economics Approach to Sustainability.* Washington DC: Island Press.

Johannes, R.E. 1978. Traditional marine conservation methods in Oceania and their demise. *Annual Review of Ecology and Systematics* 9: 349–64.

Kauffman, S.A. 1993. *The Origins of Order.* New York: Oxford University Press.

Larkin, P.A. 1977. An epitaph for the concept of maximum sustained yield. *Transactions of the American Fisheries Society* 106: 1–11.

Lee, K.N. 1993. *Compass and Gyroscope: Integrating Science and Politics for the Environment.* Washington D.C: Island Press.

Ludwig, D. 1993. Environmental sustainability: Magic, science and religion in natural resource management. *Ecological Applications* 3: 555–8.

Ludwig, D., Hilborn, R., and Walters, C. 1993. Uncertainty, resource exploitation and conservation: lessons from history. *Science* 260: 17, 36.

McCay, B.J. and Acheson, J.M., eds. 1987. *The Question of the Commons. The Culture and Ecology of Communal Resources.* Tucson: University of Arizona Press.

Norgaard, R.B. 1994. *Development Betrayed. The End of progress and a Coevolutionary Revision of the Future.* London and New York: Routledge.

Odum, H.T. 1971. *Environment, Power, and Society.* New York: John Wiley.

Ophuls, W. 1992. *Ecology and the Politics of Scarcity Revisited.* New York: Freeman.

Ostrom, E. 1990. *Governing the Commons: The Evolution of Institutions for Collective Action.* Cambridge: Cambridge University Press.

Perrings, C., Turner, R.K. and Folke, C. 1995. *Ecological Economics: The study of Interdependent Economic and Ecological Systems, Beijer Discussion Paper Series No. 55.* Stockholm: Beijer Institute.

Polanyi, K. 1980. *The Great Transformation.* Boston: Beacon Press.

Robinson, J., Francis, G., Legge, R. and Lerner, S. 1990. Defining a sustainable society: Values, principles and definitions. *Alternatives* 17: 36–46.

Simon, J., and Kahn, H. 1984. *The Resourceful Earth: A Response to Global 2000.* Oxford: Basil Blackwell.

362   *C.S. Holling, Fikret Berkes & Carl Folke*

Walters, C.J. 1986. *Adaptive Management of Renewable Resources*. New York: McGraw-Hill.
WCED 1987. *Our Common Future. World Commission on Environment and Development*. Oxford: Oxford University Press.
Wilson, J.A., Acheson, J.M., Metcalfe, M. and Kleban, P. 1994. Chaos, complexity and community management of fisheries. *Marine Policy* 18: 291–305.
Worster, D. 1977. *Nature's Economy. A History of Ecological Ideas* Cambridge: Cambridge University Press.

14

Integrated management of a temperate montane forest ecosystem through wholistic forestry: a British Columbia example

EVELYN PINKERTON

Introduction: the problem and the response

By the early 1990s, a 20-year critique of poor management of the North American northwestern temperate rainforest and associated transition zone had gained momentum and focus. This area of Canada (British Columbia) and the US Pacific Northwest (Oregon, Washington, southeast Alaska) had been drawing increasing attention from tribal, environmental and public interest groups concerned about the unsustainable rate of logging, often called 'overcutting'. Although systems of land ownership and tenure differed in these jurisdictions, they all faced two common problems. Timber companies were liquidating the remaining original old growth forest with dispatch – faster than a second-growth forest could replace it. Just as controversial were 'forest practices': the way logging and other tree-farming methods affected forest ecosystems – including fish, wildlife, birds, soils, water, plants and microorganisms.

In most cases where logging had occurred extensively (40–80 hectare clearcuts), critics believed that ecosystem resilience had been lost (Maser, 1988; Hammond, 1991). One measure of this loss was that the area replanted or regenerated after such large clearcuts did not support the volume of timber and little or none of its former associated animal and plant life in the second 'crop rotation'. Most other forest values had been sacrificed to timber production, but even timber production was compromised in the long run. The conventional 'industrial forestry' model was seen as single purpose, short-sighted, and benefiting only one sector (Marchak, 1979, 1989; Drushka and Mahood, 1990; Drushka, Nixon and Travers, 1993).

This chapter describes a creative and promising response to this problem by an Amerindian group, the Eagle Clan ('Lax'skiik') of the Gitksan people of northern British Columbia. The Lax'skiik used a combination of

363

traditional knowledge and Western landscape ecology approaches to make a sustainable logging plan which would not radically disrupt the forest ecosystem in a 25000-hectare watershed that is one of their traditional territories. They struggle to implement this plan through the assertion of aboriginal rights and an alternative economic development strategy for the entire region, which would include non-aboriginal people in forestry jobs.

The ultimate goal of the struggle – to take on a significant share of the decision-making about forest use in their area – would logically result in a co-management agreement between the Gitksan and the Ministry of Forests, and possibly also the current holder of forest tenure (lease) rights. Effective forestry co-management agreements are typically difficult to achieve, because timber rights are often held by powerful interests which have captured the regulatory agency, the scientific discourse, and the political ideology which justifies the current allocation of rights. This analysis therefore focuses on the development of leadership and successful teamwork in forging a new paradigm, both for who should participate in forest management and also for which principles should guide that management. By focusing on the 'launching' stage of a co-management initiative, this chapter contributes to the identification of factors permitting effective forestry co-management agreements to emerge (c.f. Pinkerton, 1992; Benidickson, 1992; Matakala, 1995). The analysis is innovative in examining the interaction between the development of local leadership and the capacity of the proposed management paradigm to address issues of sustainability and ecosystem resilience.

The chapter analyses how leadership and an ecosystem resilience oriented plan developed by looking at several components of the process: (1) the development of local skills and capacity through training in mapping watersheds and landscape-level forestry planning; (2) the integration of forestry planning with Gitksan traditional knowledge, laws and customary use; (3) the development of political will and vision by a leadership; and (4) the political process of asserting rights to plan for ecosystem values. These components are analysed in the context of interaction between the cultural setting – Gitksan understandings of the relationships between the forest and human society – and the natural setting– ecosystem resilience and its limits.

Political dimensions of the problem

For at least four decades, timber policy in British Columbia has been dominated by major timber companies with a single focus on short-term maximum timber and woodfibre extraction. Although over 90% of British Columbia forests are on Crown land (owned by the state), lease agreements with these major companies have given them both the overwhelming majority of timber supply and a privileged position in decision making. This condition emerged from an initial desire by government to induce the entry of major firms and thus rapid economic development (Scott, 1991; Drushka, 1993). The distribution of revenues from the exploitation of forests that favoured large firms (as opposed to society as a whole, smaller firms, or resource dependent communities) became part of the status quo. This status quo was then defended as these large firms, which had captured policy making, continued to shape policy in their own interests, as predicted by Peltzman's framework (1976).

Critics of this situation up to the present, including governmental commissions of inquiry (Pearse, 1976; Peel, 1991; Tripp, 1994), have noted three critical facts.

(1) Timber supply is being exhausted by a non-sustainable rate of cut in pure timber volume terms. Travers (1993) notes that 50% of public timber cut has been felled in the last 13 years. As the first growth or original forest is liquidated, the second-growth forest is not maturing fast enough to make the same volume of timber available in time for continuous cutting (as British Columbia's timber policy theoretically requires). Some regions would have to wait up to 30 years without logging before there is sufficient timber to log again.

(2) There is inadequate compliance with forest practices guidelines or rules (i.e. how logging and silviculture should be done). These guidelines and regulations are intended to protect non-timber resources, such as fish and wildlife, which use the forest (British Columbia Ministry of Forests, 1993a, 1993b; British Columbia Parliament, 1994). For example, a recent audit of compliance with British Columbia's Coastal Fisheries Forestry Guidelines (Tripp, 1994) – guidelines intended to reduce the impact of clearcut logging on fish-bearing streams – found high rates of non-compliance with the guidelines and high levels of logging impact on fish streams. British Columbia's new Forest Practices Code lays out ambitious intentions of addressing these problems, but critics feel implementation will be seriously hampered by the existing institutional arrangements (Sierra Legal Defense Fund, 1995).

(3) Communities affected by forest management through their dependence on forest resources have little power in policy making and rulemaking, although they suffer most of the consequences of poor management

decisions (Pinkerton, 1992, 1993). The content and timing of land-use planning exercises in British Columbia are negotiated in multi-stakeholder processes, in some cases without benefit of regulation (Haddock, 1995).

These three aspects of the situation have been well documented and are examples of problems in forest management which may exist on a global level. They are the backdrop rather than the centre of this discussion, however. The main focus here is to analyse how an attempt to reverse this situation has met with some success.

An important dimension of the political problem in Canada is that aboriginal rights of First Nations, as defined by the Canadian courts in the 1990s, have also begun to affect resource management policy to some degree. *R. v. Sparrow* [(1990)1 S.C.R. 1075] ruled that First Nations have a right to a priority allocation of fish for sustenance, social and ceremonial purposes, based in section 35 of the *Constitution Act* of 1982. This right is thus protected at a higher (federal constitutional) level than rights granted under provincial laws or contracts. *Delgamuukw v. B.C.* [(1993) 5 W.W.R. 97 (B.C.C.A.)] held that aboriginal rights were not extinguished (as a 1991 lower court decision had held), but that ownership and jurisdiction over land and resources had not been established by the Gitksan in this case. It then fell to the British Columbia Ministry of Forests (MOF) to develop a policy to implement this judgment as it applied to forest management decisions. The MOF has jurisdiction over all forest land in the province of British Columbia, although how this can mesh with aboriginal rights is contested by First Nations.

Since the nature of the aboriginal right is only broadly and vaguely defined by both these court cases, ministerial policy is likewise vague. The Gitksan hereditary chiefs (who brought the *Delgamuukw* case to trial) are taking the initiative in defining what the right means in forest management. The leadership role played by the Lax'skiik, one of the four Gitksan clans, in this regard is in fact a subject central to this discussion.

The draft policy developed by the MOF following the 1993 *Delgamuukw* case defined aboriginal rights in the following ways: 'aboriginal rights arise from activities which were integral to the distinctive culture of an aboriginal society prior to sovereignty' (1846). The MOF interpreted aboriginal rights as site specific, depending upon patterns of historical occupancy and use of land, exercised by the collective First Nation, possibly practised in a modernized form, including the rights to fish, hunt, trap and berry pick for sustenance, social, spiritual and ceremonial purposes. The MOF's policy was 'to involve First Nations in forest management planning processes; and to accommodate aboriginal rights and prevent or minimize infringement of these rights by proposed forest management activities. . . The MOF will give priority to an aboriginal activity where the forest management activity clearly limits, impedes or denies the right'.

The policy was also fairly specific about the need for a process to implement these general policy goals. In consultation with the MOF, when a First Nation agreed to participate in planning, the MOF was to develop 'a mutually agreeable process which includes consultation: during strategic planning processes; prior to awarding a license; prior to authorizing a management plan, development plan, or range plan or during annual development plan review'.

In this case, the Lax'skiik's ability to assert these rights rides on their ability to demonstrate how the fish and game which sustain the people use the watershed as habitat, as well as how the Lax'skiik themselves used the watershed, through encampments and trails in their pursuit of fish, game, berries, medicinal plants, etc. It further depends on the Lax'skiik ability to show that they can balance timber extraction with these other forest values.

While there has always been some science informing how non-timber values can be protected – and recently there is much more science than previously about how to do this better – this science has not been applied by timber companies and the ministry because there was too much political pressure to ignore other values if they significantly lessened timber extraction. The Gitksan rights, then, put them in a unique position to press this political point, where many other community-based groups had so far failed to have a significant impact on practice.

Scientific dimensions of the problem

Industrial foresters trained in timber management and silviculture (tree planting and tending) generally have a limited understanding of forest ecosystem function. In popular parlance, the Faculty of Forestry at the University of British Columbia is known as 'the Faculty of Logging' because it, like the major companies, is oriented toward maximizing woodfibre yield. This limited perspective has implications both for accurately predicting timber supply and for protecting other forest values. Forest ecologists believe that industrial foresters' predictions of future timber supply in second and third rotations are based on naive projections which ignore soil depletion, erosion, and the effects on timber growth rates and health of the loss of biodiversity in the second-growth forest causing, for example, higher rates of insect infestations and fire (Maser, 1988; Hammond, 1991). The 'falldown effect' (the difference between timber volumes in old-growth forests, several hundred years, old that are currently being logged and the timber volumes in the second-growth forests, less than 100 years old, that must sustain the industry in the near future) was identified by a Royal Commission (Pearse, 1976). Despite this finding, the rate of cut has not decreased. As noted above, the political situation will not allow a paradigm shift toward forest ecosystem management as a policy objective, since the timber companies appear to be solely interested in short-term timber values. Such a shift would require a rethinking of the British Columbia 'sustained' or 'sustainable' yield policy, which had been defined since the 1921 Forest Act in terms of a relatively even flow of timber, and since the 1978 Forest Act in terms of both timber flow and immediate economic needs as perceived by the minister. Thus the current scientific paradigm is an impediment to a larger definition of the problem.

The 'new forestry', new professionals and woodlot managers

Innovators in the Pacific Northwest who began to change this situation came both from inside and outside conventional forestry. The insiders, Jerry Franklin and his colleagues at the University of Washington and Oregon State University (Jim Sedell, Fred Swanson), introduced the concept that logging plans should allow key structural features of the forest to be preserved. For example, instead of leaving isolated forested 'islands' in large clearcut areas, logging should not remove forested corridors which allow wildlife to travel between such islands. Such wildlife movement over a normal range of habitat allows survival of species which play an important role in maintaining forest ecosystem function. Likewise, older decaying trees should be left as wildlife habitat instead of being burned. Chris Maser's research in forest ecology revealed the critical role of plant communities, mycorrhizal fungi, and lichens on decaying tree bodies and in soils in fixing nitrogen and making nutrients and water available to trees (Maser, 1988). Furthermore, the forest was viewed in this model as a living, interconnected web of functions, both above and below ground. Conventional clearcutting and silvicultural techniques of removing debris and burning logged areas often destroy critical organisms and thus disturb key functions maintaining forest and tree health.

However, new forestry concepts as applied in Washington were more mitigative of massive clearcutting than they were actually protective of ecosystem structure and function. Wildlife experts debated about how large the patches of remaining forest should be, and tended to assume these would be surrounded by large clearcuts.

At the same time, a handful of professional 'wholistic' (sometimes spelled 'holistic') foresters have become more active and vocal in British Columbia in the last decade (Hammond, 1991; Drushka *et al.*, 1993), and their perspective became better known with the increasing circulation of *Forest Planning Canada*, now the *International Journal of Ecoforestry*, published in Victoria, British Columbia. These individuals had greater understanding and caution about what is still not known, as well as what is known but still not applied. They were largely ignored by industrial foresters and worked for communities and First Nations attempting to do their own long-term forest-use plans with multiple goals. Perhaps because more old-growth forests, and therefore more options, remained in parts of British Columbia, wholistic foresters there tended to be more visionary. They asked: 'What kind of forest do you want in the future?' and thought in terms of maintaining the basic structure and function of the *entire* forest.

A few 'natural selection' forest farmers (owning small to medium-size woodlots) have also received considerable attention in the last decade, most notably Merv Wilkinson in British Columbia (Loomis and Wilkinson, 1990) and Orville Camp in Oregon (Camp, 1984). They harvest their wood-lots by a gradual, continuous thinning and removal of the weaker individuals of a species, including some smaller trees and some trees which have reached the end of their natural lives. This allows the stronger members of a species to grow more vigorously, with less competition. Thinning the forest to keep it productive without altering its structure may be compared to the fishing strategy of the James Bay Cree (see Chapter 5). In both situations, resilience is maintained by keeping some of the larger, older individuals, as well as individuals in a diversity of age classes. This harvesting strategy stands in contrast to that of industrial forestry, which either clearcuts all the trees simultaneously and replants a monoculture of same age and same species trees, and to that of conventional woodlot owners, who take out all of the larger trees at one time.

Both natural selection woodlot owners and wholistic foresters now use the term 'ecoforestry' to denote ecosystem-based planning for a variety of forest values. Because of their ownership of smaller parcels, however, woodlot owners usually lack opportunities to practise this strategy on a scale consistent with the needs of many forest species to move across different parts of the landscape. Land stewardship programmes in Ontario have enjoyed some success in persuading neighbouring landowners to co-ordinate their forest use to allow continguous forests to support more diverse species (Hilts, 1994), and government agencies in the Pacific Northwest are also attempting such voluntary co-ordination.

Regulatory reforms

Political pressure opposing conventional forest practices and overcutting in the 1980s eventually resulted in some reforms in the regulation of federal, state and private forest management in the US. Much of the pressure came from treaty tribes with rights to protect fish habitat, which have worked with a growing and well-funded environmental movement (Pinkerton, 1992). Similar reform efforts in British Columbia have had far more limited results.

In 1993, the Clinton Plan for the US Forest Service adopted an ecosystem policy for regulating federal forests, a dramatic change in emphasis. The US Forest Service, through the Presidents' Forest Plan, now requires a much wider (91 metre) riparian zone (unlogged buffer zone along fish-bearing streams) in federal forest lands. Even upland and intermittent streams without fish are given 30–45-metre protection, because of their impact on fish downstream. This can be contrasted with the 10–30-metre riparian zone in the new British Columbia Forest Practices Code, and the little or no protection for other streams.

The US Forest Service is also piloting landscape-level planning and joint watershed planning with Washington State forest lands, and with private forest landowners. The introduction of landscape-level planning was a significant improvement over earlier reforms at the state level affecting only individual cutting permits, because the movements of wildlife, fish and water operate at a (broad) landscape level. For example, species such as moose require low-elevation forest cover in winter and a connected high-elevation brousing area in summer. Coho salmon may swim far upstream to spawn, but use other areas of a stream for rearing. Planning for the long-term cumulative impacts of logging – conventionally approved through individual permits to cut one small patch at a time – requires a landscape perspective. In 1994, Washington State began several forms of landscape-type forest planning, although it is difficult to know how successful these will be in a timber-dominated state with about half the forests under private ownership. British Columbia should have better opportunities, since over 90% of forests are under the same form of public ownership.

Values and institutions shaping Lax'skiik wholistic forestry

The Lax'skiik forest planning which took shape between 1989 and 1995 benefited from recent advances in Western wholistic forestry thinking, but gave these its own particular form. To appreciate this form, it is necessary to understand more about the organization and values which shape Gitksan society. The Gitksan form the larger political, social and linguistic group to which the Lax'skiik belong.

Benefits from natural resource use do not simply accrue to individuals in Gitksan society. Each geographic territory is the responsibility of a distinct House, the corporate kin grouping which owns and manages resources in Gitksan society (Duff, 1959; Garfield, 1966; Cove, 1982; Gottesfeld, 1994b). The House carries out its responsibilities through its chief, who is also the conduit for allocating resource access and the economic benefits of the House's work on its territories to House members. The chief gives his or her permission for House members and certain in-laws to access resources, and must be able to provide adequately for those members in order to maintain the status of chief. Participation in feasts provides a major occasion for chiefs to carry out social and political obligations to House members and other Houses. To do this properly, thus validating one's status and position in society, requires that a chief demonstrate an ability to take care of the territory and provide resource access to members and relatives. A chief whose territory has been stripped of forest resources will eventually not be able to 'feed his people', and will be faced with a problem. This occurrence is not absent from Gitksan territory, where many chiefs and Houses have been, or have felt themselves to be, powerless to resist the government and timber industry's plans.

The importance of the Lax'skiik story lies partly in showing how the leadership and commitment of one clan and its leaders can demonstrate alternatives, set a direction, and hold up a standard for the whole of Gitksan society and others. Since the traditional system ideally requires such responsible behaviour of chiefs – and since the Gitksan know that industrial logging is destructive – the traditional system can be used to show others a more responsible approach. This has in fact been an important effect of the Lax'skiik activities.

The Lax'skiik perspective thus differs from the wholistic forestry perspective previously mentioned in at least four important ways. The differences illustrate the significance and potential power of the Lax'skiik example for the larger society.

1. *Adequate planning scale.* The Lax'skiik House territories, probably orig-
   inally formed around fishing sites, tend to be isomorphic with watersheds
   and to cross-cut riverine, mid-slope and alpine zones, facilitating a year-
   round cycle of activities which included fishing, trapping, hunting, berry
   picking and logging (Cove, 1982: 5). The Lax'skiik thus have an opportunity to plan for whole ecosystems on a genuinely landscape level
   because of the size and configuration of their territories. In addition,
   they are likely to get co-operation in planning from other Gitksan
   Houses with adjacent territory. This larger scale of planning creates
   important opportunities to conserve a broader range of forest values.
   Many forest species (e.g. moose) range broadly over large territories in
   different seasons. Logging impacts on one part of the system may accumulate and affect other parts of the system (e.g. debris and silt entering
   small upland streams wash downstream and add to siltation there).

2. *Traditional knowledge.* The Lax'skiik – like most aboriginal peoples –
   have a long historical relationship with their territory. Since they have
   used the same territory for generations, they can bring to planning a
   long historical memory and a complex understanding of the impact of
   human activities on a particular geographic area and all its resources.
   For example, controlled burning of small forest areas to create berry
   patches, often in montane sub-alpine areas 'half way up the mountain',
   was an important activity in pre-contact times, although the practice
   was suppressed by the Ministry of Forests by the 1940s (Gottesfeld,
   1994a). Knowledge of forest succession, and of the behaviour and abundance of animals hunted and trapped in specific areas, was multi-generational and complex.

3. *Personal identification with territory.* The Lax'skiik and Gitksan in
   general have a personal and spiritual identification with their territories
   and resources, which form the basis of their cultural and economic life.
   'A territory was a House's sacred space which it shared with other beings
   fundamentally no different in kind from humans; all having similar
   underlying form, consciousness, and varying degrees of power.
   Relations to them were not seen as unilateral and exploitative, but rather
   reciprocal and moral' (Cove, 1982). Hence the territory cannot be sold
   or alienated from them (as it can from woodlot owners). In a spiritual
   as well as an economic sense, the fate of the territory parallels their own
   fate. ('I will not be anybody unless I can live in the land. It's what makes
   us exist'.) Therefore, they perceive a parallel between the abuse suffered
   by the land and resources through industrial forestry, and the abuse
   suffered by their society and people through domination by the

European political and industrial system, drugs and other addictions. Both the land and the people need to be healed, and their healing is linked. A healthy territory and a healthy people go together. The Lax'skiik sense of their relationship to territory is the basis for their management of it. Such a relationship has much to teach others who wish to develop and support a multi-generational stewardship ethic between local populations and their resources.

4. *Economic and cultural importance of multiple forest uses.* Forest values besides timber have considerable economic and cultural importance to the Lax'skiik and other Gitksan. They fish salmon for both subsistence and commercial purposes; the Gitksan have a planning process for the protection and rehabilitation of fish habitat and depressed stocks (Morrell, 1989; Pinkerton and Weinstein, 1995). They also obtain part of their food supply from hunting and berry picking, and additional economic benefits from trapping fur-bearing, forest-dwelling animals. Some Gitksan also collect forest plants for nutritional and medicinal purposes (Gottesfeld, 1994a; 1994b), and observe that the animals they hunt also eat certain plants which heal the animals' wounds or illness. Finally, the forest is used by some Gitksan as a place of spiritual retreat, renewal and education of younger people.

The next section illustrates how traditional Gitksan values and institutions interacted with the political context of modern logging and the opportunities afforded by the wholistic forestry paradigm.

The development of Lax'skiik forestry

Individual co-evolution with social and natural systems

How the Lax'skiik developed their wholistic forestry plan and political strategy is illustrated through the story of how a clan wing chief, Art Loring, developed and pursued his vision of the future Gitksan forest. Telling one person's story is not to subtract from, or discount the roles of, other important players. It is merely an effective way of tracking the parallel progress of industrial logging in the area, and Gitksan response to it. The mounting degree of disturbance caused by industrial logging as it intensified in the area eventually provoked the Gitksan to generate an alternative. It is significant that this alternative emerged from some Gitksans' intimate experience with early selective logging, then with industrial logging, and also with subsistence activities on the land before, during and after industrial logging had transformed it.

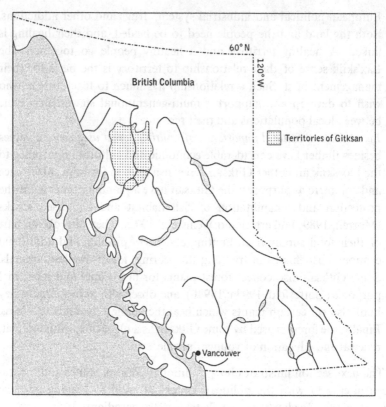

Figure 14.1. Gitskan traditional territories.

Gitksan political organization and social control

First it is necessary to situate the story. The Gitksan are a Tshimshanic-speaking people, numbering about 6000, of which some 5000 live in six reserve communities along about 200 km of river or elsewhere on the 28 160 km² of the upper Skeena, lower Bulkley and Nass Rivers in northern British Columbia (Figure 14.1).

In 1977, the Gitksan chiefs jointly laid formal claim to this area as their traditional territory, according to the claims settlement policy adopted by Canada in 1973 for aboriginal peoples. Aboriginal claims then involved establishing a use right to land and resources which had continuously supported a particular aboriginal group. The Gitksan felt that the Canadian aboriginal claims framework did not allow adequate expression of their rights and responsibilities towards the land, so in 1987 the joint chiefs also filed a court case claiming jurisdiction and ownership of the land

[*Delgamuukw v. B.C. (1991) W.W.R. 97 (B.C.S.C.)*], as discussed below. For a limited period (July 1994 to July 1995), they and the province agreed to adjourn the appeal of the *Delgamuukw* case in the Supreme Court of Canada while they worked collectively towards a treaty with British Columbia and Canada. In July 1995, the adjournment was extended for an agreed period as treaty negotiations proceeded. The Gitksan now operate a co-ordinating body called the Gitksan Treaty Office, with some 15 staff.

Individual Gitksan house chiefs mediate the resource use of their members within their own territories, as discussed above. They also participate in the broader Gitksan society through a shared understanding of Gitksan law. Attempts to codify Gitksan law have been unsuccessful. In the abstract, it consists of a central understanding about the need to respect and acknowledge the spirit within all things and people. In practice, it consists of social norms, rules and customs as interpreted by the chiefs through their recollection of past cases and how they were dealt with.

One important way the Gitksan sanction law-breakers is through questioning their right to their Gitksan name and social status. People who hold names and status come under public scrutiny through the names used to address them in formal interactions. If their Gitksan name is not used, it is implied that they have done something which makes them unworthy of the name and status. Public censure may also occur through speeches or actions at public ceremonial feasting at the local or regional level, or through political meetings at which decisions are made by consensus. Status may be reclaimed by putting on a feast for injured parties and redressing one's behaviour. All these mechanisms of social control may be used strategically in developing resource management policy.

The individual and his historical era

Art Loring, born in 1955, grew up in the early days of cedar pole and pine (railway) tie logging in Gitksan territory in the 1950s and early 1960s. Much of this work was done by Gitksan Houses with short-term timber leases. Unknown to the government of the day, House groups who ran the cutting camps on adjacent territories worked together and respected one another's territories in their activities. Logging was done selectively with horses and did not create a major disturbance in the forest ecosystem. Loring's mother carried him in a backpack as she worked at logging alongside her husband in their camp in the forest. She began this work at age 16, and continued it while raising her children. The younger children helped at the small sawmill they also ran as a family.

By the time Loring was 14 and began his own logging career, the MOF had granted cutting rights on Gitksan territory to major pulp companies using clearcut methods on up to 500-hectare openings. Loring logged for the company, and by the age of 21 had become a faller, the most danger-ous, highly skilled, highly paid, and high-status job in the woods. For the next 12 years, he participated fully in the aggressive and competitive male culture and lifestyle which accompanied it, including heavy alcohol consumption, drugs and parties. But through the hopeful vision of his grandmother, and the perseverance of his wife Mathilda, 'who remained strong and steadfast', Loring gradually came to think of this work and life-style as 'the destruction that was beginning to hurt my life and my kids'. He reflects: 'My grandmother said: "That woman [Mathilda] is going to save your life", and she was right. Everything evolves from the woman in our society. The guidance you get from mothers and grandmothers comes back when you start to heal'.

The 'destruction' also became more and more apparent as Loring went back to hunt every winter in one territory being logged. Comparing it to the area where he himself logged, he recognized how rapidly the entire ecosystem was unravelling. 'When we first broke into these areas, you could see 50 moose in a day. The coho [salmon] would be skittering around – it was a major spawning area'. But over time as the area was 'logged right to the creek and river banks, nothing over three metres high left, skid trails going up side hills, machines digging into the mud until they couldn't move, leaving 3 to 4 foot deep skid trails, . . . there was intense erosion . . . the water table was hammered'. Wildlife and fish disappeared from the area: 'steel-head [salmon] are almost completely gone from lake areas at the top of the system; coho still came up the mainstem for awhile, but had nowhere to spawn; now there are few left. . . . You rarely see grizzly bear there now; they need roots that you only find in wet habitats in early spring. . . '.

The growing awareness of the destructive forces in his life and environ-ment also coincided with Loring being asked to join the Eagle Clan (Lax'skiik), his mother's original clan, to help protect their territory. The Gitksan are a matrilineal society which traces inheritance through the female line. Loring's mother had lived away from her natal clan and been adopted at a young age by another clan. A Lax'skiik elder, carrying out the traditional role of handpicking leaders, claimed Loring back into the clan. There are four Lax'skiik Houses, all resident at Gitwangak (Kitwanga), which together make up the Lax'skiik (Eagle) clan. Loring was asked into a House when he and his wife were invited to use a fishing site by its elderly House chief owner. One day Loring quit his logging job and began to discuss with the other Lax'skiik the importance of protecting their territory from industrial logging. Soon Loring was given a Gitksan name, and was made a Lax'skiik wing chief. The role of this position is to bring the concerns of the clan members to the four head chiefs of the clan. These occurrences laid the groundwork. Loring then moved from strength to strength, acting with greater and greater confidence and articulateness as he developed the knowledge, vision, leadership and support to carry out the innovations which followed.

Some of this support also came from the Gitksan leadership. Gitksan society in general was intensely involved during the 1980s in preparations for their land claims negotiations and their court case. Research, policy development, training, interviewing elders, and developing their own Gitksan language immersion school were all part of a broad effort to revitalize and enlarge traditional institutions to take on modern problems. The slow pace of mainstream institutional response was a constant irritant to the Gitksan. No aboriginal claims had been resolved since negotiations began in 1973, and the courts had not decided yet what aboriginal rights really meant. While these discussions continued, industrial logging had only increased its pace. So many Gitksan searched for alternatives, as it became evident that they had only themselves to rely on to protect their resources.

In 1988, the Gitksan chiefs gave support to a number of blockades led by the Lax'skiik on various logging roads in response to the increased rate of logging on their territories. Some of these blockades were successful. In 1989, the Lax'skiik blockaded construction of a logging road being pushed into their territory in the Fiddler Creek watershed. They subsequently rejected the company's offers of logging contracts, and decided to develop the territory themselves for selective logging, tourism, sportsfishing guiding, and other activities. Working with Richard Overstall, a 20–year veteran forest policy analyst then at the Gitksan Office of Hereditary Chiefs, the Lax'skiik contacted wholistic forester Herb Hammond, and put together a proposal for a forest-mapping and training programme to be conducted by Hammond for clan members. The Lax'skiik received some start-up funds from the Gitksan Government Commission (the combined elected band councils) and hosted a feast to explain their proposal to the other Gitksan chiefs, and to seek their support. These actions launched the Lax'skiik leadership in developing alternative forest policy, planning, and training. Loring was asked to lead the effort.

As in many cases of successful collective action (Popkin, 1979; Feeny, 1983), leaders can be effective when they are perceived as being motivated by more than a narrow self-interest. Loring abandoned a highly paid job and took risks in exploring new ground. He worked with Lax'skiik and Gitksan elders, chiefs and political leaders of high status from the less-elevated position of wing chief. His competence in the woods, his knowledge of logging and of the land, his choice to work on the land in a fairly traditional manner (unlike many others), his ability to master the concepts and language of wholistic forestry and meld it with Gitksan traditional values, his fearlessness in taking a public stand, and his action orientation – all these qualities were valuable and relatively scarce resources contributed to Gitksan society at large, in a manner which did not threaten the status hierarchy of Gitksan society. Loring was thus able to be a credible and effective leader of a larger political effort.

Integrating wholistic forestry with local knowledge

The Lax'skiik began with a four-year training programme (1989–92) in wholistic forestry in which eight Lax'skiik enrolled. Forester Herb Hammond spent two weeks of each year working with them in the field and classroom on forest mapping, inventory and wholistic forestry principles. They found these compatible with their own values, goals and understandings of natural systems. In addition, they were able to teach Hammond something about the spiritual dimensions of what Hammond called 'wholism'. The Lax'skiik called this 'living in harmony with the forest and the plants' and gave importance to interactions of people and animals with medicinal plants. These interactions suggested reasons for some animal movements across the landscape. For example, moose plagued by insects would find relief by walking through certain dense shrubs which release an insect-repellent sap.

Principles in wholistic forestry also provided a language which could bridge Gitksan spiritual understandings about the interconnectedness of the natural world and the Western science of landscape ecology. It provided tools for Loring to articulate his vision in a scientific language and framework which was not intrinsically foreign to a Gitksan worldview.

The forestry mapping, training, and the watershed plan

Richard Overstall and Geographic Information System (GIS) expert Marvin George assisted in the process of putting those who took the training in a position to teach other Gitksan how to map and do forest inventory. At the same time, Loring took out a few Lax'skiik logging each winter to apply the new skills to low-impact logging with horses. Loring also organized a course by the Fur Institute of British Columbia on fur-bearing animals and their habitats.

The training with Herb Hammond culminated in the development of a wholistic forestry plan for the unlogged 25000-hectare Fiddler Creek watershed in Lax'skiik territory. This involved a plan for first removing from the cut the zones most sensitive to disturbance (because of steep slopes). Then, those zones of critical habitat for fish and wildlife were removed. Gitksan fisheries experts mapped coho salmon spawning and rearing areas. Hammond zoned very wide – 600–1000 yards (550–900 metres) – riparian corridors along each side of the banks of the stream (at least five times as wide as those required by the US Forest Service) and connecting corridors across the valleys. Art Loring eventually mapped the habitat of the most important wildlife species: moose, grizzly bear, mountain goat, pine martin, and bald eagle. Several of these used the riparian zone heavily for travel, feeding, mating and producing young. In this case, Hammond also considered economic opportunities for tourism and set aside a tourism zone.

The forested areas left after these zones were removed could be logged, but not by clearcutting. Hammond estimated that 20% of the forest in the 'wholistic timber zones' would be left on the site to replenish soil and provide structures for forest ecosystem processes.

The wholistic timber zones would thus yield 20% less than the conventional MOF estimate of timber yield in these zones. The wholistic plan also classified about 40% less of the drainage as suitable for logging than MOF standards would require. So, if there were no other considerations, wholistic forestry methods would produce about 40% of the timber from this drainage that conventional forestry would produce. The tourism zone would remove another 15%, meaning the entire wholistic forest plan would produce about 25% of the timber that an MOF scenario would produce.

Of course, other aspects of the ecology/economy calculus must be considered to grasp the trade-off being made here. Economic benefits to the logger are greater per unit of timber logged when big machinery is not used. Logging by low-impact and little-capitalized methods, such as horses, funnels about 75% of the gross costs of falling and yarding directly to workers. More mechanized logging methods funnel only 13–26% of these costs to workers, because machines are so costly to acquire, to finance and to run (Allen, 1989; Silva Ecosystems Consultants, 1993). In other words, machines only make sense in very rapid and high-volume logging which is concentrated in one area which is not sensitive. Or smaller, light machines may be suitable for less sensitive sites.

Overall cost per unit of production is higher with wholistic logging, but higher prices could be obtained if logs were sold on an open log market instead of through the current system. In addition, certified selectively harvested logs command a premium price in the UK and Japan.

A complete analysis of benefits is beyond the scope of this chapter. It would involve assessing the direct and indirect economic benefits of tourism, fishing, hunting, trapping, non-cash food (game), berries and medicinal inputs. It is more difficult to put a price tag on the value of a stable and sustainable stream of benefits, because some parts of the ecosystem are impossible to restore at any cost after conventional industrial logging. For example, some species of salmon adapted to the conditions of a particular watershed cannot be recovered once that gene pool is lost (Helle, 1981; Withler, 1982). Thin soils on steep slopes cannot be rebuilt in the short term once they are lost.

The stage following the mapping of the Fiddler Creek watershed was larger-scale landscape-level analysis and planning. This required the coordination of local plans with planning on adjacent jurisdictions, in order to connect wildlife corridors into the next valley for species such as grizzly bear. Other chiefs asked Loring to help in their own planning, and to train them in low-impact logging. During the planning of the Fiddler Creek watershed, negotiations with the MOF and with the tenure holder, Repap (a large Montreal-based paper company), were pursued, as discussed below.

The Hobenshield agreement: trade-offs in sharing an area

No agreement has been reached about Fiddler Creek, and no logging occurs there yet. Art Loring currently logs, and conducts some training, in another Lax'skiik territory, where an agreement was reached with the local family-owned Kitwanga Lumber Company (the Hobenshields), who held timber rights and a local mill. The agreement followed a 1992 Lax'skiik blockade of this area (three years after the original Lax'skiik blockade of the Fiddler Creek watershed), and specified that the Hobenshields and the Lax'skiik would work together on planning silviculture, restoration, and some limited logging of the area (c.f. Wild, 1992).

The Lax'skiik had received $30000 from the Sustainable Environment Fund of the British Columbia Ministry of Aboriginal Affairs in 1992 to train its members to conduct landscape analysis and planning in the larger territory using a PAMAP GIS environment. Hammond provided initial instruction in setting up the system, but the rest was conducted by the Lax'skiik, who had developed the mapping and inventory skills. The Hobenshield forester participated occasionally in the wholistic forestry course and found considerable common ground with the Lax'skiik. For example, the Hobenshield letter of agreement states: 'We agree that the progressive clearcut logging of the Nash Y area had proceeded at too rapid a rate, and that soil degradation, loss of habitat and biodiversity, and damage to wetlands and water quality has occurred, and that a long period of rest and healing is the appropriate treatment'. The company expressed a willingness for the Lax'skiik to log, by subcontract, some or all of the areas on Lax'skiik territory within the company's tenure area. 'We are testing how Native and non-Native can work together'.

Loring is also conducting some adaptive management experiments in how to log so that the forest is less vulnerable to 'blowdown' (shallow-rooted trees next to a logged opening being blown over by high winds). In one area he takes out the entire overstory, because there is a 20-year-old understory with canopy closure which may give adequate protection against blowdown; in another area, he leaves some of the older trees. Over time, he will be able to see which technique is most effective for preventing blowdown.

Management rights asserted by the Lax'skiik

The 1993 *Delgamuukw* decision and the MOF policy devolving from it required that government work to consult and plan co-operatively with First Nations so that their aboriginal rights to hunt, trap and fish would be protected. The planning of the Fiddler Creek watershed is a test case of whether government and First Nations can agree about an adequate level of protection of aboriginal rights. The Lax'skiik assert that the mapping of fish and game habitat shows what kind of protection is required for the animals and fish which they can lawfully access. They also hold that the cabins, trails and resource-gathering sites must be preserved, as they are needed to hunt, fish, trap and pick berries.

The Lax'skiik understanding of their rights stands in sharp contrast to that of government. The province has viewed aboriginal rights from a far narrower perspective, e.g. the right to use traditional hunting trails and fishing sites, or the right of *access*. The Lax'skiik argue that if the trails merely pass through clearcuts and destroyed ecosystems, their access is meaningless, and their rights will be violated. This amounts to an argument that rights of access are meaningless without rights to protect ecosystem function. Furthermore, they have offered a definition of ecosystem function, while the province has no definition so far. The judgment in *Saanichton Marina Ltd.* v. *Claxton* [1989 36 B.C.L.R.(2d.)79 (B.B.C.A.)] affirmed that the right of access to fish by First Nations carries with it the right to protect fish habitat, lending force to the Lax'skiik position, insofar as fish habitat protection is part of the wholistic forestry plan.

Art Loring notes that the Lax'skiik are willing to share the logging of Fiddler Creek with the company which holds logging rights in the area, as long as the Lax'skiik forestry plan for the area is used. The Lax'skiik might, for example, log the more sensitive areas in the timber zone while the company logged the less sensitive. The Lax'skiik have now mapped all the human and animal trails, campsites, old village sites at three levels of elevation in the Fiddler Creek drainage. If an agreement cannot be reached eventually, a court action will probably occur.

Forest ecosystem function

If the Lax'skiik claim the right to protect ecosystem function, how is the forest ecosystem understood in wholistic forestry thinking? This particular ecosystem is at the boundary of the the coastal rainforest and the coastal/interior transition Cedar–Hemlock Biogeoclimatic Zone. Coniferous species – western hemlock, mountain hemlock, subalpine fir, western red cedar, amabilis fir, and spruce – are mixed with hardwoods – mostly black cottonwood, paper birch, and alder. Much of the forest is in steep terrain, where the mountains rise 1500 m above the valley floor, and where many sites have shallow soils less than 50 cm deep, over an impermeable layer. The upper parts of the tributary drainages contain old-growth stands of large 500-year-old trees. Closer to the river the stands are of mixed age and succession.

Wholistic forest planners recognize the role played by some key species, processes, areas, or sub-ecosystems in preserving forest ecosystem function. Some of these are as follows:

1. *The speed and direction of water flow through the system.* Logging roads, which alter the direction of water flow, and large clearcuts, which change the rate of flow, have profound impacts on soil loss, nutrient transport and water retention in dry seasons. Wholistic planning includes the use of horses in very sensitive areas, while small equipment

and elevated log removal systems are used in less sensitive areas. These techniques minimize the disruption of water-flow patterns.

2. *The forest canopy* (the 'roof' of foliage in tree tops) regulates temperature, rain impact and snowmelt. Wholistic forestry planning minimizes
   disruption of these moderating functions by keeping clearcut openings
   which disrupt the canopy to 5 hectares or less in most cases. The forest
   planners prefer to plan most timber harvest simply as 'thinning' operations which leave the canopy relatively intact.

3. *The riparian zone* (the wetted zone along streams) is a key sub-ecosys-
   tem which functions as a key corridor and use area for animals, a stabi-
   lizer of water flow during intense rain, and a buffer for upland and
   streambank erosion. The larger riparian zone of influence may be
   equally important in certain circumstances in mediating the terres-
   trial/aquatic interface. Both these zones serve such multiple functions
   that wholistic planning preserves them from logging.

4. *Cross-valley forested corridors* allow wildlife to travel under forest cover
   over the landscape to summer and winter range and reach adequate
   forage and shelter in season. Non-retention of these key forested corridors results in animals being unable to access their habitat in season, or
   suffering exposure and stress leading to reduced survival. Lower survival
   rates of animals result in dramatic changes in browsing patterns and
   hence plant species composition.

5. *A diversity of tree species* reduces the level of insect damage. For
   example, the spruce weevil attacks young spruce in an open plantation.
   But beneath a canopy of cottonwoods shading the young spruce, the
   temperature is far less favourable for spruce weevil, which does not
   become a threat in these conditions. When the spruce grow taller and
   break through the canopy of the cottonwoods, they are far less subject
   to damage by the weevil. Ant colonies in ancient decaying cedar are the
   main attractor for birds, which then remain in the area to devour insects
   attacking other conifers.

6. *A mix of hardwood and coniferous tree species* allows deciduous hardwood trees such as alder to fix nitrogen (making it available to conifers),
   to build soil and reduce acidity (through leaf decay and decay of
   downed tree bodies), and to provide shade for early growth of conifers.
   Deciduous trees are often called the 'nurses' of conifers, because they
   rapidly colonize small disturbed areas and provide more favourable
   conditions for their growth. As the conifers outgrow the hardwoods, the
   latter are eventually choked out and their decaying bodies then nourish
   the conifers. The decaying bodies of trees which hold nutrients for

Evelyn Pinkerton

growing trees are particularly important where shallow soils can less easily hold nutrients. Small disturbances which result in the introduction of deciduous trees thus eventually serve to 'feed' surrounding trees. Lightening strikes or aboriginal burning patterns which resulted in small contained fires (Gottesfeld, 1994a) provided small to moderate levels of disturbance, which resulted in the introduction of deciduous hard-woods. The MOF no longer allows this type of controlled burning (orig-inally done to produce berry patches), but small-patch logging can serve the same function without the major ecosystem disruption associated with large clearcuts.

7. *Nutrient transport* to trees through fish swimming into upper reaches of the ecosystem. After spawning, dying salmon are retrieved from the river by bears, otters and eagles, and partially consumed. Decaying salmon carcasses on land, in the river, or salmon recycled through animal diges-tive tracks, appear to contribute critical nitrogen and carbon to tree growth, according to recent research – at least where salmon are still abundant. For example, 15–20% of the nitrogen in the new needles of hemlock trees in the riparian zone was found to be of marine origin (Bilby, Fransen and Bisson, 1996). Salmon are thus the functional equiv-alents of hardwoods as nitrogen suppliers. As salmon (particularly coho) are depleted (as they are in the Skeena River system), this functional redundancy is diminished, leaving a less-resilient forest ecosystem.

8. *Keystone terrestrial species* such as rodents eat subterranean fungi on tree roots containing nutrients, water, fungal spores, nitrogen-fixing bac-teria, and yeast. Rodent pellets distributed throughout the forest spread the spores and the associated package which serve to inoculate new areas with mycorrhizal fungi, which in turn make nutrients and water avail-able to their host trees (Maser, 1988; Hammond, 1991). Rodents can thus help to recolonize disturbed areas with nutrient transfer and possi-bly nitrogen-fixing capacity. (It is not known whether or not mycorrhizal fungi in this area fix nitrogen). However, rodents seek protection from predators such as eagles or hawks, and will not travel far from protective forest structures such as downed trees. Therefore, they can rebuild the nutrient-supplying function of the forest floor only where the dis-turbance is not too great.

9. *Avalanche chutes* provide habitat and transportation routes for grizzly bears, mountain goats, small fur-bearers, and songbirds, as long as the chutes are associated with old-growth forest (Herb Hammond, personal communication, 1995). Some tree species such as mountain ash take root in such habitats, and provide excellent songbird feed. Avalanche

chutes are another example the the role of small to moderate disturbances increasing or supporting diversity, in this case by increasing habitat for some species.

In summary, a fragile ecosystem with thin soils and steep, unstable slopes requires a balance of processes. On the one hand, processes which cycle water and nutrients through the system at a rate and in a manner which makes them available to trees through gradual release provide stability. On the other hand, the system is rendered more resilient through the periodic introduction of new nutrient sources and animal habitats as old ones are depleted or systems evolve. Decaying bodies of salmon on an annual basis, and nitrogen-fixing hardwoods or nutrient-transferring mycorrhizal fungi spread by rodents on a multi-year basis, supply these functions in alternative ways. Conventional logging methods in this ecosystem clearly cause disruptions to ecosystem function on a scale which will allow reconstruction of a forest ecosystem only very slowly or not at all. Wholistic logging methods mimic natural disturbance and aboriginal burning patterns, creating minor to moderate canopy gaps at the patch level, which allow the introduction of critical new nutrients, moisture and light sources or leave structures for the re-establishment of former sources.

Lax'skiik wholistic forestry looks at ecosystem resilience as the necessary product of sustainable management. The function of all forest plant and animal species may not be known at present, but their diversity is likely to be important in ecosystem function. Therefore, the Lax'skiik seek to maintain viable populations of all species and both stand and landscape-level forest structure characteristic of undisturbed forests, and consider 'sustainable' management to mean management which conserves a rough approximation of the species mix before logging. At the same time, they are experimenting with small-scale logging which creates different levels of disturbance and selectively removes some species or ages in small areas (as in Loring's adaptive management experiment to minimize blowdown).

Conclusion

This case study exemplifies a new paradigm for managing forests of the north-west coast of North America on an ecosystem basis, and shows ways in which political and ideological impediments to implementing this paradigm can be overcome. The wholistic forestry paradigm is an extension of work in 'new forestry', woodlot management, and related reforms in the US It goes beyond these innovations in several ways, which are summarized below. This summary could also be considered a series of propositions about the conditions under which wholistic co-management of local forests is most likely to occur. The most important aspects of the first three propositions are common to other documented forestry co-management arrangements. The remaining propositions are largely new in this context, particularly the last three.

1. *A local wholistic management plan.* A community oriented towards and identified with its particular geographic place insists on sharing management with government by producing a local forest ecosystem management plan, incorporating the sustainable and integrated management of multiple forest values, including wildlife, fish, soils, water, and various forest plants, in addition to timber. The direct involvement in planning at the community level allows far greater government accountability to be built into planning.

2. *Planning combines traditional local knowledge and values and the science* of landscape ecology. Art Loring brought his experience of ecosystem change through the historical evolution of low-impact to high-impact logging in geographic areas he knew intimately. This recombination of local knowledge, scientific ecological knowledge, and Gitksan traditional understandings allowed a new management paradigm to break through the conventional industrial scientific paradigm for how to manage forests.

3. *The definition of sustainability involves small-scale disturbance* by low-impact logging at the patch level, or partial removal through thinning. Both methods allow new inputs of species or nutrients while conserving basic forest structure and function. The degree of disturbance which can be tolerated (and can play a positive role in fostering diversity and hence ecosystem resilience) is not great in this fragile montane ecosystem, and the limits are to be discovered by adaptive management.

4. *Landscape scale of planning.* The size of Gitksan House territories, their tendency to coincide with watersheds, and the ability of Gitksan chiefs to work together, create the possibility of planning on a large landscape scale in a manner which seldom occurs under current public or private ownership of forest lands.

5. *Management rights of local communities.* The aboriginal rights, established in Canadian courts or by treaty of some local communities, to fish, wildlife and berries give them a unique opportunity, as well as economic and cultural incentives, to protect forest values other than timber. While there is usually broad public concern for the protection of these

values, a dispersed public often lacks mechanisms to express this concern effectively, or the resources to act upon it.

6. *Development of local skills and capacity.* Training in mapping, inventory, planning, and alternative low-impact logging methods allowed the community to integrate the new paradigm with the practical requirements of the local landscape and local manpower.

7. *Development of vision, leadership, and political will.* A leader with the experience and ability to work with a team of elders, chiefs, Gitksan leadership, and policy and technical experts was essential to the development of a vision. The leader's willingness to forgo some personal advantages and to take personal risks lent credibility to the effort. By developing the active support and co-operation of these parties, the leader was able to translate his courage to innovate into action. This was particularly critical in the context of a captive agency.

8. *The ability of leadership to call upon traditional local values.* The development of wholistic forestry principles based on traditional Gitksan values and worldview allows the leader to set an example as a moral as well as a political call on chiefs and elders to fulfill traditional obligations to protect their territories and the resource access to fish, game and other resources of House members.

Acknowledgements

Three anonymous reviewers, and Richard Overstall, Fikret Berkes, Carl Folke, Art Loring, and Herb Hammond provided very helpful comments which aided in revisions of drafts of this chapter. The author is grateful to the Social Sciences and Humanities Research Council of Canada for funding parts of this research related to the US, and to the Beijer International Institute of Ecological Economics for logistical assistance in part of the Canadian research.

References

Allen, R. 1989. Is it time to recycle the economic process?, *Forest Planning Canada 5*: 16–18.

Benidickson, J. 1992. Co-management issues in the forest wilderness: a stewardship council for Temagami. In *Growing Demands on a Shrinking Heritage: Managing Resource Use Conflicts*, pp.256–75, ed. M. Ross and J.O Saunders. Calgary: Canadian Institute for Resources Law.

Bilby, R., Fransen, B. and Bisson, P. 1996. Incorporation of nitrogen and carbon from spawning coho salmon into the trophic system of small streams:

Evidence from stable isotopes. *Canadian Journal of Fisheries and Aquatic Sciences* 53(1): 164–73.

British Columbia Ministry of Forests 1988. British Columbia Ministry of Environment, Federal Department of Fisheries and Oceans, Council of Forest Industries. *Coastal Fisheries Forestry Guidelines*, 2nd. edn. Victoria, BC: Ministry of Forests.

British Columbia Ministry of Forests 1993a. *British Columbia Forest Practices Code. Changing the Way We Manage Our Forests*. Discussion Paper. Victoria, BC: British Columbia Ministry of Forests.

British Columbia Ministry of Forests 1993b. *British Columbia Forest Practices Code. Rules*. Victoria, BC: British Columbia Ministry of Forests.

British Columbia Parliament 1994. *Forest Practices Code of British Columbia Act*, Victoria, BC.

Camp, O. 1984. *The Forest Farmer's Handbook: A Guide to Natural Selection Forest Management*. Ashland, Oregon: Sky River Press.

Cove, J.J. 1982. The Gitksan traditional concept of land ownership. *Anthropologica* XXIV(1): 3–18.

Drushka, K. 1993. Forest tenure: forest ownership and the case for diversification. In *Touch Wood. BC Forests at the Crossroads*, pp.1–22, ed. K. Drushka, B. Nixon and R. Travers. Madeira Park BC: Harbour Publishing.

Drushka, K., Nixon, B. and Travers, R., eds. 1993. *Touch Wood. BC Forests at the Crossroads*. Madeira Park, BC: Harbour Publishing.

Duff, W. 1959. *Histories, Territories, and Laws of the Kitwancool*. Victoria, BC: British Columbia Provincial Museum.

Feeny, D. 1983. The moral or the rational peasant? Competing hypotheses of collective action. *Journal of Asian Studies* 42: 769–89.

Forest Resource Commission 1991. *The Future of Forests*. Victoria, BC: Forest Resource Commission.

Garfield, V. 1966. The Tsimshian and their neighbors. In *The Tsimshian Indians and their Arts*, pp.3–70, ed. V. Garfield and P.S. Wingert. Seattle: University of Washington Press.

Gottesfeld, L.M.J. 1994a. Aboriginal burning for vegetation management in northwest British Columbia. *Human Ecology* 22: 171–88.

Gottesfeld, L.M.J. 1994b. Conservation, territory, and traditional beliefs: an analysis of Gitksan and Wet'suwet'en subsistence, northwest British Columbia, Canada. *Human Ecology* 22: 443–65.

Haddock, M. 1995. *Forests on the Line. Comparing the Rules for Logging in British Columbia and Washington State*. Vancouver: Sierra Legal Defence Fund; New York: Natural Resources Defense Council.

Hammond, H. 1991. *Seeing the Forest Among the Trees: the Case for Wholistic Forest Use*. Vancouver, BC: Polestar Press.

Helle, J. 1981. Significance of the stock concept in artificial propagation of salmonids in Alaska, *Canadian Journal of Fisheries and Aquatic Sciences*. 38: 1665–71.

Hilts, S. 1994. The natural heritage stewardship program. In *Environmental Stewardship: History, Theory, and Practice*, pp.109–17, ed. M.A. Beavis, Workshop Proceedings, Institute of Urban Studies. Winnipeg: University of Winnipeg.

Loomis, R. and Wilkinson, M. 1990. *Wildwood: a Forest for the Future*. Gabriola, BC: Reflections Press.

Mahood, I. and Drushka, K. 1990. *Three Men and a Forester*. Madeira Park, BC: Harbour Publishing.

Marchak, P. 1979. *In Whose Interests? An Essay on Multinationals in Canada.* Toronto: McClelland and Stewart.

Marchak, P. 1989. History of a resource industry. In *A History of British Columbia*, pp.108–28, ed. P. Roy. Toronto:Copp Clark Pitman.

Maser, C. 1988. *The Redesigned Forest.* San Pedro: R.& E. Miles.

Matakala, P. 1995. *Decision-making and conflict resolution in co-management: two cases from Temagami, northeastern Ontario.* Unpublished PhD thesis, University of British Columbia.

Morrell, M. 1989. The struggle to integrate traditional Indian systems and state management in the salmon fisheries of the Skeena River, British Columbia. In *Cooperative management of local fisheries: new directions for improved management and community development*, pp.231–48, ed. E. Pinkerton. Vancouver: University of British Columbia Press.

Pearse, P. 1976. *Timber Rights and Forest Policy in British Columbia.* Report of the Royal Commission on Forest Resources, Vol.1. Victoria, BC: Ministry of Forests.

Peel, A.L. 1991. *The Future of Our Forests,* Victoria, BC: Forest Resources Commission.

Peltzman, S. 1976. Towards a more general theory of regulation. *Journal of Law and Economics* 19: 211–40.

Pinkerton, E. 1992. Translating legal rights into management practice: overcoming barriers to the exercise of co-management. *Human Organization* 51: 330–41.

Pinkerton, E. 1993. Co-management evorts as social movements: the Tin Wis coalition and the drive for forest practices legislation in British Columbia. *Alternatives.* 19: 33–8.

Pinkerton, E. and Weinstein, M. 1995. *Fisheries that Work: Sustainability through Community-based Management.* A report to the David Suzuki Foundation, No. 219, 2211 West 4th Avenue, Vancouver, BC.

Popkin, S.L. 1979. *The Rational Peasant: the Political Economy of Rural Society in Vietnam,* Berkeley: University of California Press.

Scott, A. 1991. *Individual rights to use the forest.* Unpublished MSS, Department of Economics, University of British Columbia, Vancouver, BC.

Sierra Legal Defense Fund. 1995. *The Forest Practices Code of British Columbia Act. A Critical Analysis of its Provisions.* Vancouver, BC: Sierra Legal Defence Fund.

Silva Ecosystems Consultants 1993. Summary of Fiddler Creek initial landscape analysis. Unpublished report for the Lax'skiik.

Travers, R. 1993. Forest policy: rhetoric and reality. In *Touch Wood. BC Forests at the Crossroads*, pp.171–224, ed. K. Drushka, B. Nixon, and R. Travers. Madeira Park, BC: Harbour Publishing.

Tripp, D. 1994. *The Use and Effectiveness of the Coastal Fisheries Forestry Guidelines in Selected Forest Districts of Coastal British Columbia.* Victoria, BC: Ministry of Forests.

Wild, N. 1992. *Blockade.* One-hour documentary film. National Film Board of Canada and Canada Wild, Vancouver.

Withler, F.C. 1982. Transplanting Pacific Salmon. *Canadian Technical Reports on Fisheries and Aquatic Sciences* 1079.

15

Managing chaotic fisheries

JAMES M. ACHESON, JAMES A. WILSON & ROBERT S. STENECK

Introduction

The marine fisheries of the world are in trouble. Since 1970, total catches of fish in the world have plateaued despite great increases in fishing effort (FAO, 1977, 1987). Even worse, the past few decades have seen the decline of some of the world's most important fisheries. In countries in the industrialized West, including the United States, Canada and Northern Europe, virtually every major fishery has witnessed stock depletion and reduced catches (Sullivan, 1987: 2; McGoodwin, 1990: 1–4; Bencivenga, 1991; Johnson, 1991; Rosen, 1991; Walsh, 1991; Austin, 1992; *New York Times,* 1992a, 1992b).

As one important fishery after another declines, there is a clear consensus that current fisheries management practices are not working well. However, there is no consensus on the solution to the problem. Most administrators and scientists employed in fisheries management agencies tend to believe that all would be well if politicians would effectively enforce the regulations they recommend rather than catering to special interests in the fishing industry. However, there is a growing conviction, which we share, that the science itself is seriously flawed and that fisheries management will not improve until some radical changes are made.

In this chapter, we propose adopting another way to manage fisheries, which we call *parametric management.* The idea for this approach came about during our efforts to model fish populations.[1] Essentially, it calls for managing the ecological variables which set limits on the population levels observed. These variables are the parameters of the model: hence the name parametric management. It is important to note that parametric management is not completely new. Rather, it extends and modifies solutions to the problems of fisheries suggested by other scientists advocating an 'ecosystem' approach (the integrated ecological components influencing marine populations). Most important, this approach to management is one which has been tried in numerous other maritime societies. In advocating parametric management, we are essentially arguing that the solution to our own fisheries problems is to be found in the management practices of many tribal and peasant societies.

The chapter begins with a review of the scientific underpinning of the approach to management advocated by most fisheries biologists; it then describes the ways in which societies in Third World countries manage their fisheries, and the reasons for these management practices, which involves a discussion of the theory of chaos. Finally, parametric governance is discussed.

Management of fisheries in the industrial world

In virtually all modern industrialized countries, marine fisheries are managed by a centralized government through the use of quotas or other controls designed to limit fish mortality over the entire range of a single population (or discrete stock). Reliance on top-down management techniques stems in part from the assumption that those using natural resources are strongly motivated to overexploit them. This assumption is derived from the theory of 'common property' resources, which is perhaps the most influential body of theory guiding efforts to manage resources in modern Western countries (Hardin, 1968; McCay and Acheson, 1987; Acheson, 1989; Feeny *et al.*, 1990; Anderson and Simmons, 1993). According to this theory, all resources which are held in 'common' – such as rivers, oceans, forests, rangelands – will be subjected to escalating overexploitation, which ceases only when the resource is depleted. Why should the skipper of a fishing boat curb his own effort when the fish he leaves will only be taken by someone else, perhaps in a matter of hours? Under these circumstances, it is only logical for those using 'common property' resources to take as much of the resource as fast as possible. A closely related assumption is that those using natural resources cannot or will not conserve the resources on which their livelihood depends. Under these circumstances, management of these resources must be accomplished by the government.[2]

For those trained in scientific management, it is also an anathema to manage a species over only part of its range. From the view of fisheries scientists and administrators, it is not rational to protect a species in one zone only to have it migrate into another area where it can be taken by other people due to a difference in regulations. As a result, the units to be managed range along hundreds of miles of coast and can only be managed by central governments with jurisdiction over the entire area. Lobsters, for example, extend from Newfoundland to the Carolinas; swordfish migrate from the Caribbean to Newfoundland and Iceland. From the point of view of the National Marine Fisheries Service, it makes sense to have a set of uniform regulations for the entire US coast rather than one for each state (Sherman and Laughlin, 1992).

Perhaps most important, the basis of current management efforts are stock/recruitment models, which are concerned with the relationship between exploitive effort and the size of populations of fish. The central idea is that the long-term abundance or sustainability of a species is strongly linked to the amount of exploitive effort on that stock. The theory assumes that the larger the parent stock, the larger the number of eggs in the water, which in turn results in an increase in future additions to the population (i.e. recruitment).

The model also assumes that the size of that parent stock is a function of exploitive effort. At a constant level of effort, the population size of a stock will tend towards a stable size. At low levels of effort, the population is reduced only slightly. At high levels of effort, which typically occur in open-access fisheries, populations will be low, and the reproductive output of the stock (i.e. recruitment) will be low as well. It follows from this assumed relationship that low stock levels are the result of overfishing; the cure is to lower fishing effort.[3]

In the past several decades, a good deal of work has been done to refine this model. However, empirical research has not been able to demonstrate that these stock/recruitment relationships exist in most cases.[4] Most commonly, there is no correlation between the amount of fishing effort and recruitment (Daan *et al.*, 1990: 382). Nevertheless, fisheries managers have steadfastly maintained their faith in stock/recruitment models and have attempted to conserve and maintain the size of fish populations by manipulation of the amount of fishing effort.

Three aspects of these models need to be stressed. First, they maintain that the size of the population of a fish species is a function of fishing effort. This relationship can be described mathematically. Second, the objective of management is to limit effort to achieve a total allowable catch (TAC), an amount of fish which will result in a maximum sustainable yield (MSY, or the maximum difference between gross income and the cost of fishing) in the long run. A favourite management technique is direct quotas, which specifies the maximum amount of fish that can be taken (Cushing, 1988: 276). At present, administrators, economists and scientists are much taken with the idea of individual transferable quotas (ITQs), which promise not only to control the amount of fishing, but also to achieve economic efficiency (Anderson, 1992). Third, the objective of numerical management is to control the tonnage of fish caught. There is no concern for the effects of fishing on the broader environment or on factors sustaining the life cycle of fish species.

Although fisheries managers and administrators are imbued with the theory of common property and stock/recruitment models, actual regulations vary considerably from fishery to fishery, and have met with varying degrees of success. Some examples are in order.

Between 1977 and 1980, the groundfish of the Gulf of Maine were managed by the Federal Government of the United States working through the New England Regional Fisheries Management Council. A mesh-size regulation was established, but the primary management tools were trip quotas and three-month quotas. That is, a quota was established for three size classes of vessels and a total quota on each of the managed species was established for each three-month period (e.g. January through March). When the TAC was reached, fishing was banned for the remainder of the three-month period (Dewar, 1983: 176–7; Acheson, 1984: 321). This effort failed due to massive political agitation and the fact that the self-reporting system led to widespread cheating and misreporting of catches.

In the shrimp fishery of Sinaloa in northwest Mexico, the Federal Government of Mexico controlled exploitation of shrimp primarily by establishing fishing areas and controlling entry into the fishery (McGoodwin, 1987: 222–5). The government established co-operatives in the 1930s and granted each co-operative exclusive access to certain lagoons and other coastal areas. Over the course of time, it became impossible for additional people to join the co-operatives. Moreover, only co-operatives had the right to sell shrimp to packing plants, which paid a price set by the bureaucracy. Last, the Mexican fisheries bureaucracy created a closed season – a period many months long during which no shrimp fishing was allowed at all.

In the New Zealand fisheries, licences to fish are limited and the government is reducing the number of vessels in the fishery through a vessel-buy-back programme. However, the most important managerial tool in use is that of ITQs. That is, a TAC is set for each species to be managed. Rights to a portion of the catch are allocated to each of the firms in the fishery. The ITQs can be sold to other firms, which is designed to ensure that 'ownership should eventually rest with the most efficient harvesters' (Sissenwine and Mace, 1991: 149). Although the programme is so new that it is difficult to assess, Sissenwine and Mace conclude it has not led to any improvement

394   *James M. Acheson, James A. Wilson & Robert S. Steneck*

in the fishery, either from a biological or an economic standpoint (1991: 152–3).

Modelling chaos

If standard stock/recruitment models do not have much predictive value, what factors do influence changes in stock levels and influence catches? In an effort to answer this question, our research group constructed a computer simulation model of fish stocks between 1987 and 1990. The work of this group strongly suggests that even simple communities of fish may exhibit chaotic population patterns. That is, the population level of individual species varies unpredictably within definite limits (Wilson et al., 1991a; 1991b).

In order to understand the implications of these findings for management, more detailed information on our model and the theory of chaos is necessary. The model approximates the qualitative characteristics of fish and ocean ecosystems, using parameters typical of ocean and fisheries systems. The assumptions on which the model was constructed are the following.

1. We are dealing with an age-structured community of five groundfish species in temperate waters such as the Gulf of Maine. The figures we used for such variables as spawning, growth, and mortality are typical of such populations.
2. A large amount of niche overlap exists.
3. The size of the community as a whole is limited by food constraints that set an overall carrying capacity.
4. Interactions among the five species are marked by 'community predation' – big fish of all species eat little fish of all species indiscriminately (Sissenwine, 1984; 1986).

The output of the model produces results with several characteristics observed in fisheries. There is compensation among species so that when the population of one declines, another increases to take its place. Thus the biomass of individual species can vary in unpredictable ways, but the total biomass of the community of fish is relatively stable. In addition, there is no relationship between the size of the spawning stock and recruitment. Stock/recruitment models aside, no relationship has been found between spawning stock and the number of adult fish resulting in the future (i.e. recruitment) (Hall, 1988; Daan et al., 1990).

Most importantly, the output of our model mirrors the kind of unpredictable population changes observed in actual populations of fish over time. There are many possible sources of variation. One of the most important factors producing chaotic variations in population is community predation. In short, as the community approaches its carrying capacity, the scarcity of food results in more small fish being eaten by larger fish. Cushing (1977) has demonstrated that cannibalism produces swings in the population of a single species. Our model is a multiple species version of the same phenomenon. The result is chaotic variation in stock sizes.

Others, such as Hastings and Higgins (1994), also argue that marine populations may be chaotic. However, there is certainly no consensus on this matter yet, and none will probably arise for a period of years. We cannot regard the case as proven by any means.[5]

If oceans and fish stocks are indeed chaotic systems, management is possible, but it must be based on principles different from those stemming from stock/recruitment models, which assume equilibrium conditions and long-run, predictable relationships between fishing effort and future populations of fish. Understanding the implications for management necessitates some understanding of two aspects of chaotic systems.

First, chaotic systems are not marked by a complete lack of order. Clear cause-and-effect relationships exist, as physicists point out, but the complexity and non-linearity of these relationships make output of these systems highly unpredictable. Moreover, the outcomes of chaotic systems take place within well-defined limits determined by the initial parameters of the model. In models of biological systems, these parameters represent such things as spawning potential, growth rates, habitat, migration, and the way they are organized both spatially and temporally. In nature, these basic input factors (i.e. parameters) remain relatively stable over the course of time. Their stability is the reason that fish populations can be expected to vary within certain limits, even though population changes will remain unpredictable.

In addition, chaotic systems are extremely sensitive to 'initial conditions' (Gleick, 1987: 22-3). That is, small changes in current circumstances can produce huge changes in the future state of a chaotic system. For example, a storm near a groundfish spawning area might result in a small year-class; whereas favourable water temperatures and amounts of food might result in high recruitment several years later.

Given the deterministic nature of chaotic systems, it should be possible, theoretically, to predict the effects of initial conditions on fish populations. As a practical matter, this is impossible. Small variations or inaccuracies in measurement would probably result in huge errors in prediction, given the sensitivity of such models to 'initial conditions' (Gleick, 1987: 8). Given the large number of relationships in such systems and the complicated feedback mechanisms, errors in measurements are almost certain. In the real world, it would be impossible to obtain the huge amount of accurate, fine-grained, continuously updated data that would be necessary to make accurate prediction possible (Wilson *et al.*, 1991a, 1991b; Wilson and Kleban, 1992).

In marine fisheries, we are fortunate to be able to assess stocks within 30% of their actual size; we can only guess at community predation inter-relationships; and great variations exist in estimates of factors such as natural mortality. Thus, it is virtually impossible to know the likely outcome of management quotas given our state of knowledge and our real measurement capabilities.

However, this approach suggests that the stock depletion observed today is the result of degradation of the parameters of the system. More importantly, the fact that the populations of fish vary within specifiable limits if the system parameters are undisturbed suggests that fisheries can be managed by maintaining those parameters. If this approach to management is taken, the goal of regulation would be to maintain such parameters as breeding grounds, migration routes, nursery areas, and spawning and other critical life processes. This can be accomplished by rules concerning fishing locations, fishing area and techniques. We call this *parametric management* since control over fish populations is affected by influencing the parameters of the model.

As shall be seen in the next section, many tribal and peasant societies use parametric strategies in managing their fish stocks. In these societies, the emphasis is on maintaining the variables that affect fisheries systems – not the amount of fish taken.

Fisheries management techniques in traditional societies

Fisheries management is scarcely new or confined to the industrialized countries of the world. Management of marine fisheries is found throughout the world, and there is a growing body of literature on these fisheries and the way they are regulated at the local level.[6] (see Johannes, 1981; McCay and Acheson, 1987; Acheson, 1988; Berkes, 1989; McGoodwin, 1990; Bromley, 1992; Anderson and Simmons, 1993; Dyer and McGoodwin, 1994). However, there can be little question that practices in modern Western societies where 'scientific' management is the rule are different from those used in peasant and tribal societies. We have reviewed in some detail the management practices found in 30 traditional societies. An extensive analysis of our findings has been reported in another publication, along with the bibliography (Wilson *et al.*, 1994). Analysis of these cases shows three essential differences.

First, the rules and techniques used in traditional societies are the product of very different cultures, and are buttressed by ideologies which have no analogues in the modern West. In these societies, rules are often enforced by informal community pressure or religious sanctions rather than by officials of a formally constituted government. Both can be very effective in ensuring compliance.

Second, in all of the societies studied, resources are managed by local-level communities that have riparian or coastal rights over small areas which people know intimately (Johannes, 1981: 11–84). The small size of the management units is consistent both with the small scale of the political units of tribal and peasant societies and with the low level of technology, which limits access to distant areas.

We believe that some kind of riparian or coastal rights is a necessary precondition for any other kind of resource management regulation.[7] A rule or regulation cannot apply generally. It can only apply within the territory of a group willing and able to enforce the rule.[8]

Third, and most important, the techniques on which traditional societies place primary dependence are different from those in our own society, and they are used in different combinations.

All of the rules and practices we found in these 30 societies regulate 'how' fishing is done. That is, they limit the times fish may be caught, the locations where fishing is allowed, the technology permitted, and the stage of the life cycle during which fish may be taken. None of these societies limits the 'amount' of various species that can be caught. Quotas – the single most important concept and tool of scientific management – is conspicuous by its absence. In the 30 traditional societies studied, we found no instance of the use of quotas in managing any fisheries. An idea of the combinations of management techniques employed in some of these traditional societies can be seen in Table 15.1.[9]

These three principles can be seen in the ethnography of any number of societies. Some examples are in order. In India, for example, Gadgil (1985) reports that there was a system of caste fishing territories, which lasted until recent times (Gadgil, 1985: 139). The Hindu month of *Sravana* (August) was a closed season when no fish or meat could be harvested; and the use of fish poisons was limited to a few days coinciding with a communal festival. There were religious prohibitions on fishing during the mating season

Table 15.1. *Management techniques used in ten traditional societies around the world*

Case number	Society	Reference	Resource	Areas	Limited access	Seasonal limits	Technology	Protect breeding stock	Protect young	Size limits	Conservation ethic	Protect overcrowding	Other	Quota
1	India	Gadgil (1985)	Fish, animals, plants	X	X	X		X					X	
2	Oceania	Baines (1989)/Johannes (1978)	Fish	X	X	X		X	X	X	X			
3	Japan (Tokugawa period)	Kalland (1984)	Fish	X	X	X	X		X		X			
4	Nile Delta	Rowntree *et al.* (1984)	Fish	X		X								
5	Titicaca	Levéil and Orlove (1990)	Game and totora	X										
6	Cree	Berkes (1987)	Fish	X			X				X			
7	Sri Lanka	Alexander (1980)	Fin fish									X		
8	Equador	Southon (1989)	Fish						X			X		
9	Brazil	Robben (1994; personal communication)	Fin fish	X			X	X	X					
10	Turkey	Berkes (1986)	Fish	X		X	X	X		X				

to protect the breeding stock. Sacred ponds were protected, providing a sanctuary (Gadgil, 1985: 143). All resources, including fish, were the property of the gods and not of the caste, and permission needed to be asked from the gods before they could be harvested. Despite the lack of enforcement by a government, Gadgil argues that these practices maintained populations at sustainable levels (1985: 135).

In Oceania, inshore fishing grounds were considered the property of local villages, and access to them was allocated by the local village chief. Individuals from other communities were generally allowed to fish in these waters if they paid a fee. Some species were protected by supernatural sanctions. Johannes also reports that turtles were thought to be owned by God, and neither the adult turtles nor turtle eggs could be taken (1981: 64–7). Throughout Oceania, species of fish considered an emergency food supply could not be caught in good weather when other species were available.

In Tokugawa Japan, both land and sea were part of the fiefs of feudal lords. The people of fishing villages were allowed to fish in coastal waters only by paying the lord. After a time, village fishing territories evolved, but ownership still resided with the lord (Kalland, 1984: 11–12). Permission to use fish did not give unlimited rights of exploitation. Fishing was often restricted to certain seasons. Licences needed to be obtained from the feudal authorities to take abalone and whales, and to use very large nets (Kalland, 1984: 22–4). Fisheries sanctuaries were also recognized and protected. However, no effort was made to limit the number of boats or the number of people involved in the fishery (Kalland, 1984: 23). Many of the latter techniques are in use in Japan today.

Does parametric management work? Evidence from the Maine lobster industry

It is very difficult to demonstrate convincingly that any effort to manage fisheries preserves stocks, given the large number of ecological, biological and cultural variables involved. For this reason, it is very difficult to demonstrate the superiority of parametric management over any other approach. In a few cases, enough data exist to buttress the case for parametric management. One is the Maine lobster industry, with which we have been working for over two decades.

Over the course of the past 120 years, Maine lobster catches have varied greatly. Between 1880 and 1900, they ranged from 14 million to a maximum of 24.5 million pounds (6.35 to 11.1 million kilograms). In the years between World War I and World War II, catches dropped precipitously, a period which we call the 'lobster bust'. From 1919 to 1938, catches were only 5 to 7 million pounds (2.25 to 3.2 million kilograms) per year (Dow 1967: 5). After World War II, catches recovered and remained relatively stable – between 16 million and 24 million pounds (7.25 and 10.8 million kilograms) for 40 years (Maine Department of Marine Resources, 1995). Since 1989, however, lobster catches have increased tremendously. In 1994 the catch was 38 million pounds, which is an all-time record. We call the period from 1989 to 1995 the 'lobster boom'.

These changes in catches cannot be explained in terms of basic changes in regulations. The lobster industry of Maine has been managed parametrically throughout its entire history. There are two kinds of rules influencing 'how' people fish: the informal territorial system and formal laws enforced by the wardens.

The Maine lobster fishery has long been organized into communal territories. In order to go fishing at all, one must become a member of a 'harbor gang', the group of fishermen who go lobstering from a single harbour. Once one has gained admittance into such a group, one can only set traps in the traditional territory of that particular harbor gang. Members of harbor gangs are expected to obey the rules of their gang concerning fishing practices, which vary somewhat from one part of the coast to another. In all areas, a person who gains a reputation for molesting others' gear or for violating conservation laws will be severely sanctioned (Acheson, 1988: 48–83). Incursions into the territory of one gang by fishers from another are ordinarily punished by surreptitious destruction of lobster gear. There is strong statistical evidence that the territorial system, which operates to limit the number of fishers exploiting lobsters in each territory, helps to conserve the lobster resource (Acheson, 1975; 1988: 153–9; 1993).

Since the 1880s, Maine's formal system of lobster management has depended on laws to protect the breeding stock and to increase the number of eggs in the water. These have taken the form of size regulations and prohibitions on taking 'berried' lobsters (i.e. those with eggs). At present, the most important conservation laws are minimum and maximum size measures, a prohibition against catching lobsters with eggs, and a law to prohibit the taking of lobsters which once had eggs and were marked – i.e. the 'V-notch' law (Acheson, 1988: 138–41; 1989; 1993). All of these rules are designed to protect the breeding stock. There have never been any laws to limit the amount of lobster that can be caught by an individual or in aggregate, nor are there any limits on the number of boats or licences.

In an effort to understand the complex of factors influencing lobster catches, we examined all of the many hypotheses that biologists and lobster

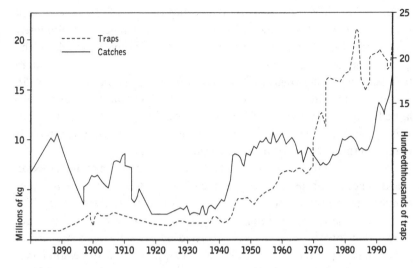

Figure 15.1. Catch and effort data for the Maine lobster industry, 1880–1994. (Data obtained from the Maine Department of Marine Resources, 1995.)

fishers have suggested cause changes in catches. We have concentrated on the 'bust' and the 'boom' periods because we thought it would be easier to observe variables underlying catches in these periods of unusual catches. The results of this study have been reported in great detail in another publication (see Acheson and Steneck, 1997).

For our purposes, there are four important conclusions. First, virtually none of the explanations for the bust and the boom advanced by members of the fishing industry is supported by the evidence. Hypotheses concerning changes in prices, predation by large groundfish, changes in the legal measure, changes in food supply for lobsters, etc. are not buttressed by the data available.

Second, historic catch-and-effort data from the lobster industry of Maine do not support the stock/recruitment model at all. The calamity of the 1920s and 1930s cannot be explained in terms of high fishing pressure. As can be seen from Figure 15.1, during this period, catches were low, but numbers of traps and licence holders were low as well. We see no way that lobster stocks could have been depleted by fishing pressure alone during these years – particularly in view of the relatively inefficient gear in use. Moreover, in recent decades, fishing effort has increased dramatically. Since the early 1970s the number of licence holders has doubled. More importantly, since 1950 the number of traps in use has gone from 400 000 to over 2 000 000. Stock/recruitment models would lead us to believe that such extremely high levels of effort should reduce the size of the parent stock substantially, result in a low number of eggs in the water, and ultimately lower recruitment and catches. Since the 1970s biologists employed by government agencies have predicted that such increases in effort would, in fact, lead to disaster (Bell, 1972; Krouse, 1972: 172; Thomas, 1973: 54–5). Nothing of the kind has occurred. However, the fact that the predicted collapse has not happened does not seem to have shaken the faith of the biologists concerned with lobster management. Stock/recruitment models remain in the forefront of their thinking.

Third, we are convinced that two factors have had a primary influence on the changes in lobster catches observed from 1880 to the present: water temperature and the growth of a conservation ethic in the industry. The effect of these factors is interrelated and complex. Recent studies have shown that lobster abundance is regulated by the settlement strength of lobster larvae (Wahle and Steneck, 1991; 1992), and that successful larval settlement is triggered by a thermal threshold at temperatures of 15°C and above. These high water temperatures would occur in late summer if they occur at all (Boudreau, Simard and Bourget, 1991; 1992). Since it takes seven to ten years for larval lobsters to mature to a size at which they can be recruited into the fishery, the years when the August temperature was below 15°C should have resulted in lower than average landings seven to ten years later; and there should be high catches seven to ten years after a year in which the temperature was over 15°C.

A regression analysis of the relationship between water temperature and landings from 1920 to 1994 reveals another interesting and important pattern related to temperature. The years from 1946 to 1986 fall on or close to the thermal-mediated settlement success (TMSS) line, which indicates the years when settlement should and should not be successful. These results suggest that water temperature is closely related to landings during those years (Figure 15.2). However, the landings during the 'bust' and 'boom' years do not fall on the TMSS line. Landings during these periods cannot be explained simply in terms of larval settlement success. It is possible that the 'bust' resulted mainly from low settlement success since temperatures in the early (1920s) bust years were on the low side. However, the 'boom' clearly is not the result of warmer August temperatures. As can be seen in Figure 15.2, temperatures during the boom were not the highest recorded and were similar to those in many years between 1920 and 1986, and even several years in the 'bust'. However, the fact that neither the 'bust' of the 1920s and 1930s nor the 'boom' of the 1990s falls on the TMSS line indicates that factors other than water temperature are

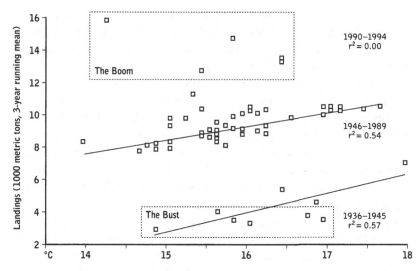

Figure 15.2. Regression analysis of August temperature (seven-year lag) and lobster catches, 1936–1994. (The 'bust' went from 1919 to 1939. Our data on the 'bust', however, begin with catch data from 1936. Given the seven-year lag, the 1936 catches are correlated with temperature conditions in 1929; those from 1945 reflect temperature conditions in 1939. Catch from earlier years could not be analysed because of discontinuous temperature records before 1929.)

involved in depressing catches in the 'bust' and elevating them to record levels in the 'boom'.

We believe that one of the most important factors producing the 'bust' and the 'boom' is the absence or presence of a strong conservation ethic. In the 1920s and 1930s, violations of the conservation laws were at epidemic proportions.[10] The law protected small lobsters and females with eggs, but both of these rules were widely flouted. Some fishermen made a practice of smashing up short lobsters caught in their traps to serve as bait (Clifford, 1961: 124). There was also a massive trade in illegal lobsters, which found their way to restaurants in neighbouring New England States and beyond (Judd, 1988: 612). In the early 1920s violations of the legal size limits were so severe that the Commissioner of the Maine Sea and Shore Fisheries Department resorted to drastic measures, at one point closing the entire lobster fishery for a period of time (Clifford, 1961: 204).

Over the course of the past 70 years, support for conservation has grown in the lobster industry, and the massive violations of the law which marked the earlier decades of the century became a thing of the past. By the middle of the 1980s, lobster regulations had become largely self-enforcing, and fishermen themselves sanction people who take short lobsters, molest other people's traps or engage in other violations (Acheson, 1988: 90). More importantly, the industry began to support critically important conservation regulations. In 1933, the double gauge law was passed, resulting in minimum and maximum size measures.

Efforts to conserve the lobster were advanced by the advent of the 'V-notch' law in 1948. Fishermen who catch an egg-bearing female may voluntarily cut a notch in her tail before they return her to the water. That female may never be legally taken as long as the notch is visible. This programme has received almost universal support from fishermen, with the result that there are literally millions of 'V-notched' lobsters in the Gulf of Maine (Bayer, Daniels and Waltz, 1989; Acheson, 1993: 76). In 1979, a law was passed making it mandatory to place escape vents in all lobster traps to allow undersize lobsters to escape from traps, thereby reducing mortality on sublegal lobsters. In 1995, the leaders of the lobster industry lobbied successfully for a law dividing the coast into small zones, which will allow people in each local area to set a trap limit. Knowledgeable biologists generally concede that these legal measures have helped conserve the lobster resource, although the magnitude of their effect is still being debated (Acheson and Steneck, 1997; M. Fogarty 1995, personal communication).

What produced this change in attitudes toward conservation? We believe that the experiences of the disastrous interwar period produced a kind of social learning that has influenced the industry to the present. By the 1930s, catches had fallen to the point where large numbers of fishermen left the industry. In 1928, there were 4000 fishermen; by 1933, there were 2800 (Judd, 1988: 619). There were no other jobs. The need to protect the breeding stock was impressed on the memory of many fishermen, along with the idea that the 'thieves' were doing damage to everyone.

We conclude that changes in lobster catches are strongly affected both by water temperature and by parametric fishing practices. The effects of low temperatures in the 1920s and 1930s were exacerbated by violations of conservation laws to produce the 'bust'. However, the 'boom' of the 1990s, we believe, is due less to water temperature and probably more to conservation efforts (Acheson and Steneck, 1997).

The case of the Maine lobster fishery underlines the importance of environmental factors in influencing changes in stocks. This case study also affirms the idea that preserving basic biological processes is the secret to maintaining the health of stocks. When regulations designed to maintain the breeding stock were violated widely, catches fell dramatically. In addition, lobster traps are a type of gear that do not disturb the environment, and 'V-notching' and the oversize measure preserve a wider age distribution of lobsters than would exist otherwise.

Why do people manage parametrically?

Why have so many societies managed fisheries by essentially regulating 'how' people fish? We argue that these kinds of regulations reduce information and enforcement costs to manageable levels – especially if they are administered through decentralized institutions. It is relatively easy to monitor whether people are fishing along a prohibited beach, or taking a forbidden species, or using illegal gear. Numerical management involves a sizeable system to gather data and maintain records. If one is going to regulate fishing effort to achieve MSY, one needs to know how many fish of various types are being caught, when, and by whom. People in tribal and peasant societies can monitor 'how' people fish; they do not have the organizations and technology necessary to manage numerically. Thus, parametric management not only results in relatively low enforcement costs, it is also the only kind of management people of these societies can use.

In addition, rules regulating 'how' fishing is done are apt to get more support from those in the industry because they seem to be in their best interests. People in fishing communities know a great deal about the habits of various species, including feeding patterns, migration routes, spawning areas, nursery areas, predation, habitat, and seasonal rhythms (Johannes, 1981: 32–58), knowledge which is critical for fishing success. They know how to maintain these critical life-cycle processes with rules controlling technology, fishing locations and fishing times. Such rules, in their view, are based on biological reality. They are judged to be effective and sensible, which increases the willingness of people to comply with them.

In tribal and peasant societies, conservation rules are the result of political processes at the local level. They are likely to promote conservation of the resource in ways that reflect local norms, so that the goals of individuals and communities coincide. They are not imposed by a central government with little knowledge of the local area or a personal stake in the outcome. Since there is an incentive to obey conservation rules, costs of enforcement are low.

Johannes has pointed out, 'When fishermen do not understand the purposes of fishing regulations or perceive them as being imposed arbitrarily by outsiders they are not liable to look on them with favor or obey them voluntarily' (1981: 84). We are convinced that many fisheries management regimes in the United States founder on exactly such shoals. Fishermen know that fish stocks and catches fluctuate greatly and in unpredictable ways. A number of close observers of fisheries have pointed out that chaotic change is biological reality to people who live from the sea (Smith, 1990: 5; Pálsson, 1994: 918–21). Policies which ignore this reality seem ineffective and even foolish. Stock/recruitment models are in this category. Obedience to such rules imposes costs which will not result in future benefits. Costs of enforcement are further increased by the difficulties of monitoring such systems.

The problems inherent in numerical management became apparent to observers of the New England fisheries when the New England Regional Council attempted to manage the groundfish stocks of the Gulf of Maine using trip quotas and three-month quotas in the late 1970s. The plan was monumentally unpopular, and the self-reporting system led to massive cheating (Acheson, 1984). In 1980, this plan was put in abeyance and none was substituted for it until the 1990s. The basic problem was that fishermen did not believe the rules would conserve the stocks. In the words of one fisherman, they were 'stupid'. This disrespect led to escalating cheating. When some people got away with cheating, others were motivated to do the same to 'get their share' of the fish. People not only had the motivation to cheat, they were able to do so because of the difficulties involved in monitoring the numerical rules. The management plan could not survive under these conditions.

This is not to suggest that high enforcement costs alone are responsible for the demise of conservation institutions. Many factors are necessary for institutions to be generated.[11] But among the most essential conditions, as Elinor Ostrom (1992: 55) makes clear, is the ability to monitor and punish those who violate rules. Numerical management, which demands a huge amount of data on catches, makes monitoring prohibitively costly. In this regard, parametric management has a clear advantage.

In summary, when the rules are costly to enforce, no effective institutions will be generated. When people do not believe rules are effective, when rules cannot be monitored, and when they conflict with local norms, enforcement costs grow rapidly, numbers of people become free riders on the system, and management plans fail.

Discussion

Numerical management, as it is currently practised, is not working well. We are convinced that the major problem lies in the concepts used, and not in the inability to enforce regulations due to the political activities of the fishing industry.

Data from historical fisheries, our modelling exercise and the Maine lobster industry all support the idea that fish populations vary chaotically. These unpredictable changes in populations appear to be due primarily to complex changes in environmental factors. Some of these factors can be affected by human beings; others cannot be. We maintain that if there were no effort on fish stocks, populations would fluctuate chaotically due to environmental factors alone. In many cases, fishing effort has a relatively small effect on catches. In the Maine lobster industry, for example, one of the primary factors influencing catches is water temperature, a variable well beyond human control. In this fishery, changes in rules, enforcement, and fishing practices appear to have a secondary effect on catches.

We suggest that numerical approaches to management be replaced with parametric management which is more appropriate for fisheries which are chaotic in nature. That is, we are proposing that fisheries be managed through rules on 'how' fishing is done in order to maintain regular biological processes, rather than by attempting to control 'how much' is caught. Many environmental factors remain relatively constant over time, including: growth rates, spawning potential, migration routes, predation patterns, and nursery grounds. These are the parameters of our computer model, and as long as they remain stable, populations of fish fluctuate only within certain limits. These parameters can be maintained by rules on technology, and by protecting animals in certain periods of their life cycles. We do not claim that parametric management would result in stable or even predictable catches. However, we believe it would avoid the kinds of stock failures and disasters being experienced in so many fisheries at present (see Holling, 1987). That may be all we can reasonably expect to accomplish.

Several aspects of the parametric approach should be stressed. First, parametric rules appear to be effective when they are properly implemented. In the lobster industry, regulations have always been parametric in nature. But the existence of parametric rules alone did not prevent the 'bust' of the 1920s and 1930s. Those rules only became effective when the 'bust' produced a kind of social learning that resulted both in better self-enforcement and in additional parametric regulations.

Second, regular biological processes can be observed by human beings, and do not change much, if at all. This means that knowledge about these biological processes can be gathered at reasonable cost and that knowledge, once learned, should last a considerable time. The numerical approach to management demands such a large amount of continuously updated information that we doubt if it can be implemented, except at tremendous cost.

Third, if effective managerial institutions are going to be implemented at reasonable costs, fishermen must be convinced that rules are working so that today's sacrifice will result in benefits in the future. If rules cannot be monitored, people will have strong incentive to become 'free riders', resulting in rules that will be very costly or impossible to enforce. Rules on 'how' people fish are much more easily monitored than 'quotas'. They also seem realistic and effective to fishers, who know a good deal about the life cycle of their target species. Both of these factors should help to produce institutions that people in the industry will support at less cost.

Fourth, in tribal and peasant societies, management units are relatively small. Local level management may be impractical in an industrial society. We suggest a hierarchical approach to regulation may be more effective. That is, some biological factors are characteristic of the species over its entire range (e.g. spawning potential); others are very localized (e.g. habitat). Thus some regulations will be required to affect the fisheries system on a large scale, while others can be tailored to affect smaller-scale events and processes.

Where possible, management should be handled by small-scale units, which have two advantages: (1) management of small units eases the problems of obtaining information on biological processes (it is easier to learn the intricacies of a small zone as opposed to a large one); and (2) local-level management would presumably be more able to frame rules to fit local conditions in ways that take into account local practices and norms. This, in turn, should result in a higher degree of political support. However, this is not to suggest that there are not biological phenomena that demand management on a very large scale.

Fifth, the primary advantage of the parametric approach is that it seeks to preserve those variables important to maintaining the entire biological system. Stock/recruitment models lead those managing by the numerical method to play down the importance of habitat or other ecological factors which are critically important to preserve the system in the long run. The objective of numerical management is to control mortality on target species with no attention to the environment. This, we maintain, is the fatal flaw in the numerical approach.

Last, there are a number of fisheries that have been managed by local-level political units using a parametric approach (see Table 15.1). Controlling 'how' people fish appears to have had beneficial results. The approach we are advocating is scarcely untried. In stark contrast to what most fisheries biologists believe, we believe it is the only kind of management regime that is likely to be effective in the long run.[12]

Notes

1 This model was completed between 1988 and 1991 in the course of a project entitled 'University of Maine Chaos Project', financed by the University of Maine Sea Grant Office. Special thanks are due to Dr. Ann Acheson who copy-edited numerous drafts of this chapter.
2 Garrett Hardin, perhaps the most widely read common property theorist, argues that governments might have to act in highly autocratic fashion to protect resources (Hardin, 1968; 1977).
3 Economists argue that effort should be limited to a point where maximum economic yield occurs; biologists argue that effort should be restricted to produce maximum sustainable yield (Wieland, 1992).
4 There are a few cases that buttress the stock/ recruitment model of fisheries. During World War I and World War II, when fishing effort decreased the stocks went up. We might mention the case of the striped bass and the Fraser River salmon stocks as well. However, a large number of factors other than effort could explain an increase in these stocks during these periods.
5 Even if fisheries do not prove to be chaotic, they are certainly complex. This complexity alone will make prediction very difficult.
6 Berkes goes so far as to say that management is virtually ubiquitous in tribal and peasant societies. In this regard, he says that fisheries 'are almost never truly open-access' (1985: 204).
7 In worldwide perspective, riparian rights are very common (Acheson, 1981: 280–81; Berkes, 1985).
8 Territoriality should not automatically be equated with limited entry. Groups often have the right to devise and enforce rules in specific zones; they may not have the right to exclude all others from using that area.
9 All of the cases in Table 15.1 are stable and sustainable fisheries. In all cases the ethnographer reported that regulations are devised to conserve the resource.
10 The factors producing the low catches of lobsters in the interwar period have been discussed in some detail in another publication (see Acheson, 1992: 152–7).
11 Olsen (1965) defines a 'public' or 'collective good' as any set of rules that helps a community to achieve a goal which cannot be achieved by individual effort. He points out that individuals will not automatically support the effort to create public goods, even when those goods benefit everyone in the community. 'Selective incentives' (i.e. sanctions, rewards, etc.) are necessary to ensure compliance.
12 Most fisheries biologists are scarcely convinced of the utility of the parametric approach to fisheries management. For a traditional response to the idea of parametric management, see Michael Fogarty's (1995) critique of one of our earlier papers on this topic.

References

Acheson, J.M. 1975. The lobster fiefs: economic and ecological evects of territoriality in the Maine lobster industry. *Human Ecology* 3(3): 183–207.

Acheson, J.M. 1981. Anthropology of fishing, *Annual Review of Anthropology* 10: 319–29.

Acheson, J.M. 1984. Government regulation and exploitive capacity: the case of the New England groundfishery. *Human Organization* 43(4): 319–29.

Acheson, J.M. 1988. *The Lobster Gangs of Maine.* Hanover, New Hampshire: New England University Press.

Acheson, J.M. 1989. Management of common-property resources. In *Economic Anthropology*, pp.351–78, ed. S Plattner. Stanford: Stanford University Press.

Acheson, J.M. 1992. Maine lobster industry. In *Climate Variability, Climate Change and Fisheries*, pp.147–65, ed. M. Glantz. Cambridge: Cambridge University Press.

Acheson, J.M. 1993. Capturing the commons: legal and illegal strategies. In *The Political Economy of Customs and Culture:Informal Solutions to the Commons Problem,* pp.69–83, ed. T.L. Anderson, and R.T. Simmons. Lanham, Maryland: Rowman and Littlefield Publishers.

Acheson, J.M. and Steneck, R. 1997. Bust and then boom in the Maine lobster industry: the perspectives of fishermen and biologists on the efficacy of management. *North American Journal of Fisheries Management*, November issue.

Alexander, P. 1980. Sea tenure in southern Sri Lanka. In *Maritime Adaptations: Essays on Contemporary Fishing Communities*, pp.91–112, ed. A. Spoehr. Pittsburgh: University of Pittsburgh Press.

Anderson, L.G. 1992. Consideration of the potential use of individual transferable quotas in US fisheries. *The National ITQ Study Report* 1: 1–71.

Anderson, T.L. and Simmons, R. 1993. *The Political Economy of Customs and Culture.* Lanham, Maryland: Rowman and Littlefield.

Austin, Phyllis 1992. Who killed Maine's multi-million dollar fishery? *Maine Times* 14:2

Baines, G.B.K. 1989. Traditional resource management in the Melanesian South Pacific: a development dilemma. In *Common Property Resources: Ecology and Community-Based Sustainable Development,* pp.273–95, ed. F. Berkes, London: Belhaven.

Bayer, R., Daniels, P. and Waltz, C. 1989. Egg production of V-notch American lobster along coastal Maine. *Journal of Crustacean Biology* 9: 77–82.

Bell, F. 1972. Technological externalities in common-property resources: an empirical study of the U.S. northern lobster fishery. *Journal of Political Economy* 80(1): 148–58.

Bencivenga, J. 1991. New England's fishery in decline. *Christian Science Monitor* July 26, V.83: 3.

Berkes, F. 1985. Fishermen and the tragedy of the commons. *Environmental Conservation* 12(5): 199–205.

Berkes, F. 1986. Marine inshore fishery management in Turkey. In *Proceedings of the Conference on Common Property Resource Management*, pp.63–83. Washington DC: National Research Council, National Academy Press.

Berkes, F. 1987. Common property resource management and Cree Indian fisheries in subarctic Canada. In *The Question of the Commons,* pp.66–91, ed. B.J. McCay and J.M. Acheson. Tucson: University of Arizona Press.

Berkes, F., ed. 1989. *Common Property Resources: Ecology and Community-based Sustainable Development*. London: Belhaven Press.

Boudreau, B., Simard, Y. and Bourget, E. 1991. Behavioral responses of the planktonic stages of the American lobster *Homarus americanus* to thermal gradients, and ecological implications. *Marine Ecology Progress Series* 76: 12–23.

Boudreau, B., Simard, Y. and Bourget, E. 1992. Influence of a thermocline on vertical distribution and settlement of post-larvae of the American lobster (*Homarus americanus*. Milne-Edwards). *Journal of Experimental Biological Ecology* 162: 35–49.

Bromley, D.W., ed. 1992. *Making the Commons Work: Theory, Practice and Policy*. San Francisco: ICS Press.

Clifford, H. 1961. *The Boothbay Region: 1906–1960*. The Cumberland Press, Freeport.

Cushing, D.H. 1977. The study of of stock and recruitment. In *Fish Population Dynamics*, pp.105–28, ed. J.A. Gulland. London: John Wiley and Sons.

Cushing, D.H. 1988. *The Provident Sea*. Cambridge: Cambridge University Press.

Daan, N., Bromley, P.J., Hislop, J.R.G. and Nielsen, N.A. 1990. Ecology of North Sea fish. *Netherlands Journal of Sea Research* 26: 343–86.

Dewar, M. 1983. *Industry in Trouble: The Federal Government and the New England Fisheries*. Philadelphia: Temple University Press

Dow, R.L. 1967. Temperatures, fishing effort point up growing problem. In *The Influence of Temperature on Maine Lobster Supply*. Sea and Shore Fisheries Research Bulletin No. 30, Augusta.

Dyer, C.L. and McGoodwin, J.R., eds. 1994. *Folk Management in the World's Fisheries: Lessons for Modern Fisheries Management*. Niwot: University of Colorado Press.

FAO 1977. *FAO Yearbook of Fisheries Statistics*, Vol. 45. Rome: Food and Agricultural Organization of the United Nations.

FAO 1987. *FAO Yearbook of Fisheries Statistics*, Vol. 65. Rome: Food and Agricultural Organization of the United Nations.

Feeny, D., Berkes, F., McCay B.J. and Acheson, J.M. 1990. The tragedy of the commons: twenty-two years later. *Human Ecology* 18(1): 1–19.

Fogarty, M. 1995. Chaos, complexity and community management of fisheries: an appraisal. *Marine Policy* 19(5): 437–44.

Gadgil, M. 1985. Social restraints on resource utilization: the Indian experience. In *Culture and Conservation: The Human Dimension in Environmental Planning*, pp.135–54, ed. J.A. McNeeley and D. Pitt. Dublin: Croom Helm.

Gleick, J. 1987. *Chaos: Making a New Science*. New York: Viking.

Hall, C. 1988. An assessment of several of the historically most influential theoretical models used in ecology and of the data provided in their support. *Ecological Modelling* 43: 5–31.

Hardin, G. 1968. The tragedy of the commons. *Science*: 162: 1243–8.

Hardin, G. 1977. Living on a lifeboat. In *Managing the Commons*, pp.261–79, ed. G. Hardin and J. Baden. San Francisco: W.H. Freeman.

Hastings, A. and Higgins, K. 1994. Persistence of transients in spatially structured ecological models. *Science* 263: 1133–6.

Holling, C.S. 1987. Simplifying the complex: paradigms of ecological function and structure. *European Journal of Operations Research* 30: 139–46.

Johannes, R.E. 1978. Traditional marine conservation methods in Oceania. In *Annual Review of Ecology and Systematics 9*, pp.349–64, ed. R. Johnson. Palo Alto: Annual Reviews Inc.

Johannes, R.E. 1981. *Worlds of the Lagoon: Fishing and Marine Life in Palau District of Micronesia*. Berkeley: University of California Press.

Johnson, G. 1991. Scarce tuna, scant profits. *Los Angeles Times* Aug 26, 110:D1.

Judd, R.W. 1988. Saving the fisherman as well as the fish: conservation and commercial rivalry in Maine's lobster industry, 1872–1933. *Business History Review* 62: 596–625.

Kalland, A. 1984. Sea tenure in Tokugawa Japan: the case of the Fukuoka domain. In *Maritime Institutions in the Western Pacific*, pp.11–36, ed. K. Ruddle and T. Akimichi. Osaka: National Museum of Ethnology.

Krouse, J.S. 1972. *Size at First Sexual Maturity for Male and Female Lobsters Found Along the Maine Coast*. Lobster Information Leaflet No. 2. Augusta, Maine: Department of Sea and Shore Fisheries.

Leveil, D.P. and Orlove, B. 1990. Local control of aquatic resources: community and ecology in Lake Titicaca, Peru. *American Anthropologist* 92: 362–82.

Maine Department of Marine Resources. 1995. *Summary of Maine Lobster Fishery Statistics*. Augusta, Maine: Department of Marine Resources.

McCay, B.J. and Acheson, J.M., eds. 1987. *The Question of the Commons: The Culture and Ecology of Communal Resources*, pp.1–34. Tucson: University of Arizona Press.

McGoodwin, J.R. 1987. Mexico's conflictual inshore Pacific fisheries: problem analysis and policy recommendations. *Human Organization* 46: 221–31.

McGoodwin, J.R. 1990. *Crisis in the World's Fisheries*. Stanford: Stanford University Press.

New York Times. 1992a. Plenty of fish in the sea? Not any more. March 15.

New York Times. 1992b. Salmon in decline off west coast. March 20, 141: B6.

Olsen, M. 1965. *The Logic of Collective Action*. Cambridge: Cambridge University Press.

Ostrom, E. 1992. *Crafting Institutions for Self-Governing Irrigation Systems*. San Francisco: ICS Press.

Pálsson, G. 1994. Enskillment at sea. *Man* 29(4): 901–27.

Robben, A.C.G.M. 1994. Conflicting discourses of economy and society in coastal Brazil. *Man* 29(4): 875–900.

Rosen, Y. 1991. Alaska fisheries depleted by foreign fleets. *Christian Science Monitor* 83: 7.

Rowntree, J. et al. 1984. Fishery management in the northern Nile Delta lakes of Egypt: the case of Hosha. In *Studies and Reviews* 61(2), UNESCO: 542–55.

Sherman, K. and Laughlin, T., eds. 1992. *NOAA Technical Memorandum NMFS-F/NEC 91. The Large Marine Ecosystem (LME) Concept and Its Application to Regional Marine Resource Management*. Woods Hole: Northeast Fisheries Science Center.

Sissenwine, M.P. 1984. Why do fish populations vary? In *Explorations of Marine Communities*, pp.59–94, ed. R.M. May. Berlin: Springer-Verlag.

Sissenwine, M.P. 1986. Perturbation of a predator – controlled continental shelf eco-system. In *AAAS Selected Symposium 99. Variability and Management of Large Marine Eco-systems*, pp.55–85, ed. K. Sherman, and L.M. Alexander, Boulder: Westview.

Sissenwine, M.P. and Mace, P. 1991. ITQs in New Zealand: The era of fixed quotas in perpetuity. *Fishery Bulletin* 90: 147–60.

Smith, M.E. 1990. Chaos in fisheries management. *MAST* 3: 1–13.

Southon, M. 1989. Competition and conflict in an Ecuadorian beachseine Fishery. *Human Organization* 48: 365–9.

Sullivan, K. 1987. Open entry fishery = tragedy of the commons. *Fisheries* 12: 2.

Thomas, J. 1973. *An Analysis of the Commercial Lobster (Homarus americanus)*

Along the Coast of Maine, August 1966 Through December 1970. NOAA
   Technical Report NMFS SSRF-667. Washington DC: US Department of
   Commerce, National Marine Fisheries Service.
Wahle, R.A. and Steneck, R.S. 1991. Recruitment habitats and nursery grounds
   of the American lobster (*Homarus americanus* Milne Edwards): a
   demographic bottleneck? *Marine Ecology Progress Series* 69: 231–43.
Wahle, R.A. and. Steneck, R.S. 1992. Habitat restrictions in early benthic life:
   experiments on habitat selection and in situ predation with the American
   lobster. *Journal of Experimental Marine Biology and Ecology* 157: 91–114.
Walsh, M.W. 1991. Peril for a fish – and a way of life. *Los Angeles Times* July 20
   110:A1.
Wieland, R. 1992. *Why People Catch Too Many Fish: A Discussion of Fishing and
   Economic Incentives.* Washington DC: Center for Marine Conservation.
Wilson, J.A., Acheson, J.M., Kleban, P. and Metcalfe, M. 1994. Chaos,
   complexity and community management of fisheries. *Marine Policy* 18:
   291–305.
Wilson, J.A., French, J., Kleban, P., McKay, S. and Townsend, R. 1991a. Chaotic
   dynamics in a multiple species fishery: a model of community predation.
   Ecological Modelling 58: 303–22.
Wilson, J.A., French, J., Kleban, P., McKay, S. and Townsend, R. 1991b. The
   management of chaotic fisheries: A bio-economic model. In *Proceedings
   From The Symposium on Multiple Species Fisheries,* pp.287–300, ed. M.
   Sissenwine, and N. Daan. Copenhagen: International Council for the
   Exploration of the Sea.
Wilson, J.A. and Kleban, P. 1992. Practical implication of chaos in fisheries.
   MAST 5: 67–75.

16

Ecological practices and social mechanisms for building resilience and sustainability

CARL FOLKE, FIKRET BERKES & JOHAN COLDING

Introduction

Several previous studies have shown that conventional scientific and technological approaches to resource and ecosystem management are not always working, and may indeed make the problem worse (e.g. Regier and Baskerville, 1986; Ludwig, Hilborn and Walters, 1993; Gunderson Holling and Light, 1995). Why this is so is not entirely clear, but may have to do with the focus on the wrong kinds of sustainability and on narrow types of scientific practice (see Chapter 13). The combination of powerful centralized institutions and functionally specialized divisions of labour is part of the ideology of conventional scientific practice. Not only does this scientific ideology hinder resource management reform, but the accompanying political and economic ideologies make adaptive social change difficult as well. It is not surprising that there is a widespread search for new approaches, with visions of smaller-scale, more environmentally sound and more democratic resource management systems which are more responsive, adaptive and resilient.

When we designed the project that led to the present volume, we knew that there was already considerable evidence of cultural capital pertaining to sustainable resource use and the maintenance of resilient ecosystems (Berkes and Folke, 1994). However, the evidence was scattered through a broad stream of multidisciplinary literature, and there was a need for systematic treatment. Hence, the major objective of the project and the book was to create a transdisciplinary framework through which we could evaluate examples of socially and culturally evolved management practices based on ecological knowledge and understanding, and the social mechanisms behind them.

Many of the cases in the book investigate resource management systems that incorporate both Western science and traditional or local practice, and their change over time. In casting our net wide to mobilize a wider range of considerations and sources of information than those used in conventional resource management, our broader long-term objective was to improve the management of selected ecosystems by learning from a diversity of management systems and their dynamics.

To focus discussion, we formulated three hypotheses in the introductory chapter. The first part of this synthesis chapter deals with the hypotheses, followed by two sections that deal with the question of how adaptiveness and resilience can be built into institutions so that they are capable of responding to the processes that contribute to ecosystem resilience. The two sections cover management practices and the social mechanisms behind these practices, respectively. These are followed by a section in which we ask if there are similarities, general patterns and principles that can be drawn from the cases. Based on these principles, some conclusions are then formulated and an attempt is made to identify some productive lines of inquiry to assist in the designing of more resilient and sustainable resource management systems.

The first hypothesis proposed that maintaining resilience was important for both resources and social institutions, and that the well-being of social and ecological systems is thus closely linked. Chapters in the volume provide examples of deep crises in conventional resource management systems (see Chapter 11 and 12). But the bulk of the chapters analyse social – ecological linkages, mainly in local community-based institutions. These chapters illustrate how adaptiveness and resilience have been built into institutions so that they are capable of responding to and managing processes, functions, dynamics and changes in a fashion that contribute to ecosystem resilience. Thus, the well-being of social and ecological systems seems to be closely linked. The fact that such linked social – ecological systems are found so widely, and have a track record often over a long period, suggests that they are highly adaptive.

In the introductory chapter it was also hypothesized that successful knowledge and resource management systems will allow disturbance to enter at a scale which does not disrupt the structure and functional performance of the ecosystem and the services it provides. Conventional resource management is predisposed to block out disturbance, which may be 'efficient' in a limited sense in the short term. But since disturbance is endogenous to the cyclic processes of ecosystem renewal (Holling *et al.*, 1995), conventional resource management tends to increase the potential for larger-scale disturbances and even less predictable and less manageable feedbacks from the environment. These feedbacks, or surprises, can have devastating effects on ecosystems and on societies that depend on the resources and services that ecosystems generate (see Chapter 12). As resilience or the buffering capacity of the system gradually declines, flexibility is lost, and the linked social–ecological system becomes more vulnerable to surprise and crisis (see Chapter 13).

Although crisis may be a necessary condition to provide the understanding and impetus for change (Gunderson *et al.*, 1995), it should not be allowed to build up to a level where it challenges the survival of the community, the region, or society as a whole. As we learned from the collapse of the Newfoundland cod fisheries (see Chapter 12), the cost of learning from such crises may be extremely high, and social and economic consequences severe.

Several of the management practices and associated social mechanisms identified in this volume prevent the build-up of large-scale crises. They allow disturbance to enter at a lower level and they build resilience. Thus, disturbance does not accumulate to challenge the existence of the whole social–ecological system. It seems that such social–ecological systems allow for internal renewal while maintaining overall structure (see Chapter 13). We believe that such adaptations have been made possible through management practices based on ecological understanding, and generated through a trial-and-error learning process. As was hypothesized in the introductory chapter, and as shown throughout this volume, there will be social mechanisms behind these management practices based on local ecological knowledge, as evidence of a co-evolutionary relationship between local institutions and the ecosystem in which they are located. The practices and mechanisms provide a reservoir of active adaptations in the real world which may be of universal importance in designing for sustainability.

However, for the social–ecological system to persist, the integrity of locally adapted systems in which the practices and mechansism are embedded needs to be protected (but not isolated) from external driving forces such as misguided macro-economic policies or trade opportunities that lead to unsustainable resource exploitation. Such support is often provided by umbrella institutions such as the tenurial shell for local community forest management created by the Mexican state (see Chapter 9), or through nested sets of institutions (Ostrom, 1990; Hanna, Folke and Mäler, 1996). Such nested systems of governance are best accomplished through a federal, not a hierarchical, approach (Ostrom and Schlager, 1996), as illustrated by the co-management process in Maine's soft shell clam fishery for the sharing of rights and responsibilities between the State of Maine and the local community (see Chapter 8). The results of the systematic search for such practices and mechanisms of local and nested institutions are presented in the next section.

Most of the social–ecological linkages synthesized below are drawn from the chapters of the book. They are derived from empirical observations through case studies from both contemporary and traditional resource management systems, and combinations thereof. The practices and mechanisms discussed are not separate phenomena, but interlinked with one another and co-evolving. For example, monitoring change in dryland ecosystems is closely related to the rotation of grazing areas and linked with the role of leaders or stewards and the use of ceremonies as a cultural code. Many social–ecological linkages, such as those discussed in this chapter have not previously been identified specifically for their role in the management of resources and ecosystems.

Management practices based on ecological knowledge

The practices based on ecological knowledge are presented in a sequence in Table 16.1, from monitoring and management of specific resources to management of landscape and watersheds as well as dynamic process and renewal of ecosystems at multiple scales. The table is by no means exhaustive. It is merely a starting point for the further identification of social – ecological linkages and their contribution to resilience and sustainability.

Monitoring the state of resources is a common practice among resource users, and virtually every case study in this volume provides local recipes and examples for monitoring. For example, the *caiçaras* of the Brazilian Amazon monitor the maturity of the trees to start the tapping of rubber (see Chapter 6). Icelandic fishermen communicate about the abundance of fish stocks (see Chapter 3), coastal communities in Maine monitor clam populations to help determine the areas needing enhancement (see Chapter 8), and herders of the Sahel monitor grazing pressure and the state of the pasture to make decisions about rotating or relocating herds (see Chapter 10).

Total protection of certain species is common among traditional societies. The reasons for protection varies from avoiding species that are poisionous or are used for medicinal purposes (see Chapter 6) to preserving keystone species in the ecosystem (Colding and Folke, 1997).

Several practices involve the protection of *vulnerable life-history stages* of a variety of species (Johannes, 1978). For example, there are prohibitions on catching lobsters with eggs in the Maine fisheries (see Chapter 15). In south India, many wading birds are hunted outside the breeding season,

Table 16.1. *Social–ecological practices and mechanisms for resilience and sustainability*

1. Management practices based on ecological knowledge
Monitoring change in ecosystems and in resource abundance
Total protection of certain species
Protection of vulnerable stages in the life-history of species
Protection of specific habitats
Temporal restrictions of harvest
Multiple species and integrated management
Resource rotation
Management of succession
Management of landscape patchiness
Watershed management
Managing ecological processes at multiple scales
Responding to and managing pulses and surprises
Nurturing sources of renewal

2. Social mechanisms behind management practices

a) Generation, accumulation and transmission of ecological knowledge
    Re-interpreting signals for learning
    Revival of local knowledge
    Knowledge carriers/folklore
    Integration of knowledge
    Intergenerational transmission of knowledge
    Geographical transfer of knowledge

b) Structure and dynamics of institutions
    Role of stewards/wise people
    Community assessments
    Cross-scale institutions
    Taboos and regulations
    Social and cultural sanctions
    Coping mechanisms; short-term responses to surprises
    Ability to re-organize under changing circumstances
    Incipient institutions

c) Mechanisms for cultural internalization
    Rituals, ceremonies and other traditions
    Coding or scripts as a cultural blueprint

d) Worldview and cultural values
    Sharing, generosity, reciprocity, redistribution, respect, patience, humility

but not at heronaries, which may be on trees in the middle of a village (Gadgil, Berkes and Folke, 1993).

Sacred groves, often a small part of a forest set aside for spiritual or religious purposes, were once widely used for the *protection of habitat,* and continue to be important in many areas. The tribal state of Mizoram in northeastern India has re-established such habitats, now calling them 'safety forests' (see Chapter 2). Even small sacred groves may be surprisingly effective in conserving biodiversity. A botanical survey in a Nigerian sacred grove yielded 330 plant species as compared to only 23 in the non-protected areas (see Chapter 7). Sacred groves are not the only example of culturally protected habitats. Chapter 10 describes the use of buffer areas of rangelands normally protected from grazing except in the case of emergencies. Gadgil *et al.* suggest that traditional conservation practices in relation to refugia might have originated to serve secular functions although they were implemented through systems of social sanctions based on religious beliefs and social conventions. A version of a refugia are the extractive reserves which include concerns to biological and cultural diversity, as well as access and equitability in the use of resources by the commoners (see Chapter 6).

Temporal restrictions of harvest is a common practice of many modern systems of management (see Chapters 8 and 15). It is also used in some traditional management systems, for example among many groups of African herders (see Chapter 10), and groups of Canadian Amerindian hunters, where hunting areas are 'rested' so that the animals can replenish themselves (see Chapter 5). In the Hindu-Kush Himalayas, there are seasonal and periodic restrictions on product gathering from the village commons (see Chapter 11).

Many traditional and contemporary systems apply *multiple species management,* including integrated farming and cultivation systems. The Nigerian case study (see Chapter 7) identifies an agroforestry system combining food crops and domesticated trees as the oldest farming practice. Since the beginning of the twentieth century, this agroforestry system has become a perennial mixed plantation that includes cash crops such as cocoa, oil palm and coffee. Chapters 9 and 11 both describe mixed cultivation systems in which some of the species maintain ecosystem structure and function. The *milpa* system, for example, uses tools and techniques that support the processes and functions of the agroforest ecosystem. In the Himalayas, the reduced use of deep-rooted species is resulting in an increase in the frequency of mud slides on unstable slopes. The wholistic forestry of British Columbia aims to conserve the structure and function

420   Carl Folke, Fikret Berkes & Johan Colding

of forest ecosystems by a variety of means, including maintenance of hardwood and coniferous tree species (so that species such as alder can help fix
nitrogen for the conifers), and management of a diversity of tree species
and age-groups that reduces insect outbreaks (see Chapter 14).

Rotation is another management practice. Chisasibi Cree hunters rotate
trapping areas (ideally) on a four-year cycle to allow populations of beaver
to recover. They use a similar rotational technique for fishing areas, using
the declining catch per unit of effort as the feedback that informs decision-
making (see Chapter 5). In semi-arid regions such as the fringe of Sahel,
plant productivity is seasonal and follows the rains. Many of the larger herbivores, as well the traditional cattle herders, have adapted to this pattern
by migrating seasonally. The yearly cycle of nomads and their cattle provides a rotational management system, enabling the recovery of heavily
grazed rangelands. Throughout arid and semi-arid Africa, grazing land
was rotated seasonally and, in some cases, adjacent grazing areas were
rotated in the same season as well (see Chapter 10). Local communities
enforce grazing rotation in parts of the Hindu-Kush Himalayas (see
Chapter 11).

Management of succession is examplified by the shifting cultivation
system (milpa), as used in tropical Mexico. This system is well adapted to
make ecologically sound use of tropical moist forest. While crops are
growing, the regenerating vegetation is renewing the site for the next milpa
cycle, and many of the regrowth species in old cornfields will eventually
become trees that provide firewood, construction materials, dyes, craft
materials, canoe bodies, medicine, and other resources. Agriculture
becomes a sequential cropping of crops and non-crops (see Chapter 9).
There are many similar agroforestry systems elsewhere that actively manage
succession (see, for example, Berkes, Folke and Gadgil, 1995).

The use of landscape patchiness is described in Chapter 10. The small-
scale movements of Sahelian herders were designed to mimic the variability and unpredictability of the landscape. Pastoralists diversified by having
an appropriate mix of animal species in the herd to utilize different vegetation types and patches. In a contemporary adaptation of traditional
herding rules, the Maasai of Kenya progressively widen the radius of
grazing around wells as the wet season advances, so as to leave enough
forage around the wells for the dry season. Chapter 4 hypothesizes that the
distribution of small-sized agricultural parcels is an adaptation for the
utilization of multiple ecological zones in the landscape. A similar situation
may also exist in the Himalayas, where ecological patchiness exist by elevation zones (see Chapter 11).

Southeast Asia and Oceania had, and to some extent still have, a wealth of pre-scientific *watershed management* systems. Examples include ancient Hawaiian *ahupua'a* (Costa-Pierce, 1987), the Yap *tabinau*, the Fijian *vanua*, and the Solomon Islands *puava* (Ruddle, Hviding and Johannes, 1992). In ancient Hawaii, whole river valleys were managed as integrated farming systems, from the upland forest all the way to the coral reef. Similarly, in the Solomon Islands, a *puava* in the widest sense includes all resources and ecosystems in a watershed, from the top of the mainland mountains to the open sea outside the barrier reef (Hviding, 1990).

There is some evidence that locally devised systems may be *managing ecological processes at multiple scales*. *Milpa* succession, as described in Chapter 9, may be seen as managing food crops on a one- to three-year scale, while managing other crops and products on a 30-year scale. The James Bay Cree hunters in Chapter 5 seemed to be managing simultaneously beaver populations on a 4–6-year scale, fish on a 5–10-year scale, and caribou on a 80–100-year scale. Chapters 10 and 14 also describe management at multiple spatial scales. The wholistic forestry in Chapter 14 is concerned not only with the production of fibre over several square kilometres, but also with the maintenance of ecological processes involving soil bacteria at the spatial scale of a few square metres. In the case of African herders, Chapter 10 recognizes two different sets of practices and rules for the larger-scale movements (macro-mobility) and the smaller-scale movements (micro-mobility).

An example of *responding to and managing disturbances and surprises* is the establishment of range reserves within the annual grazing areas of African herders. These reserves provide a 'savings bank' of forage, maintaining the resilience of both the ecosystem and the social system of the herders. These areas serve as buffers when disturbance, such as drought, challenges the processes and functions of the system (see Chapter 10). Sacred groves in India absorb disturbance by serving as firebreaks for cultivated areas and villages (see Chapter 2). There are other examples in the literature about the ability of traditional peoples to manage disturbances. Lewis and Ferguson's (1988) work from Canada, Australia and California showed that aboriginal peoples possessed detailed technical knowledge of fire, and used it effectively to improve feeding habitats for game and to assist in the hunt itself.

Disturbances triggered by events like fire, wind and herbivores are an inherent part of the internal dynamics of ecosystems, and often set the timing of successional stages (Holling *et al.*, 1995). They create the opportunity for renewal. Many traditional societies *nurture sources of ecosystem renewal* by creating small-scale disturbances. Methods of traditional agroforestry, like those of the *milpa* script, enable people to produce a crop of maize without seriously disrupting the renewal of the forest (see Chapter 9). Some, like the nomads in Chapter 14, behave like a disturbance by following the migratory cycles of the herbivores from one area to another. The pulses of grazing by herbivores contribute to the capacity of the semi-arid grasslands of Africa to function under a wide range of climatic conditions. If this capacity is reduced, an event that previously could be absorbed can flip the grassland ecosystem into a state which is dominated and controlled by woody shrubs (e.g. Walker, 1993).

Social mechanisms behind management practices

We have organized the sequence of social mechanisms as a hierarchy that proceeds from ecological knowledge to worldviews. Institutions, in the sense of rules-in-use, provide the means by which societies can act on their knowledge and use it to produce a livelihood from the resources in their environment. Both knowledge and institutions require mechanisms for cultural internalization, so that learning can be encoded and remembered by the social group. Worldview or cosmology gives shape to cultural values, ethics, and basic norms and rules of a society.

Generation, accumulation, and transmission of ecological knowledge

The evolution of the Cree caribou hunting system, following a resource crisis, illustrates how a society can *reinterpret signals for learning.* The crisis caused by the disappearance of caribou in the 1910s triggered learning, and the redesign of the management system was encoded in ethical and cultural beliefs of the Cree (see Chapter 5). Another example is provided in Chapter 12 about Newfoundland cod fisheries. Inshore fishers, who had traditionally seen failures in the fishery as natural and transient, began to realize, when the offshore fishery escalated, that failure could be caused by fishing itself. The irony of the case is that the inshore fishers were unable to convince the managers of the impending crisis; the managers were preoccupied with the offshore fishery and completely missed the signals that the inshore fishers were learning from – until the entire stock collapsed.

The *revival of ecological knowledge* requires strong traditions (Chapin, 1991), and resource management institutions. The re-establishment of beaver management rules in the 1950s by the Cree (see Chapter 5) provides an example of the revival of local ecological knowledge for restoring a resource population. In the absense of strong traditions, the redevelopment of refugia in many parts of India, based on traditional ecological knowledge, may require monetary or other incentives (see Chapter 2). Traditional local values and ecological knowledge are being revived through the wholistic forestry among the Gitksan of British Columbia (see Chapter 14).

Folklore and knowledge carriers help maintain ecologically sound management practices. For example, tales of a maize culture hero are associated with all stages of the *milpa* agroforestry system: the hero warns people of impending doom if they stop making *milpa* properly (see Chapter 10). Among the Cree, the elders' words have a profound effect on the younger hunters, as illustrated in the caribou story (see Chapter 5). The hunters' guild of the Yoruba functions as a knowledge carrier to maintain ancient traditions and indigenous ecological knowledge (see Chapter 7), and the Icelandic fishermen serve as carriers of practical knowledge (see Chapter 3).

Several of the chapters give examples of the *integration of knowledge*. The Maine's soft shell clam fishery is characterized by the integration of informal local knowledge and formal scientific information, generated locally (see Chapter 8). In Chapter 6, Begossi argues that the mix of traditional and new knowledge of the *caiçaras* and *caboclos* increases the resilience of their social–ecological systems. A two-way knowledge transfer between fishermen and biologists in Iceland has the potential to improve the management of the fish stocks (see Chapter 3).

Intergenerational transmission of ecological knowledge is common among traditional societies and is embedded in the social and cultural system. An example of such a transmission is the *milpa* script, which is passed on to children and sustained by cultural beliefs, mythologies and yearly festivals (see Chapter 9). Cases presented in Chapters 3, 5, 6 and 7, and others, also involve intergenerational transmission of ecological knowledge.

Free information exchange on the productivity and potential of the rangeland ecosystem between different groups and tribes of African pastoralists (see Chapter 10) illustrates the existence of *geographical transfer of ecological knowledge* between local communities. There is good evidence that learning and redesign of beaver management diffused across the Canadian subarctic in the 1800s (see Chapter 5). The similarity of parametric management in some 30 fishing societies throughout the world indicates geographic transfer of knowledge (see Chapter 15). Johannes' (1978) detailed study of reef and lagoon fishery management in Oceania shows the pervasiveness of knowledge diffusion inferred through striking similarities across island groups in the basic design of the management system. It is the elaboration of the basic design that apparently has given rise to diversity.

Structure and dynamics of institutions

The story in Chapter 14 of how a clan wing chief developed and pursued his vision of the future Gitksan forest is a telling case of the *role of stewards/wise people*. The collective leadership of stewards is the key common-property resource management institution among the Cree. It is common that a hunt leader acts as a steward of the resources on behalf of the community, as well as a social and, in the old days, also a spiritual leader (see Chapter 5). Similarly, the Ara hunters are custodians for their sacred areas as well as for many communal property areas. The traditional guild of Ara hunters is headed by the chief of the hunters, and guided by an Ogun priest who performs ritual duties (see Chapter 7).

Community assessments are common in both traditional and contemporary society. Keesing (1981) gives the name action or task group to people who gather in some more organized fashion to perform a common task. Many tribal task groups engage once a year in a large-scale communal hunt, a group-level assessment. Such a group exercise may have served the purpose of evaluating the status of prey populations and their habitats. This in turn may have helped in continually adjusting resource harvest practices for resilience (Gadgil *et al.*, 1993). In the Maine clam fishery time, and effort to develop and implement management plans are proportionally shared by the major beneficiaries of the resource through inclusion of users in resource surveys and other assessments and through rotating membership on shellfish conservation committees (see Chapter 8).

The chapters provide several examples of *cross-scale institutions* – those that operate at more than one level. The tenurial shell created by the Mexican state supports the traditional belief structure of the Huastec, which in turn supports ecologically sustainable land use (see Chapter 9). The re-establishment by the Cree of resource management institutions for beaver was backed by government regulations (see Chapter 5). In the Maine soft shell clam fisheries, the bundle of rights moving from the citizen to state level is nested in ascending levels of authority (see Chapter 8). The territorial rights of tribes, sub-tribes and clans in south-central Sudan is another example of nested institutions (see Chapter 10).

Taboos and other regulations are critical social mechanisms. Various kinds of taboos have been recognized to be ecologically functional and to have the potential of building resilience in ecosystems (e.g. Rappaport,

1967; Harris, 1971; 1979; Johannes, 1978; 1981; Chapman, 1985; 1987; Gadgil, 1987). Food taboos on game and fish are part of *caboclos* and *caiçaras* cultures, where species are avoided due to toxic, medicinal or ecological reasons (see Chapter 6). In pre-colonial Ara, there were sacred forests and sacred trees of various types, just like in India (see Chapter 2) and elsewhere. The forests were believed to be occupied by spirits and the trees were presumed to be inhabited by ghosts, and their use was forbidden by taboos (see Chapter 7). A characteristic regulation in many traditional and many contemporary communities is to manage resources not by managing numbers but by managing the resource and ecosystem through social conduct (see Chapter 15). For example, pastoral herders in Africa manipulate the stocking rate through rules that track ecological dynamics and control micro-mobility (see Chapter 10). Chapter 12 shows the mismatch between regulations and proper management of ecosystem processes and functions, a mismatch that is common in many parts of modern society (see also Ludwig *et al.*, 1993; Gunderson *et al.*, 1995).

Social and religious sanctions are addressed in several of the chapters of this book. The Gitksan of British Columbia sanction those who do not follow the norms and rules of the community by questioning their right to their Gitksan name and social status (see Chapter 14). In the Maine lobster fishery, one must become a member of a 'harbor gang' to participate in the fishery. Members are expected to obey the rules of their group concerning fishing practices, and a person who violates the rules will be severely sanctioned. The territorial system, which operates to limit the number of fishers exploiting lobsters in each territory, helps conserve the lobster resource (see Chapter 15). Other social sanctions include accusations of witchcraft against those who attempt to appropriate community resources for private gain. In the Yoruba religion, supernatural forces provided the basis upon which the Alara and his chiefs sanctioned the undesired behaviour of inhabitants (see Chapter 7). Among Mesoamericans, traditional curers reinforce socially appropriate behaviour during their interactions with patients, by relating illness to misuse of resources and other misbehaviour (see Chapter 9). Sanctions related to religious beliefs are common in this context (see Chapter 6).

There are *coping mechanisms* to respond to surprises. Coping strategies are differentiated from adaptive strategies on the basis of the time-scale of response, the level of vulnerability, and the type of risk faced by households and communities. Coping strategies tend to be short-term responses in abnormal periods of stress. Berkes, Duffield and Ham (1996) report from Himalayan livelihood systems that a household which is able to turn to a number of coping strategies in times of stress (e.g. the number of alternative short-term agricultural opportunities, handicraft manufacturing, and tourist-guiding jobs that young men took on following the failure of a major crop) will weather contingencies more successfully than those which have only one option to pursue (e.g. selling the only cow of the household). Further, the continued availability of a range of coping strategies may be necessary for livelihood strategies to remain adaptive in the long term.

In both traditional societies and contemporary resource management agencies, there seems to be an *ability to reorganize under changing circumstances*. For example, tighter management rules seem to emerge in some cases following an intensification of resource use or decrease in resource supply. Examples include the emergence of tightly regulated, family-based hunting territories in the last 200 years in James Bay in place of loosely regulated community-based territories (Berkes, 1989), and the emergence of a system of 'cascading property rights' for water use in Florida in the droughts of the 1970s and 1981–82 (Light, 1983).

There are also *incipient institutions* with rules that seem to 'kick in' following certain kinds of stresses. The 'sleeping territoriality' of some Pacific Islands, which serves a blueprint of property rights that is activated when fisheries resources are becoming scarce, provides an example of an incipient institution (Edvard Hviding, University of Bergen, personal communication).

Mechanisms for cultural internalization

Rituals, ceremonies and other traditions provide examples of mechanisms for cultural internalization. Rituals help people remember the rules and interpret signals from the environment appropriately (see Chapter 5). Chapin (1991) states that where traditions remain strong, people see no need to preserve esoteric knowledge, they simply practise their culture. Chapter 9 discusses how religious institutions reinforce community cohesion in both indigenous and mestizo rural communities across Mexico. Ritual obligations, rights to community resources, and obligations to manage those resources are linked. Other examples of resource management systems with interlinked rituals are the shifting agriculture, or *jhum* system, of north-east Indian tribal societies (Ramakrishnan and Patnaik, 1992), and the 'first-salmon ceremony' of several northern California tribes through Oregon, Washington and British Columbia (Child and Child, 1993).

Coding of scripts can function *as cultural blueprints*. The *milpa* system is a cultural script, an internalized plan consisting of a series of routine steps with alternative subroutines, decision nodes, and room for experimentation. Ecological knowledge is encoded in the local variation of the *milpa* script, derived from experiences and experiments of farmers over generations. The making of *milpa* is the central, most sacred act, one which binds together the family, the community, the ecosystems, the universe (see Chapter 9). The authors argue that the strong cultural support buffers the script from disruption by new economic demands, introduction of new technologies, or other changes. The partial inheritance system of Dalecarlia, Sweden, that embedded ecological knowledge and social equality, also had a strong cultural support and survived external pressures for several centuries (see Chapter 4).

Worldview and cultural values

It is often assumed that 'our' system of acquiring knowledge, through the application of Western scientific principles, is a universal epistemology (Redclift, 1994). In fact, we know from the cases in the book that there are *multiple epistemologies*, different ethical positions with respect to the environment, and different cultural traditions in the perception of ecosystems and resources. To extend our consideration of the range of resource management alternatives requires an openness to different epistemologies and cultural traditions, and the worldview behind them. Here we concentrate on cultural values as expressions of worldview.

Cultural values, such as sharing, generosity, reciprocity and redistribution of resources, are characteristics of traditional societies and important also in contemporary common-property systems. For example, stewards among the Cree can regulate access but cannot exclude other community members from using the land for their basic needs. High-status hunters are expected to be generous and to share the products of that land (see Chapter 5). Similarly among the Huastec, members of one family have the right to ask another family to borrow land or harvest forest products to meet their subsistence needs. They also share rights to medicinal plants. Equitable distribution of community land and water resources prevents overexploitation, while communally shared core values and institutions maintain the community's overall resource use within acceptable bounds (see Chapter 9). The system of fictive kinship among Mesoamericans reinforces bonds of reciprocity. Among pastoralists, rights to use territories belonging to others are exercised when tribes know that in the future they can return the favour (see Chapter 10).

428   *Carl Folke, Fikret Berkes & Johan Colding*

Reciprocity is also found in contemporary communities. The institutional structure and the co-management process of the Maine soft shell clam fishery are based on co-operative interactions and mutual give-and-take. The values of co-operation and reciprocity, rooted in tradition, are at the heart of negative attitudes toward private leasing of the intertidal area or limiting access of community residents. Just as in several of the local systems described in Chapter 15 the use of the clam resource is managed through indirect measures such as open and closed areas, open and closed seasons, and enhancement of production, a preference that appears to be based in the traditional desire to keep resource access open to all members of the community (see Chapter 8). Another example of reciprocity is the co-operation between small-scale fishers, biologists and local politicians on the Vestman Islands, Iceland, united in an effort to remove destructive fishing technology and reduce catches of overexploited fish stocks (see Chapter 3).

Expression of worldview through respect, patience and humility; and people being viewed as a part of nature are common in traditional communities. The Lax'skiik and Gitksan of British Columbia, in general, have a personal and spiritual identification with their territories and resources, which form the basis of their cultural and economic life. The fate of the land parallels their own fate, and this relationship forms the basis for their management of the land and its resources (see Chapter 14). The *caiçaras* in Brazil believe in forest guardians; in spirits that protect animals and spirits of the water who punish those who fish too much (see Chapter 6). Among the Huastec of Mexico, the real owners of the land and forest are divine beings and spirits (including ancestors), and the Earth is a member of the community. The community has the obligation to treat the Earth and all community members with respect and concern for their continued well-being. In the Cree view of the natural world, human–animal relationships continue to be a spiritual and religious matter, and there is reciprocity between the hunter and the animal. As seen in the caribou and beaver cases described in Chapter 5, limiting the harvest and avoidance of waste are two of the main ways, along with various rituals, in which respect is shown.

General patterns and principles

The basic assumption behind the work in this volume is that resource and ecosystem management is necessary but that it requires fundamentally different approaches, not mere tinkering with current models and practices.

As was made clear in the introductory chapter, the volume seeks to integrate two streams of resource management thought that differ from the classic utilitarian approach. The first involves rethinking resource management science in a world of complex systems with non-linear relationships, thresholds, uncertainty and surprise, using systems approach and adaptive management (Holling, 1978; Walters, 1986). The second involves rethinking resource management social science by focusing on property rights institutions (Hanna *et al.*, 1996), and in particular common-property systems (McCay and Acheson, 1987; Berkes, 1989; Ostrom, 1990; Bromley, 1992).

The combination of those two streams of resource management, and the systematic treatment of practices and their social mechanisms, have resulted in new insights concerning social–ecological linkages that contribute to building resilience (see Table 16.1). First we identified a diversity of management practices based on ecological knowledge. The practices ranged from monitoring and managing specific resources, to ecologically sophisticated practices that respond to and manage disturbance and build resilience across scales. Many previous studies of indigenous knowledge have emphasized its value, for example for medicinal purposes, and for biodiversity conservation through genetic diversity and biotechnology. The emphasis in this book has been on the role of indigenous knowledge in responding to and managing processes and functions of complex systems. Ecosystems and biological diversity are managed to secure a flow of natural resources and ecological services on which the local social–ecological system depends.

Secondly, a number of social mechanisms behind these practices, were identified and organized in a sequence from the generation, accumulation and transmission of ecological knowledge to the underlying values of a given culture in which that knowledge is embedded. Presumably, the social mechanisms identified here represent only a tiny fraction of the reservoir of human–environmental adaptations. In Table 16.2, a set of new or 'rediscovered' principles of resource and environmental management is proposed, emerging from the chapters of this book. The following generalizations appear to be conducive for building resilience and sustainability.

Using management practices based on local ecological knowledge. Such management practices include protection of certain species, stages in their life history, and habitats; multiple species and multiple scale management of ecosystem processes; temporal restriction of harvest,

Table 16.2. *Principles drawn from local social–ecological systems for building resilience*

Using management practices based on local ecological knowledge
Designing management systems that 'flow with nature'
Developing local ecological knowledge for understanding cycles of natural and
  unpredictable events.
Enhancing social mechanisms for building resilience
Promoting conditions for self-organization and institutional learning
Re-discovering adaptive management
Developing values consistent with resilient and sustainable social-ecological
  systems

---

resource switching and rotation; management of landscape patchiness; different phases of ecosystem development; and whole watershed management.

Designing management systems that 'flow with nature'. Allowing disturbances to act on the social–ecological system and managing environmental feedbacks that signal disturbance prevent the accumulation of disturbance that may otherwise lead to large-scale unpredictable events that challenge the well-being and even existence of the system.

Developing local ecological knowledge for understanding cycles of natural and unpredictable events. Society must be able to respond to and manage predictable changes (pulses) and unpredictable changes (surprises) in ecosystem processes and functions to secure a flow of natural resources and ecological services

Enhancing social mechanisms for building resilience. There must be social mechanisms in a society by which information from the environment may be received, processed and interpreted to build resilience of the linked social–ecological system. These include mechanisms for the generation, accumulatation, and transmission of ecological knowledge; the existence of a diversity of dynamic institutions; mechanisms for cultural internalization; cultural values conducive to sustainability; and modern society equivalents of such traditional mechanisms as rituals, taboos, and social and religious sanctions.

Promoting conditions for self-organization and institutional learning. All social and ecological systems have the capacity for self-organization. In linked social–ecological systems, there exists a co-evolutionary relationship between local institutions and the ecosystems in which they are located. Due to self-organization through mutual entrainment (co-evolution), many events that appear as surprises in conventional resource management may be well-known phenomena in linked social–ecological systems. Many local institutions learn how to respond to environmental feedbacks and to build resilience. The lack of such learning in some centralized conventional resource management systems results in ecological illiteracy.

Rediscovering adaptive management. Adaptive management can be seen as a rediscovery of dynamic practices and institutions already existing in some traditional systems of knowledge and management, and to some extent in contemporary local communities. Drawing on management practices based on ecological knowledge and understanding the social mechanisms behind the development of them may speed up the process of adaptive management.

Developing values consistent with resilient and sustainable social – ecological systems. A precondition for the above principles is the development of values that will facilitate sustainability. The necessary worldview underlying these values and ethics would be one which is in tune with an ecologically sustainable societal development and resilient resource management.

Conclusions

A number of social and environmental movements share the vision of a smaller-scale, community-based, more environmentally sound, and more socially sound future. These visions reflect a fundamental questioning of many current institutions and ideologies in the search for a more sustainable social–ecological system. We do not go so far as to suggest the dismantling of all centralized 'big science' and technology in a fit of post-modernism! We do, however, criticize the utilitarian slant of conventional resource management and its reductionistic excesses. Modernist forestry valued only the utilitarian kind of forestry, when, as McQuillan (1993) put it, forestry 'has in fact an entire spectrum of genres'. In addition to rediscovering those spectra of alternative approaches and uses, we also propose extending the range of 'science' to recognize the value of traditional and neo-traditional or otherwise local resource management systems.

The questions of scale, in dimensions of space and time, are related to the issue of the restructuring and decentralizing of large, centralized institutions. Cases in this book show, time and again, that local-level institutions learn and develop the capability to respond to environmental feedbacks faster than do centralized agencies. In building centralized bureaucracies for environmental management over the years, we have assumed that resource management can be scaled up; that is, we can apply the propositions or models of micro-scale systems to macro-scale ones. That does not seem to be the case. As Young (1995) pointed out, 'macro-scale systems are not merely small-scale systems writ large. Nor are micro-scale systems mere microcosms of large-scale systems'.

One can also err in the opposite direction. If we were to claim that all centralized management institutions could be replaced by community-level institutions, we would be assuming that resource management could be scaled down. The evidence from the cases support the proposition in Chapter 13 that environmental and renewable resource issues tend to be neither small scale or large scale but *cross-scale* in both space and time. It follows, therefore, that the problems have to be tackled simultaneously at several levels. Thus the power of centralized management agencies should be redistributed and balanced, not eliminated. Co-management systems and the decentralization of power in a nested framework seem to offer a good compromise for dealing with cross-scale effects. For example, the communities of clam diggers in Chapter 8 know the local environment best and can formulate their own rules. But the series of communities of clam harvesters need to be co-ordinated as well.

If management is too centralized, valuable information from the resource, in the form of feedbacks, may be delayed or lost because of the mismatch in scale (e.g. population biologists in their offices in the big city are cut off from the daily reality of Newfoundland inshore fishers). If management is decentralized too much, then the feedback between the user-groups of different resources or between adjacent areas may be lost. Environmental disturbances in one area (e.g. mountain erosion) generates feedbacks somewhere else (e.g. the lower watershed). In the contemporary world of integrated markets and international environmental problems, the need for co-ordination at the higher end of the scale is greater than ever before. For example, fishery management needs to be controlled at the local level through property rights and social sanctions, with emphasis on 'how' people fish rather than how much (see Chapter 15). The problem is that these types of restrictions are usually not enough in most modern commercial fisheries, in which the fishing units may be subject to international market pressures. Throughout the world, such fisheries tend to be too large and too efficient for the fish stock they are exploiting. Thus larger-scale controls, nationally and internationally, including quota management, are often necessary (Hilborn and Walters, 1992; Hilborn and Gunderson, 1996).

The research agenda for the next decade will need to include theoretical and empirical studies that focus on institutional needs for the management of cross-scale problems. More systematic information is needed about co-management, reasons for successes and failures, institution building, use of adaptive management, and the design of supportive policies. The research agenda will also include more work on the resilience of social systems and the avoidance of ecological surprises (King, 1995), social and institutional learning (Gunderson *et al.*, 1995), and the study of renewal cycles in social systems (Holling and Sanderson, 1996).

Holling (1986) observed that institutions, like ecosystems, can become 'brittle' over time, and resource crises can result in release and re-organization (Gunderson *et al.*, 1995). These observations can lead to new empirical and theoretical work on linkages between social and ecological systems, and to the question of just what produces adaptive capacity in linked social–ecological systems, the sources of adaptive resilience, and the conditions under which the complex process of co-evolution of the coupled social–ecological system either build (create harmony) or erode (create dissonance) resilience. Regarding the renewal cycle of Holling, interesting questions include the sources and sinks of novelty and the role of wisdom. 'Novelty', a little known process in both social and ecological systems, seems to occur in rapid phases of re-organization, in transient moments, and may play a key role in the development of adaptive capacity – and so may wisdom which is valued highly in traditional societies. These issues are not merely academic; a great deal is at stake.

A society's survival is ultimately dependent on the finite capacity of ecosystems to support it with essential resources and ecological services. To secure human well-being, there is an urgent need to design institutions that safeguard this dynamic capacity. Conventional resource management has been successful in producing yields and economic growth in the short term, but has not been very successful in safeguarding the dynamic capacity of ecosystems or in managing ecological and social systems for resilience and sustainability.

Linked social–ecological systems, such as those studied in this volume have developed the ability to respond to changes and to adapt to them in an active way, not because people in these societies were 'noble savages and possessed primitive harmony' (Edgerton, 1992), but because such adaptations were key to survival. It is well known that self-regulatory mechanisms for survival tend to evolve in societies when they are faced with resource limitations (Gadgil *et al.*, 1993).

434   *Carl Folke, Fikret Berkes & Johan Colding*

The social–ecological practices, mechanisms and principles identified in this volume have the potential to improve conventional resource management by providing: insights for designing adaptive resource management systems that flow with nature; novel approaches to forestry, agriculture, fisheries, aquaculture and freshwater management; lessons for developing systems of social sanctions and successful implementation and enforcement of sustainable practices; means to avoid surprises caused by conventional resource management; experiences in managing fluctuations and disturbance, and to build resilience for sustainability. Learning from local social–ecological systems and combining insights gained in adaptive management in Western science may counteract many of the prevailing crises of conventional resource management.

References

Berkes, F., ed. 1989. *Common Property Resources: Ecology and Community-Based Sustainable Development*. London: Belhaven.

Berkes, F., Duffield, C. and Ham, L. 1996. *Livelihood Systems, Adaptive Strategies and Sustainability Indicators in the Western Indian Himalaya*. Paper Presented at the Voices from the Commons Conference, University of California, Berkeley, June 1996.

Berkes, F. and Folke, C. 1994. Investing in cultural capital for the sustainable use of natural capital. In *Investing in Natural Capital*, pp.128–49, ed. A.M. Jansson, M. Hammer, C. Folke, and R. Costanza. Washington, DC: Island Press.

Berkes, F., Folke, C. and Gadgil, M. 1995. Traditional ecological knowledge, biodiversity, resilience and sustainability. In *Biodiversity Conservation: Policy Issues and Options*, pp.281–99, ed. C. Perrings, K.-G. Mäler, C. Folke, C.S. Holling, and B.-O. Jansson. Dordrecht: Kluwer Academic Publishers.

Bromley, D.W. 1992. *Making the Commons Work: Theory, Practice, and Policy*. San Francisco: Institute for Contemporary Studies.

Chapin, M. 1991. Losing the way of the Great Father. *New Scientist*, 10 August: 40–44.

Chapman, M. 1985. Environmental influences on the development of traditional conservation in the South Pacific region. *Environmental Conservation* 12: 217–30.

Chapman, M. 1987. Traditional political structure and conservation in Oceania. *Ambio* 16: 201–05.

Child, A.B. and Child, I.L. 1993. *Religion and Magic in the Life of Traditional Peoples*. Englewood Cliffs, NJ: Prentice Hall.

Colding, J. and Folke, C. 1997. The relation between threatened species, their protection, and taboos. *Conservation Ecology 1*, June 15.

Costa-Pierce, B.A. 1987. Aquaculture in ancient Hawaii. *BioScience* 37: 320–30.

Edgerton, R.B. 1992. *Sick Societies: Challenging the Myth of Primitive Harmony*. New York: The Free Press.

Gadgil, M. 1987. Diversity: cultural and biological. *Trends in Ecology and Evolution* 2: 369–73.

Gadgil, M., Berkes, F. and Folke, C. 1993. Indigenous knowledge for biodiversity conservation. *Ambio* 22: 151–56.

Gunderson, L., Holling, C.S. and Light, S., eds. 1995. *Barriers and Bridges to the Renewal of Ecosystems and Institutions.* New York: Columbia University Press.

Hanna, S., Folke, C. and Maler, K.-G., eds. 1996. *Rights to Nature.* Washington DC: Island Press.

Harris, M. 1971. *Culture, Man, and Nature. An introduction to General Anthropology.* New York: Thomas Y. Crowell Company.

Harris, M. 1979. *Cultural Materialism: The Struggle for a Science of Culture.* New York: Random House.

Hilborn, R. and Gunderson, D. 1996. Chaos and paradigms for fisheries management. *Marine Policy* 20: 87–9.

Hilborn, R. and Walters, C.J. 1992. *Quantitative Fisheries Stock Assessment.* New York: Chapman and Hall.

Holling, C.S. 1978. *Adaptive Environmental Assessment and Management.* London: Wiley.

Holling, C.S. 1986. The resilience of terrestrial ecosystems: local surprise and global change. In *Sustainable Development of the Biosphere,* pp.292–317, ed. W.C. Clark and R.E. Munn. Cambridge: Cambridge University Press.

Holling, C.S. and Sanderson, S. 1996. Dynamics of (dis)harmony in ecological and social systems. In *Rights to Nature,* pp.57–85, ed. S. Hanna, C. Folke, and K.-G. Mäler. Washington DC: Island Press.

Holling, C.S., Schindler, D.W., Walker, B.W. and Roughgarden, J. 1995. Biodiversity in the functioning of ecosystems: An ecological synthesis. In *Biodiversity Loss: Economic and Ecological Issues,* pp.44–83, ed. C. Perrings, K.-G. Mäler, C. Folke, C.S. Holling, and B.-O. Jansson. Cambridge: Cambridge University Press.

Hviding, E. 1990. Keeping the sea: aspects of marine tenure in Marovo lagoon, Solomon Islands. In *Traditional Marine Resource Management in the Pacific Basin: An Anthology,* pp.7–44, ed. K. Ruddle, and R.E. Johannes. Jakarta: Unesco/ROSTSEA.

Johannes, R.E. 1978. Traditional marine conservation methods in Oceania and their demise. *Annual Review of Ecology and Systematics* 9: 349–64.

Johannes, R.E. 1981. *Words of the Lagoon. Fishing and Marine Lore in the Paulau District of Micronesia.* Berkeley: University of California Press.

Keesing, R.M. 1981.*Cultural Anthropology: A Contemporary Perspective.* Holt, New York: Reinhart and Wilson.

King, A. 1995. Avoiding ecological surprise: Lessons from long-standing communities. *Academy of Management Review* 20: 961–85.

Lewis, H.T. and Ferguson T.A. 1988. Yards, corridors and mosaics: How to burn a boreal forest. *Human Ecology* 16: 57–77.

Light, S.S. 1983. Anatomy of surprise. PhD Dissertation, University of Michigan, Ann Arbor.

Ludwig, D., Hilborn, R. and Walters, C. 1993. Uncertainty, resource exploitation, and conservation: Lessons from history. *Science* 260: 17, 36.

McCay, B.J. and Acheson, J.M., eds. 1987. *The Question of the Commons: The Culture and Ecology of Communal Resources.* Tucson: University of Arizona Press.

McQuillan, A.G. 1993. Cabbages and kings: the ethics and aesthetics of New Forestry. *Environmental Values* 2: 191–221.

Ostrom. E. 1990. *Governing the Commons: The Evolution of Institutions for Collective Action*. Cambridge: Cambridge University Press.

Ostrom, E. and Schlager, E. 1996. The formation of property rights. In *Rights to Nature*, pp.127–56, ed. S. Hanna, C. Folke, and K.-G. Mäler. Washington DC: Island Press.

Ramakrishnan, P.S. and Patnaik, S. 1992. Jhum: Slash and burn cultivation. In *Indigenous Vision*, pp.215–20, ed. G. Sen. Newbury Park, London: Sage Publications.

Rappaport, R. 1967. Ritual regulation of environmental relations among a New Guinea people. *Ethnology* 6: 17–30.

Redclift, M. 1994. Reflections on the sustainable development debate. *International Journal of Sustainable Development and World Ecology* 1: 3–21.

Regier, H.A., and Baskerville G.L. 1986. Sustainable redevelopment of regional ecosystems degraded by exploitive development. In *Sustainable Development of the Biosphere*, pp.75–101, ed. W.C. Clark, and R.E. Munn. Cambridge: Cambridge University Press.

Ruddle, K., Hviding, E. and Johannes, R.E. 1992. Marine resources management in the context of customary tenure. *Marine Resource Economics* 7: 249–73.

Walker, B.H. 1993. Rangeland ecology: Understanding and managing change. *Ambio* 22: 80–7.

Walters, C.J. 1986. *Adaptive Management of Renewable Resources*. New York: McGraw-Hill.

Young, O.R. 1995. The problem of scale in human/environment relationships. In *Local Commons and Global Interdependence*, pp.27–45, ed. R.O. Keohane and E. Ostrom. London: Sage.

Index

Italic numbers refer to figures or tables; n refers to notes; B refers to boxes.

437

see also herders, Sahelian; pastoralism; rangelands, Sahelian
salinity, Maine coastline 191
salmon fisheries, managed, problems 353–4
San Juan Nuevo community
  Communal Council to oversee community projects 239
  forests divided into family patches 238
  vertically integrated forest products industry 237
sanctions
  against law-breakers, Gitksan 379
  in pastoral societies 273
  social and religious 425
  see also social sanctions
savanna, derived, Ara 170, 183
schooling, for fishers 53
science
  conventional, inadequacy of 345–6
  of the integration of the parts 346
  interdisciplinary 346
  and its centrality 321–4
  one kind of? 344–7
  of the parts 346
  not for sustainable development 347
  Western, nature and limitations of 345
science of the commons 321
'science of surprise' 11
science, sustainability and resource management 342–62
scientific ideology, a hindrance 414
scientific resource management 5
  acting to reduce ecosystem resilience 8
  failure in Canadian subarctic fisheries 115–18, *117*
  need to manage a species over complete range 391–2
scientific understanding 51
scientists
  creating climate of over-optimism 326–7
  learning by fishing 63
  relationships with policy-makers, and errors in stock assessment 327
scramble competition 203–4
script(s)
  coding of, as cultural blueprints 426–7
  defined 232
  see also milpa script
secondary forest 170, 367–8
  milpa cycle in 234
  slow-maturing 365
  tropical, cleared for farming, Ara 161
self-organization 354, 357, 430–1
self-reporting 393, 406
semi-transhumance 253, 259–60
settlements, Leksand parish 77

shellfish conservation committees 196, 199, 205
shellfish conservation plans 199, 202, 203
  absence of 207
  and feedback 204
shifting cultivation 143, 170, 420
  trees burned *in situ* to maintain soil fertility 43
shifting cultivators, Manipur State 40
single species exploitation 348
skippers
  education, importance of practical learning 55–6
  knowledge, tacit and intuitive 62
  use of fishing technology 58
slash-and-burn cultivation/techniques 144
  caiçaras 135, 136–7
  prohibited in Parks and Reserves, Brazil 141
social capital 6
social cohesion, building of 307
social control
  by Gitksan 375
  Leksand parish 91
  and individual rights 92
social and ecological systems 36–9
social equality, and partial inheritance 70–1
social learning 407
  Maine lobster industry 404
social mechanisms 20–1
  behind management practices 3, 416, *418*, 422–8, 429, *430*
  enhanced for building resilience 430
social sanctions 27, 419
  regulating resource use 302
social system–ecological system interdependence 4
social system–ecosystem links
  disrupted by complex of driving forces 298
  elements which strengthen 299
  positive 214
  possibility of restoration 299–303
  revival of in Himalayas 285–310
  vanishing 305
  weak in Newfoundland 329
social systems
  ability to link with ecosystems 305
  developing/restoring 307
  in an ecosystem perspective 9
  defined 4
  evolutionary character of 356–7
  flexible 358
  Hindukush–Himalaya region 288
  local
    of rights and responsibilities, development of 7

Printed in the United States
By Bookmasters